Towards Auctioning: The Transformation of the European Greenhouse Gas Emissions Trading System

Climate Change Law, Policy and Practice Series

VOLUME 2

Editor

Professor Kurt Deketelaere

Professor of Law, University of Leuven, Belgium
Honorary Chief of Staff, Flemish Government
Honorary Professor of Law, University of Dundee, UK
Secretary-General, League of European Research Universities (LERU), Belgium

The aim of the Editor and Publishers is to publish works of excellent quality that focus on Climate Change.

Through this series the Editor and Publishers hope:

- to contribute to combating climate change, to protecting biodiversity, to reducing the impact of pollution on health and better use of natural resources;
- to provide information on cutting use of energy-improving energy efficiency, emissions and emissions trade, renewable energy, carbon capture and geological storage policy;
- to increase the access to information on climate change issues for students, academics, non-governmental organizations, government institutions, and business;
- to facilitate cooperation between academic and non-academic communities in the field of climate change throughout the world.

The titles published in this series are listed at the end of this volume

KLUWER LAW INTERNATIONAL

Towards Auctioning: The Transformation of the European Greenhouse Gas Emissions Trading System

Present and Future Challenges to Competition Law

Stefan Weishaar

Law & Business

AUSTIN   BOSTON   CHICAGO   NEW YORK   THE NETHERLANDS

Published by:
Kluwer Law International
PO Box 316
2400 AH Alphen aan den Rijn
The Netherlands
Website: www.kluwerlaw.com

Sold and distributed in North, Central and South America by:
Aspen Publishers, Inc.
7201 McKinney Circle
Frederick, MD 21704
United States of America
Email: customer.service@aspenpublishers.com

Sold and distributed in all other countries by:
Turpin Distribution Services Ltd.
Stratton Business Park
Pegasus Drive, Biggleswade
Bedfordshire SG18 8TQ
United Kingdom
Email: kluwerlaw@turpin-distribution.com

Printed on acid-free paper.

ISBN 978-90-411-3198-0

© 2009 Kluwer Law International BV, The Netherlands

Printed in Great Britain.

To my beloved wife and family – the source of my happiness
and all inspiration.

Table of Contents

Chapter 1
Introduction   1
1   Background   2
2   Research Questions   8
3   Research Methodology and Outline of the Study   11

Chapter 2
Economic Foundations   19
1   Introduction   19
2   Introduction to Social Welfare, Efficiency and the Relevance of
    Initial Allocation   20
3   Basic Economic Intuition for Abatement and Emission Trading   29
    3.1   Why Abatement?   30
    3.2   Why Emission Trading?   31
    3.3   What is the Problem of Initial Allocation?   33
          3.3.1   Why Do We Need Prices?   33
          3.3.2   Allocation   34
          3.3.3   Environmental Considerations   36
4   Overview of Industrial Economics   36
    4.1   Introduction   36
    4.2   Overview   37
    4.3   Economic Models   42
          4.3.1   A Simple Static Monopoly Model   42
          4.3.2   Abuse   44
          4.3.3   Cartels   45
          4.3.4   Merger and Acquisition   48
5   Summary   51

Chapter 3
Allocative Efficiency: A Static and Dynamic Perspective   **53**
1   Introduction   53
2   Emission Trading Systems   56
3   Static Closed Economy   59
    3.1   Static Closed Economy Model   59
    3.2   Comparison of Allocation Mechanisms and Allocative Efficiency   60
          3.2.1   Auctions as an Allocation Mechanism   60
                  3.2.1.1   Auction Formats   62
                  3.2.1.2   Multiunit Auctions   63
          3.2.2   Administrative Allocation Mechanisms   69
                  3.2.2.1   Financial Administrative Allocation Mechanisms   69
                  3.2.2.2   Free Administrative Allocation Mechanisms   70
                            3.2.2.2.1   Grandfathering   70
                            3.2.2.2.2   Relative Standard Base Mechanisms   71
    3.3. Environmental Aspects   73
    3.4   Summary   76
4   Dynamic Open Economic Setting   77
    4.1   The Dynamic Open Economy   77
    4.2   Comparison of Allocation Mechanisms and Allocative Efficiency   79
          4.2.1   Auctions as an Allocation Mechanism   79
          4.2.2   Administrative Allocation Mechanism   80
                  4.2.2.1   Financial Administrative Allocation Mechanisms   80
                  4.2.2.2   Free Administrative Allocation Mechanisms   81
                            4.2.2.2.1   Grandfathering   81
                            4.2.2.2.2   Relative Standard Base Mechanisms   81
    4.3   Environmental Aspects   82
          4.3.1   Dynamic Interaction of Firms and the Environment   82
                  4.3.1.1   Outsourcing Abatement   85
                  4.3.1.2   Outsourcing of Production through
                            Movements of Factors of Production   86
                  4.3.1.3   Substitute Domestic Goods for Imported Goods   87
          4.3.2   Initial Allocation Mechanisms and the Environment   88
                  4.3.2.1   Auctions   88
                  4.3.2.2   Administrative Allocation Mechanisms   89
                            4.3.2.2.1   Financial Administrative Allocation
                                        Mechanisms   90
                            4.3.2.2.2   Free Administrative Allocation
                                        Mechanisms   90
                                        4.3.2.2.2.1   Grandfathering 90
                                        4.3.2.2.2.2   Relative standard base
                                                      mechanisms   91
    4.4   Summary   92
5   Conclusion   93

Chapter 4
Compatibility of Allocation Formats and Directive 2003/87/EC **97**
1 Introduction 97
2 Allocation Formats under Directive 2003/87/EC 98
3 Auctions 101
4 Grandfathering 103
5 PSR 104
6 The EU ETS Amendment 107
7 Conclusion 108

Chapter 5
EU Emissions Trading System and Articles 81 and 82 **111**
1 Introduction 111
2 Economic Problem 112
   2.1 Barriers to Entry 112
   2.2 Entrenched Market Shares 115
3 Legal Analysis 116
   3.1 Development of the Jurisprudence 117
   3.2 Joined Application Jurisprudence Today under Article 81 127
      3.2.1 Analysis: The EU ETS and Article 81 EC Treaty 130
   3.3 Joined Application Jurisprudence Today under Article 82 132
      3.3.1 Application 134
   3.4 Applicability of Articles 82 and 86 EC Treaty against State
      Measures 136
      3.4.1 Application 138
4 Economic Appraisal and Summary 139
5 Joint Application Post 2012? 142

Chapter 6
EU Emissions Trading System and State Aid **145**
1 Introduction 145
2 European State Aid Regulation Applicable to Emission
   Trading Systems 146
   2.1 The Existence of State Aid: Article 87(1) 146
      2.1.1 Transfer of a Benefit or an Advantage
         (Notion of Aid) 147
      2.1.2 Aid Favouring a Certain Undertaking over Others
         (Selectivity Principle) 149
      2.1.3 Granted by the State or through State Resources 150
      2.1.4 It Should be an Undertaking or . . . Production 152
      2.1.5 Distorts or Threatens to Distort Competition 152
      2.1.6 Community Dimension: Aid Which is Capable of
         Affecting Trade between Member States 153
   2.2 Derogations of Article 87(1) 154
      2.2.1 Article 87(3)(b) 157

	2.2.2	Article 87(3)(c)	158
3	State Aid Assessment of Free Allocation Mechanisms		161
	3.1	State Aid Assessment of Grandfathering	161
		3.1.1 Transfer of a Benefit or an Advantage (Notion of Aid)	161
		3.1.2 Aid Favouring a Certain Undertaking over others (Selectivity Principle)	163
		3.1.3 Granted by the State or through State Resources	164
		3.1.3.1 By the State	164
		3.1.3.2 Through Member State Resources	165
		3.1.4 It Should be an Undertaking or . . . Production	167
		3.1.5 Distorts or Threatens to Distort Competition	167
		3.1.5.1 Firstly, Competitive Distortions	167
		3.1.5.1.1 Between Incumbent and New Entering Firms	168
		3.1.5.1.2 Trading Sectors and Non-trading Sectors	169
		3.1.5.1.3 Anticompetitive Distortions between Competing Firms of the Same Member State	170
		3.1.5.1.4 Between Trading Sectors	171
		3.1.5.2 Secondly, Does It Exceed the *de minimis* Threshold?	171
		3.1.6 Community Dimension: Aid Which is Capable of Affecting Trade between Member States	172
		3.1.6.1 Compatible with the Common Market	173
	3.2	State Aid Assessment of a PSR System	174
		3.2.1 Transfer of a Benefit or an Advantage (Notion of Aid)	175
		3.2.2 Aid Favouring a Certain Undertaking over Others (Selectivity Principle)	175
		3.2.3 Granted by the State or through State Resources	176
		3.2.3.1 By the State	176
		3.2.3.2 Through State Resources	178
		3.2.4 It Should be an Undertaking or . . . Production	181
		3.2.5 Distorts or Threatens to Distort Competition	181
		3.2.5.1 Firstly, Is There Distortion?	182
		3.2.5.1.1 Between Incumbent and New Entering Firms	182
		3.2.5.1.2 Trading Sectors and Non-trading Sectors	183
		3.2.5.1.3 Anticompetitive Distortions between Competing Firms of the Same Member State	183
		3.2.5.1.4 Between Trading Sectors	184

3.2.5.2 Secondly, Does It Exceed the *de minimis*
Threshold? 185
3.2.6 Community Dimension: Aid Which is Capable of
Affecting Trade between Member States 185
3.2.6.1 Compatible with the Common Market 186
3.3 Conclusion 187
4 Auctioning under the EU ETS Amendment 189
4.1 Auctions and Commission Involvement 190
4.2 Auctions in the Third Trading Phase 192
4.2.1 Transfer of a Benefit or an Advantage
(Notion of Aid) 192
4.2.2 Aid Favouring a Certain Undertaking over
Others (Selectivity Principle) 193
4.2.3 Granted by the State or through State Resources 195
4.2.3.1 By the State 196
4.2.3.2 Through Member State Resources 197
4.2.4 It Should be an Undertaking or . . . Production 197
4.2.5 Distorts or Threatens to Distort Competition 198
4.2.5.1 Between Incumbent and New Entering Firms 198
4.2.5.2 Trading Sectors and Non-trading Sectors 198
4.2.5.3 Anticompetitive Distortions between
Competing Firms of the Same Member State 198
4.2.5.4 Between Trading Sectors 198
4.2.5.5 Does It Exceed the *de minimis* Threshold? 199
4.2.6 Community Dimension: Aid Which is
Capable of Affecting Trade between Member States 199
5 State Aid and Distortions of Competition 200

Chapter 7
Auction Design Challenges 203
1 Introduction 203
2 Bundling 205
3 Timing 205
4 Market Structure and Transparency 207
5 Concluding Remarks 209

Chapter 8
Final Conclusion 211

Bibliography 221

List of Cases 243

Index 249

Chapter 1

Introduction

On 25 October 2003 the Directive 2003/87/EC governing the European Emissions Trading System (EU ETS) for greenhouse gas emission allowances for energy-intensive installations has entered into force. This greenhouse gas emissions trading system started with the operation of the first trading phase in January 2005. While the second trading phase (2008–2012) was subject to increased Commission scrutiny and witnessed a strong reduction of the number of emission allowances that was distributed, it did as such not see fundamental legal changes. Drastic amendments to the current legal framework championing auctioning as the ultimate mechanism of allocation for the third trading phase (2013–2020) have been proposed. In January 2008 the European Commission proposed an amendment that was in a watered down version adopted by the European Parliament on 17 December 2008. Despite the co-decision procedure that was to be employed for the legislative process, the European Parliament cooperated closely with the European Council and made the very rapid acception of an amended version of the Commission's draft possible. On 6 April 2009 the Council of the European Union has passed the amendment of the EU ETS directive. The rapid agreement that has been reached reflects the importance global climate change has in the eyes of decision makers and their desire to be able to present tangible outcomes in time for the fifteenth meeting of the United Nations Convention on Climate Change in 2009 in Copenhagen.

This book takes a Law and Economics approach to assess to which extent the Directive and the amendment restricts Member States' discretion to choose the most efficient and environmentally sound allocation mechanism.[1] Since economic theory attributes a central role to the free working of competitive market forces to

---

1. Acknowledgement: This author has been working for the research project 'Emissions trading and competitive positions of industries', conducted by the Metro Institute of the Law Faculty of Maastricht University (webpage <www.rechten.unimaas.nl/metro>).

bring about allocative efficient outcomes, anticompetitive distortions that result from the allocation mechanisms are examined in light of European Competition law provisions.

This chapter is subdivided into three interrelated parts. At first some background information is presented in section one. Given the focus of the research this section is restricted to the extent necessary. Subsequently the research questions are stated in section two. Section three describes the applied methodology and gives a concise outline of the research.

1   BACKGROUND

Since 1991 the European Commission has taken various climate related initiatives to limit greenhouse gas emissions and to improve energy efficiency. Measures include the promotion of electricity from renewable energy sources, voluntary commitments by carmakers and proposals on the taxation of energy products. In order to meet the European Greenhouse Gas reduction goals, as committed to under the Kyoto Protocol,[2] of attaining the committed average annual reduction of 8% below 1990 levels during the years 2008–2012[3] (binding upon the old fifteen EU Member States), the European Commission responded to a request from the EU Council of Environment Ministers in June 2000 by launching the European Climate Change Program (ECCP).[4] The findings of the working groups considering and giving recommendations on the most important options for reducing greenhouse gas emissions cost-effectively were summarized in a final report.[5] Following this report, the Commission presented a package of three broad

---

2. When reports indicated that mere stabilization of Greenhouse Gas emissions were insufficient to prevent climate change, members to the 'UN Framework Convention on Climate Change' (UNFCCC) committed themselves in 1997 to emission reductions, the so-called Kyoto Protocol. Greenhouse gases covered by the Kyoto Protocol are carbon dioxide (CO_2), methane (CH_4), nitrous oxide (N_2O), hydrofluorocarbons (HFCs), perflurocarbons (PFCs) and sulphur hexafluoride (SF_6). Following the ratification by Russia in November 2004, the Kyoto Protocol entered into force on 16 Feb. 2005. For a concise overview of the international and European policies towards climate change see Pallemaerts, M., and Williams, R., (2006). The decision to EC acceded the UNFCCC was taken on 15 Dec. 1993. See Council Decision 94/69/EC, (1993).
3. The Kyoto Protocol does not contain any expiry date. Yet it does also not contain any emission reduction targets for the period after 2012. Prospects for such commitments remain vague. See STEM (2005), 22. Yet at the eleventh conference of the parties (COP 11) in Montreal it has been agreed to continue to address climate change issues beyond the Kyoto compliance period. See Schipper, E. L. F., Boyd, E., (2006), 76. The continuation of the Kyoto protocol has been discussed at COP 12 in Nairobi and will be tabled in at COP 13 on Bali. Strong political support has been voiced at the G8 meeting 2007 in Heiligendam to address climate change and to achieving a comprehensive post Kyoto agreement that should include all major emitters. See G8 Summit Declaration (2007), 16.
4. COM (2000) 88.
5. ECCP (2001).

measures to tackle climate change as well as an Action Plan in the form of a Communication on the implementation of the first phase of the ECCP.[6]

These measures embraced the proposal for ratification of the Kyoto Protocol,[7] a proposal for regulating certain fluorinated gases,[8] and the proposal to establish a European Emissions Trading System.[9] The Action Plan contained, amongst other measures, a proposal for linking the Kyoto Protocol project-based mechanisms, including JI and CDM, to the European Emissions Trading System.[10] Also under the second phase of the ECCP (2002–2003) a working group worked on issues related to the linking of the European Emissions Trading System (EU ETS) with the Kyoto Protocol.[11] Following a proposal of the Commission[12] and subsequent amendments,[13] Directive 2004/101/EC linking the EU ETS with the Kyoto Protocol was adopted in October 2004.[14] Since the core field of interest of this book regards the European Emissions Trading System it suffices at this point to note that the ECCP continues to develop,[15] and that there is a wider framework in which the EU ETS is embedded.

The theoretical idea to reduce pollution by emission trading is by no means new. It was already proposed by Dales (1968) and has been discussed in a European context for years.[16] In accordance with Directive 2003/87/EC,[17] all Member States of the European Union were obliged to establish an emissions trading scheme as of 1 January 2005. Around 5000 operators with approximately 12.000 installations participate in this multi-jurisdictional attempt to reduce CO_2 emissions from four broad sectors: energy (electric power, oil refineries, etc.), the production and processing of ferrous metals (iron and steel), minerals (cement, glass, ceramics), pulp and paper.[18] The program is implemented in multiple phases: the first ranging from 2005–2007 and the second one from 2008–2012, which resembles the Kyoto Protocol compliance period. The following periods were according to the Directive

---

6. COM (2001) 580 final.
7. See COM (2001) 579 final. The Council of the European Union approved the Kyoto Protocol on 25 Apr. 2002. See Council Decision 2002/358/EC.
8. COM (2003) 492 final.
9. COM (2001) 581 final. The proposed Directive later gave way to what is today Directive 2003/87/EC, the European Emissions Trading Directive. This Directive presently allows for trading in carbon dioxide but the scheme can be extended to other greenhouse gases.
10. COM (2001) 580 final, 8.
11. Issues dealt with regarded linking, domestic offset projects and an EU Fund for JI and CDM projects.
12. COM (2003) 403 final.
13. For an examination of the linking possibilities see Langrock, T., Sterk, W., (2004).
14. See De Cendra de Larragán, J., (2006). See also Zwingmann, K., (2007), 76–81.
15. The second phase of the European Climate Change Programme (ECCP II) was launched on 24 Oct. 2005 and the amendment to Directive 2003/87/EC has been proposed on 23. Jan. 2008.
16. See Dales, J. H., (1968) and Peeters, M., (1993), 117–134.
17. The Directive entered into force on 25 Oct. 2003.
18. See Annex I of Directive 2003/87/EC.

five years long,[19] but the third trading period is extended under the amendment of the Directive till 2020.[20]

The first EU ETS trading phase was seen as a learning by doing phase.[21] It was supposed to allow Member States to get used to the system that was implemented in a very short period of time,[22] allow them to make progress towards their commitments under the Kyoto Protocol or towards meeting their particular CO_2 goals committed under the Burden Sharing Agreement[23] (binding upon the old EU fifteen Member States) with respect to the Kyoto Protocol. Not every Member State is, however, subject to emission reduction goals. At the present time two small (new) Member States of the EU are not engaged in any Kyoto Protocol commitments and thus do not have any emission reduction target.[24]

Less than five years after the adoption of the founding directive the proposal of the Commission amending the initial EU greenhouse gas emissions trading scheme has been delivered. The proposal aims at ensuring a significant reduction of greenhouse gas emissions by 2020, to be realized by a major part of the European business and power sector. The proposal includes fundamental shifts, like a far-going harmonization of allocation of the allowances, and, in this vein, auctioning as the default allocation method throughout the EU. On the 23 January 2008 the Commission introduced its proposal that has been endorsed in large parts by the Environmental Committee on the Environment, Public Health and Food Safety of the European Parliament.[25] Pursuant to the Council meeting of March 2008 an agreement on the Climate change measures should have been reached by the end of

19. See Directive 2003/87/EC, Arts 11(1) and (2).
20. See COM (2008) 16 final of 23 Jan. 2008 and European Parliament legislative resolution of 17 Dec. 2008 on the proposal for a directive of the European Parliament and of the Council amending Directive 2003/87/EC so as to improve and extend the greenhouse gas emission allowance trading system of the Community, COM (2008) 16 and the EU ETS amendment, Directive 2009/29/EC.
21. See COM (2000) 87 final, 10, as well as STEM (2005), 9 and Peeters, M., De Cendra de Largán, J. & Weishaar, S., (2007).
22. See Peeters, M., (2003). Beholding the speed of implementation the question appears to be well founded if the system was designed in an optimal way and if both environmental and economic concerns have been duely evaluated. See STEM (2005), 10.
23. The Council of the European Union agreed upon the contributions of each Member State to the overall Community reduction commitment in the Council conclusions of 16 Jun. 1998. Document 9702/98 (Annex I) of 19 Jun. 1998 of the Council of the European Union reflects the outcome of proceedings of the Environment Council of 16–17 Jun. 1998. The Burden sharing agreement is only binding upon the old EU fifteen Member States. See also COM (2001) 579 final, 23 Oct. 2001, Pallemaerts, M., & Williams, R., (2006), 39 and 43 ff. See in particular Pallemaerts, M., (2004) and Krämer, L., (2003), 303 ff. on the Burden Sharing Agreement.
24. Unlike the new ten Member States who are also parties to the Kyoto Protocol, both Malta and Cyprus were qualified as 'developing countries' within the meaning of the UNFCCC and therefore do not have any qualified greenhouse gas emission targets – they are however still obliged to participate in the EU ETS. The four National Allocation Plans are based upon business as usual scenarios. See Malta, (2004) and Malta (2006) as well as Cyprus (2004) and Cyprus (2007).
25. See COM (2008) 0016.

2008 and allowed for its adaptation at the latest early 2009.[26] In this light the severe time pressure with which the amendment has been reviewed by the European Parliament can be understood.[27] The *rapporteur* for the EU ETS amendment at the European Parliament, Avril Doyle, acknowledged that in five or six trialogues[28] taking place before the European summit the issues and parameters had been raised with the Council and that there were no surprises encountered with regard to the Council's positions. Even though the Council was only expected to adopt a common position in March 2009, it is noticeable that the text adopted by the European Parliament in its first reading appeared to be endorsing the positions of the Council expressed in its final compromise on Energy and Climate change in December 2008.[29]

For the sake of completion, below the amendment to Directive 2003/87/EC proposed by the Commission revising the original greenhouse gas emissions trading Directive is presented. This major revision would entail a dramatic change of the allocation of the tradable greenhouse gas allowances:

– Instead of a decentralized allocation by Member States, the Commission envisions that there should be an EU wide cap for the covered industries, for which auctioning will be the principal method. The Member States shall execute this auctioning, thereby following a Commission regulation governing the 'timing, administration and other aspects', in order to ensure that the auction will be done in an 'open, transparent and non-discriminatory manner'.[30] The share of allowances to be auctioned will gradually increase, resulting in full auctioning by 2020. Consequently, the free allocation shall decrease each year by equal amounts resulting in no free allocation in 2020.[31] However, in order to deal with the prevention of carbon leakage outside the EU, there will be a different regime for installations in sectors which are exposed to a significant risk of carbon leakage.[32] In 2013 and in each subsequent year up to 2020, installations in sectors which are exposed

---

26. Council of the European Union (2008), 12.
27. Comments and critical remarks from Members of the European Parliament relating to this issue were made during the debate on the proposed amendment in the session on the 16 Dec. 2008. The debate can be reviewed via the official website of the European Parliament.
28. Trialogues can be described as negotiation tools that facilitate agreements between the Council and the European Parliament and thereby speed up the legislative process. They are meetings between the Presidency of the Council, the Parliament's rapporteur for the concerning dossier and the Parliament's Chairmen of the Committee in charge and the respective Commission specialist. Such meetings are not regulated nor mentioned in the treaties and thus to be regarded as a legal gray area. The object and purpose of such trialoges are to update the various legislators of their respective positions and to inform their institutions and thus to enable them to identify common ground and controversial aspects of a legislative project and thereby facilitate the reaching of compromises. I am endebted to Dr Markus Pieper, Member of the European Parliament and his staff for clarification of the legislative process of the amendment.
29. Council of the European Union (2008b).
30. Article 10(5), COM (2008) 16 final of 23 Jan. 2008.
31. Article 10a(7), COM (2008) 16 final of 23 Jan. 2008.
32. Article 10a(8), COM (2008) 16 final of 23 Jan. 2008.

to a significant risk of carbon leakage shall be allocated allowances free of charge up to 100% of the quantity determined in accordance with the free allocation method as determined in the paragraphs 2 to 6 of Article 10a of the proposal. This means, for instance, that the maximum amount of allowances that is the basis for calculating allocations to installations which carry out activities in 2013 and received a free allocation in the period 2008 to 2012 shall not exceed, as a proportion of the annual Community-wide total quantity, the percentage of the corresponding emissions that those installations emitted in the period 2005 to 2007. A correction factor shall be applied where necessary.[33]

– At the latest by 30 June 2010 and every three years thereafter the Commission shall determine these exposed sectors, and the proposal includes some circumstances that should be reviewed when taking this decision.[34]

– Furthermore, for the *energy-intensive* sectors or sub-sectors exposed to a significant risk of carbon leakage, the Commission needs to analyse their position in the light of the outcome of international negotiations.[35] This needs to be done ultimately by June 2011, and the analysis shall be accompanied by appropriate proposals, among which the *adjustment of the proportion* of allowances received free from charge. The latter would be a specific rule within the free allocation method. Another possibility is the inclusion in the Community scheme of importers of products produced by the sectors or sub-sectors exposed to carbon leakage.

– The Commission will adopt rules for the free allocation, and will designate the sectors to be covered by the rule that no auction will apply because of the significant risk of carbon leakage.

– In any case, electricity production will be covered by the auction method.[36]

– Irrespective of the allocation method that is being employed an annual reduction of 1.74%[37] of the emission allowances allocated under the EU ETS scheme will increase the scarcity of the allowances and increase the system's environmental effectiveness.

The amendment to the EU ETS was passed on 06 April 2009 by the Council. It was previously voted upon by the European Parliament on 17 December 2008 and agreed upon by the European Commission and was published in the Official Journal as Directive 2009/29/EC on 23 April 2009. The Directive entered into force on 25 June 2009. It follows in many areas the proposal of the Commission. Most notably it also envisages a strong degree of harmonization of the allocation method.

---

33. Article 10a(4), but also 10a(5), COM (2008) 16 final of 23 Jan. 2008, for installations only included in the scheme from 2008 onwards.
34. Article 10a(9), COM (2008) 16 final of 23 Jan. 2008.
35. Article 10b, COM (2008) 16 final of 23 Jan. 2008.
36. Article 10a(6), COM (2008) 16 final of 23 Jan. 2008.
37. Article 9, COM (2008) 16 final of 23 Jan. 2008.

- The European Council endorsed the proposed linear reduction factor of 1.74% and suggested to review it by 2025.[38]
- Auctioning is to be increasingly phased in as the predominant form of allocation rising from 20% in 2013 to 70% in 2020. Full auctioning is aimed for in 2027.[39] Important derogations to auctioning in favour of free allocation are provided for by Articles 10a (transitional rules) and 10c (transitional allocation for modernization of electricity generation). While auctioning is in principle the default allocation for the power sector[40] some Member States were able to secure concessions to compensate for their poorly integrated electricity grid.[41] While 88% of the total quantity of allowances to be auctioned will be distributed to Member States,[42] 10% will be allocated for the purpose of solidarity and growth within the Community.[43] The remaining 2% is given to those Member States that were in 2005 at least 20% below their Kyoto protocol base year emissions.[44] A large share of auction revenues (50%) should (not shall) be used for emission mitigation and adaptation measures.[45]
- Sectors and sub-sectors exposed to a significant degree of carbon leakage are eligible to reveice up to 100% of allowances for free.[46] This can be reviewed when an international climate change agreement is reached.[47]
- Free allocations are in general based on *ex ante* benchmarks based upon the average of the top 10% most efficient installations in a sector or subsector during the period 2007–2008.[48]

Elements that are new regard the strong support of Carbon Capture and Storage[49] and the exclusion of smaller installations[50] and the setting of a more stringent timetable for the Commission. Undertakings may in particular benefit from the slower phasing in of auctioning and the introduction of price caps in the presence of strong price fluctuations.[51]

While these amendments are not yet implemented it is expedient to note that the current Directive 2003/87/EC obliges Member States to identify the undertakings that fall within the scope of the Directive, grant greenhouse gas emission allowances, and to establish effective monitoring and verification procedures.

---

38. Article 9, Directive 2009/29/EC.
39. Article 10a(11), Directive 2009/29/EC.
40. Article 10a(3), Directive 2009/29/EC.
41. Article 10c(1), Directive 2009/29/EC.
42. Article 10(2)(a), Directive 2009/29/EC.
43. Article 10(2)(b), Directive 2009/29/EC.
44. Article 10(2)(c), Directive 2009/29/EC.
45. Article 10(3), Directive 2009/29/EC.
46. Article 10a(12), Directive 2009/29/EC.
47. Article 10a(1), Directive 2009/29/EC.
48. Article 10a(2), Directive 2009/29/EC.
49. Article 10a(8), Directive 2009/29/EC.
50. Article 27, Directive 2009/29/EC.
51. Article 29a, Directive 2009/29/EC.

Besides obligations to report upon the application of the Directive, Member States are in particular charged with the allocation of emission allowances. Any allocation method employed in the National Allocation Plans (NAP) has to be compatible with the establishment of an absolute emissions cap and observe a number of criteria contained in Annex III of the Directive.

This book examines in which way the legal text of the Directive limits Member States' discretion to select an allocative efficient and environmentally effective allocation mechanism, and examines which role EC Competition law is currently playing in this context and how its role will be changed due to the amendment. The starting point of the analysis is the recognition that an important element for the attainment of allocative efficiency is the free working of competition and that anticompetitive distortions undermine its attainment. Such distortions are of particular relevance if an efficient allocation is not attained as a direct result of initial allocation of emission allowances but has to be established on the Emissions Trading market. Suboptimal initial allocation of emission allowances could give rise to collusion, market power and windfall profits and lead to distorts of competition on the merits. Therefore the current design challenges of the envisaged (partial) auctioning regime for the third trading period are examined.

2   RESEARCH QUESTIONS

As can already be understood from the background information given in the preceding section, climate change represents a transboundary environmental problem that requires a multi-jurisdictional problem solving approach. Such approaches are further complicated by the multitude of academic disciplines that endeavour to analyse the routes and causes of the problem and the interdependencies of the various elements, actors and decision makers. This book uses Law and Economics[52] to address climate change and only does so in the restricted framework of the European Greenhouse Gas Emissions Trading System (EU ETS). It should be emphasized, however, that the research focuses strongly upon law and discusses legal implications by reference to economic concepts.

The emphasized discipline, law, can be described as the sum of provisions, acts, ordinances, and regulations, that together with legal principles and doctrines, regulate societal relationships. It consists of various fields that seek to coherently address diverse topics related to amongst others, property, contracts, torts, crimes, institutional behaviour and extends to a myriad of aspects.[53] By regulating relationships it essentially tries to prevent undesirable behaviour of all and all kinds of

---

52. Law and Economics is an independent area that has developed and is to be distinguished from the disciplines of economics proper and law proper and is more directed to give normative policy guidance.
53. See Cooter, R., Ulen, T., (2004), Ch. 1.

members[54] of society. Where necessary the legislator has established sanctioning mechanisms in order to both punish trespass and to serve as a deterrence tool. Law thus legitimates governmental intervention in private activities, allows for the creation of policies but also protects private parties against the government by for instance guaranteeing human rights.

While law essentially tries to set incentives (rules) so that people behave in a desirable manner, economists are able to provide the framework to determine whether legislators and administrative authorities succeed in setting the incentives correctly so that they actually achieve what they are striving for in a cost efficient way. It can therefore be seen that many synergy effects can be expected to be derived from a combined Law and Economics approach.

Legislators can strive to attain a large variety of objectives such as 'social equity' for example, but for the purpose of this research it is expressly assumed that the objective is maximization of 'social welfare', that is, the maximization of net benefits to society. Firm believers in the Smithsonian invisible hand[55] trust that the market will do it right and that government is well advised not to do more than is absolutely necessary to provide for an effective and efficient functioning of the market[56] that minimizes waste to society.

Irrespective of whether one believes that government intervention is good or bad, it is beyond doubt that government is a major market player. Not only does it buy a lot through specially designed procurement systems but it also governs trading mechanisms addressing market failures. Greenhouse gas emissions can be described as negative externalities that do not have a market price and are thus created in excess of what is socially desirable. External effects can be defined as follows: an external effect or an externality exists if the consumption or production activities of one individual or firm affects another person's utility or firm's production function.[57]

Clearly, the European Emissions Trading System directly addresses market failures (externalities) on the product market and seeks to correct them, which – from an economic point of view – constitutes an area for State intervention that may be justified.[58] Justification of State intervention is a necessary requirement but knowledge regarding the optimal level and form of intervention is quite a different issue. Economic theory suggests that optimal CO_2 reduction would equate marginal abatement costs and the net present value of the (global) marginal damages of

54. That is not only natural persons are subject to law, but also legal persons as well as public entities.
55. Smith, A., (1937).
56. Prominent defenders of such a view are to be found among the ranks of the ordo-liberals but also scholars of the public choice tradition are critical of government activism.
57. See Kuik, O., (2005), 9. For an elaboration on externalities see Verhoef, E. T, (1999).
58. For a review of regulatory intervention in case of market failure the interested reader is referred to Philipsen, N. J., (2003), s. 2.2 and for a textbook on public finance and public choice see Cullis, J., Jones, P., (1998).

CO_2 emissions.[59] Damage cost estimates differ considerably and estimates are higher if one takes the higher dollar value for poorer income classes into account.[60] Estimation of climate change damage studies are subject to much reservation,[61] due to the scientific uncertainty regarding the physical changes that take place, their geographic and temporal distribution[62] and because of their sensitivity to the assumed present value attributed to damages in the future.[63]

Despite the difficulty to define an optimal greenhouse gas reduction goal government is still under the (economic) obligation to design a system in the most desirable way (i.e., efficient and least distortive). Mechanisms that can be employed to bring about the internalization of negative externalities include taxes that force private operators to recognize social costs,[64] the establishment of product standards, command and control regulation as well as trading systems. From a social welfare perspective the question can be raised if publicly orchestrated trading mechanisms such as the EU ETS are able to perform as well as competitive markets by allowing for cost effective transactions that allow market demand and supply to equalize and an appropriate allocation of initial endowments.

Since the market exchanges of public trading mechanisms can resemble transactions under privately organized exchange systems, it is not expected that there must be significant differences between them. It may, however, be assumed that the more complex a publicly designed measure is and the higher transaction costs are, the further it may be from competitive outcomes. Since transactions under the EU ETS can be executed at specialized exchanges, it is expressly assumed here that transactions costs are negligible. With regard to the second element, initial allocation, things are less clear. The equilibrium in the market will depend on the initial distribution of endowments. If Government interference generates a suboptimal initial allocation of endowments that gives rise to distortions of competition and windfall profits[65] and if

---

59. This does, however, not imply that the optimal abatement objective is to bring about a stabilization of atmospheric CO_2 concentration. On this point see Peck, S. C., Teisberg, T. J., (1994), Tol, R. S. J., (1999b). The underlying rational is that the costs of climate change largely depend upon its speed and that the damages associated with a sufficiently low change may be out weight by its benefits.
60. Tol, R. S. J., (1999a), 70. For a general welfare theoretic discussion see Fankhauser, S., Tol, R. S. J. & Pearce, D. W., (1997).
61. Tol, R. S. J., (1999a), 78, and IPCC (2001), Ch. 19.5.2.
62. See Tol, R. S. J., Downing, T. E., Kuik, O. J. & Smith, J. B., (2004).
63. Tol, R. S. J., (1999a), 71.
64. This referred to as the 'Pigou tax'. See Pigou, A.C., (1949), Ch. 8.
65. In addition it should be noticed that the position of general equilibria may vary depending on operating costs that are associated with the various allocation mechanisms. Depending on the particular preferences of society its welfare function may be best served under one allocation mechanism than under another. Since operating costs are dependent upon the design and complexity of the allocation mechanism and are best addressed within a concrete framework, this research limits itself to the examination of allocative efficiency and its impact on the general equilibrium.

initial allocations are further away from the market equilibrium, it may take longer for demand and supply to equate.[66]

It is therefore particularly lamentable that even though much has been written about the advantages and disadvantages of Emission Trading Systems, one element of crucial importance, emission allowance allocation, still needs further examination.[67] From a Law and Economics perspective, there are essentially two important elements that have to be reviewed with regard to the European Emissions Trading System in order to examine if legal rules impede or even constrain the attainment of allocative efficiency – a condition in which all possible gains from exchange are realized[68] – and the environmental objective. Firstly, whether the current set up of the EU ETS fosters allocative efficiency or whether this allocative efficiency is hindered by legal impediments or constraints. And secondly, whether Competition law can serve to remedy anticompetitive effects stemming from Member State actions taken pursuant to Directive 2003/87/EC. Anticompetitive behaviour of firms is thus not the core interest of this book because firm behaviour is directly addressed under EC Competition law rules.

In the course of this research particular emphasis is placed on allocative efficiency and distortions of competition associated with the mechanisms employed by the government in order to distribute emission allowances. More specifically the book seeks to answer the following research questions with regard to the European ETS:

> Which allocation formats are most desirable from an allocative efficiency and environmental effectiveness point of view?
>
> Is allocative efficiency safeguarded under the present European Greenhouse gas Emissions Trading System (EU ETS)?
>
> If governmental measures under the EU ETS distort competition on the merits, can Competition law be employed to contain them?
>
> What are the challenges to allocative efficiency and Competition law posed by the amendment to the Directive and its emphasis on auctioning?

3   RESEARCH METHODOLOGY AND OUTLINE OF THE STUDY

This section of the chapter briefly introduces the general framework and describes the methodology used to address the research questions. Subsequently the structure of the text is outlined.

---

66. See Weishaar, S., (2007d) for a vivid example how market adjustments may be time consuming even in the presence of efficient exchange markets.
67. Gayer, T., (2005), 1.
68. Frank, R., (1997), 350 defines allocative efficiency as a condition in which all possible gains from exchange are realized. This implies that those market participants valuing a good most have been able to attain it.

In order to examine if a publicly governed trading system such as the EU ETS can be expected to generate fully competitive outcomes, one can turn to economic research and investigate the industries and markets that are affected. Another approach is Law and Economics based and focuses on the theoretical analysis of the legislative framework which governs the trading system and the effects they are expected to give rise to. Even though it is no substitute for an economic analysis it can yield important complementary insights.

Due to problems of data availability this book follows a theory based approach with an emphasis on law that uses economic theory as a framework for reference. In order to better understand the functioning of the allocation mechanisms employed in the EU ETS simple economic concepts are used. In economics partial equilibrium models are designed to give insights to the functioning of individual markets in isolation. If the linkages between individual markets are so strong that they will affect other markets, so-called general equilibrium models are used to understand these effects. Since general equilibrium models are complex, mathematical and require a large amount of programming, this Law and Economics analysis necessarily limits itself to an examination of allocation mechanisms within a static partial setting. Thus in this book the EU ETS is examined as a publicly governed trading mechanism on the production side operating in both a closed and open economic setting.

From an economic theory point of view some qualifications of the applied approach are in order. Allocative efficiency can be defined as a condition in which all possible gains from exchange are realized.[69] This implies that those market participants valuing a good most, should be able to attain it. This is equally valid for a partial equilibrium context as it is for a general equilibrium one. In a general equilibrium framework, however, the attainment of allocative efficiency on one market is not a sufficient condition to conclude that a general equilibrium is reached. Even if all other confounding factors such as anticompetitive distortions or taxes that in the real world distort its attainment are assumed away, the attainment of one market equilibrium is not a desirable state as such since it implies that the remaining markets are in disequilibrium. Second best theory[70] therefore suggests to distort such a market to bring about the true general equilibrium.

Another point of critique that can be put forward from an economic perspective regards the use of a model that is not sufficiently capturing the intertemporal nature of the EU ETS. The European Emissions Trading System mandates allocations spanning over a multitude of successive periods and should thereby be better examined within the framework of an intertemporal general equilibrium model. While the author acknowledges that this is an interesting and valuable field of research, this work does not purport to be striving to propose

---

69. Allocative efficiency can be defined as a condition in which all possible gains from exchange are realized. See Frank, R., (1997), 350. This implies that those market participants valuing a good most have been able to attain it.
70. See Cullis, J., & Jones, P., (1998), Ch. 1.

long run optimal allocation of allowances. It is the object and purpose of this research to elucidate economic implications of legislative action in a static setting.

In addition it has to be pointed out that not each and every equilibrium will maximize social welfare. Depending on the particular preferences of society its welfare function may be served best by one set of parameters then by another. The positions of equilibria may vary depending on operating costs that are associated with the various allocation mechanisms or their underlying environmental objective. Since operating costs are dependent upon the design and complexity of the allocation mechanism and are best addressed within a concrete framework, this research is limited to the examination of allocative efficiency and its impact on a partial equilibrium.[71] It also treats the environmental objective that is being set by society as given, assuming that it is set at its optimum.[72]

As already indicated above, the determination of the optimal environmental goal is subject to a complex debate in which even the anthropogenic cause of climate change has been questioned.[73] In the presences of much scientific uncertainty, the 'precautionary principle' has been established that 'induces' action even in the absence of complete knowledge.[74] Since this research is elaborating upon allocation mechanisms, such considerations are beyond the scope of the text. For the purpose of the treatment, the environmental objective is taken as given and assumed to be set optimally.

The relevant legislative frameworks for this book are the legal sources establishing the Emissions Trading System, Directive 2003/87/EC, and European Competition law. The ongoing[75] development of the ECCP and of Member States' climate change policies, is limited to the extent necessary. The Linking Directive,

---

71. Tietenberg, T., Grubb, M., Michaelowa, A., Swift, B. & Zhang, Z. X., (1999), 34 suggest that regulatory barriers to the creation of credit trading benchmarks can be substantial. Such costs would imply high setup costs. Such costs are also assumed to be high under grandfathering yet much lower under auctioning systems. See Zwingmann, K., (2007), 288.
72. In this context one frequently differentiates between efficiency and cost effectiveness. While the former can be used to describe policies that achieve an optimal level of protection, the later is often applied to policies that reach a predetermined policy objective with least costs. On this point see also Kuik, O., (2005), 11.
73. For an objective assessments of the current literature the interested reader is referred to the IPCC reports, IPCC (2007a), IPCC (2007b), IPCC (2007c). The Intergovernmental Panel on Climate Change (IPCC), an organisation founded in 1988 by the United Nations Environmental Programme and the World Meteorological Organization, has the task to evaluate the risk associated with anthropogenic climate change by examining peer reviewed and published scientific and technical literature. The IPCC publishes special reports in support of the work of the UNFCCC. Its 2007 report regards the scientific aspects of the climate system and climate change (working group 1), the vulnerability of socio-economic and natural systems to climate change, consequences, and adaptation options (working group 2) and the options for limiting greenhouse gas emissions and otherwise mitigating climate change (working group 3). The work of the IPCC is often regarded as objective and has been accepted at the G8 Summit in Heiligendamm. See G8 Summit Declaration (2007), 15.
74. Heal, G. & Kriström, B., (2002), 11.
75. As mentioned before, the second phase of the European Climate Change Programme (ECCP II) was launched on 24 Oct. 2005.

Directive 2004/101/EC, encourages private investment in the set up of CDM and JI projects, in order to improve the liquidity of the EU ETS and to reduce the compliance costs of covered sectors.[76] Since the issue of enhanced market liquidity and the quantity of allowances to be allocated to operators is logically separable from the choice of allocation formats, the treatment of the Linking Directive is restricted. Particular attention is placed on EC Competition law. This field of law merits additional attention because – even though it has always been an important part of Community law[77] – its underlying purpose is less evident. At times conflicting objectives such as the enhancement of economic efficiency, promotion of consumer welfare and the advancement of European economic integration are cited.[78] Legal scholars such as Fairhurst and Vincenzi (2003) postulate that market integration is the primary objective of European Competition law while the creation of a level playing field and the promotion of efficiency are secondary goals.[79]

Secondary as they may be, these goals can also be quite closely related. Article 3(1)(g) EC Treaty commits the European Community to establish a system of undistorted competition in the internal market in order to attain the objectives set out in Article 2 EC Treaty. Since anticompetitive distortions can stem from public as well as private actions, judges employ a large arsenal of substantive provisions and jurisdiction to eradicate governmental barriers to trade between Member States as well as barriers to trade between private parties. Competition rules can thus be viewed to be a natural and necessary complement to the freedoms of movement of goods, persons, services and capital which are essential to the attainment of the internal market.[80] This interpretation is also expressed by the European Court of Justice (ECJ) in *Consten Grundig*, where the Court clarified that the Treaty aims at the abolition of barriers to trade between States and thus 'could not allow undertakings to reconstruct such barriers'.[81] From this, it can be seen that addressing anticompetitive distortions between the Member States will also have a positive effect upon the internal market.

Even though it does not appear that the legal use of the term 'level playing field' amounts to an acknowledged concept nor does it appear to have risen to the level of a doctrine, it is used and applied by scholars and used in European Commission documents such as the consultation document on the State aid Action Plan.[82] In its elaboration the Commission places the notion of level playing field into close neighbourhood to what economists describe as competition on the merits, that is, undistorted competition. Since this text is concerned with the

---

76. De Cendra de Larragán, J., (2006), 103, but also Zwingmann, K., (2007), 76–81.
77. Craig, P. & De Burca, G., (2003), 936.
78. Discussions about the purpose of antitrust legislation are also existing in other countries as e.g., the US or Japan. See Martin, S., (1994), 45 ff. and 61 ff.
79. Fairhurst, J. & Vincenzi, C., (2003), 441.
80. Article 14 EC Treaty. With regard to Arts 28, 29, 81 and 82 EC Treaty, see Fairhurst, J. & Vincenzi, C., (2003), 440.
81. Case 56/64 and 58/64, *Établissements Consten S.à.R.L. and Grundig-Verkaufs-GmbH v. Commission* [1966] ECR 299, 340.
82. COM (2005) 107 final, 3. See also Van der Laan, R. & Nentjes, A., (2001), 131 ff.

Law and Economics aspects of the European CO_2 Emissions Trading System, this interpretation is of particular importance.

After having introduced general aspects of the applied approach, the following paragraphs present the structure of this book. After this introductory chapter which delineates the field of reference and places the research questions into their proper context, the second chapter introduces the economic concepts which are employed in this text. The first part addresses efficiency concepts and presents a general equilibrium framework which is used to present the importance of initial allocation and adjustment processes of out of equilibria situations under an Emission Trading System. The further away an initial allocation is from an equilibrium situation the more heavily it may rely upon an effective secondary market to bring about allocative efficiency and an equilibrium. The slower the adjustment process, the more attractive allocation mechanisms become that give rise to relatively less distortions of the market equilibrium. The second part gives a short overview of the economic intuition of emission abatement and initial allocation. This is followed by an introduction to industrial economics which gives the theoretical grounding for the discussions relating to competition and Competition law. Giving a historical overview, this section reviews the Standard-Conduct-Performance paradigm, notes the existence of the Chicago school of Industrial Economics and reviews some empirics regarding the general link of industry structure, barriers to entry and profitability of enterprises. Thereafter economic concepts of monopolies, abuse, cartels and mergers and acquisition are introduced.

Chapter 3 places a coherent typology of emission allowance allocation mechanisms into an emission trading system. It examines how various assignment mechanisms deal with issues such as price determination, allocative efficiency[83] and environmental considerations in a static closed economy and subsequently an open dynamic one. It is reviewed how these mechanisms are ranked and whether they serve the attainment of a market equilibrium if one assumes that there is no global International law framework to mitigate effects stemming from countries employing lax environmental legislation (polluter havens[84]). It is not only considered how solely market-based allocation mechanisms (auctions) perform in light of the above issues but also how they compare to administrative allocation mechanisms. Two types of administrative allocation mechanisms are examined: (1) financial administrative allocation mechanisms, combining payment schemes with bureaucratic expertise, and (2) free administrative allocation mechanisms, based *inter alia* on policy considerations[85] and on historical emission records (grandfathering). In particular, the value added of relative performance standards which are for example included in the 'Performance Standard Rate' (PSR) Emission

---

83. Allocative efficiency can be defined as a condition in which all possible gains from exchange are realized. See Frank, R., (1997), 350. This implies that those market participants valuing a good most have been able to attain it.
84. On polluter havens see Antweiler W., Copeland, B. R. & Taylor, M. S., (2001).
85. Policy considerations are to be understood as social welfare considerations other than those based on economic rationale.

Trading System – as employed in the Dutch NOx Emission Trading System[86] – are elaborated upon as a means to provide allowances. Besides determining which allocation mechanism is more desirable, it is the objective of this chapter to examine the impact of these mechanisms on the declared goal of environmental effectiveness.

In order to address the question whether allocative efficiency can be safeguarded under the EU ETS, Chapter 4 examines the compatibility of prominent allocation mechanisms introduced in the preceding chapter with Directive 2003/87/EC. While it applies a strictly legal analysis in the first part, it reverts to economic theory in order to evaluate the findings of the analysis. If those allocation mechanisms that minimize market imperfections are foreclosed because they are not compatible with the legal text, it can be concluded that the EU ETS will depend more strongly upon a secondary market to attain allocative efficient outcomes.

If the allowed initial allocation mechanisms gives rise to distortions of competition and windfall profits that work to the detriment of society, it has to be examined if law can be successfully employed to mitigate such effects. EC Competition law seeks to contain distortions of competition on the merits and constitutes a relevant legal framework to analyse if distortions created through the EU ETS can be contained. The potential applicability of the four freedoms in this context is not examined in this text. Thus this research is thus interested in the examination of the incentives created under legal rules that induce undertakings to violate EC Competition law as well as indirect infringements of EC Competition law by Member States.

Chapter 5 analyses how Competition law restraints Member States to take measures that distort competition within the framework of the European Emissions Trading System. Since Competition law is particularly concerned with the action of undertakings, the ECJ has developed jurisprudence intended to broaden the scope of application of central provisions also to State actions. After describing the economic effects that have a negative effect from an industrial economics perspective, the historic evolution of the 'joint application jurisprudence' of Articles 10(2), 3(g), 81 and 82 EC Treaty is presented. Subsequently its applicability with regard to the European Emissions Trading System is examined.

Chapter 6 addresses the allocation of allowances from a State aid perspective. The State aid provision is designed to prevent Member States from granting aid that sets undertakings at a comparative advantage and affects the internal market. Following the typology developed in Chapter 3, this chapter assesses the widely applied grandfathering allocation method from both an industrial economics and legal perspective. It analyses the methods' proneness to give rise to anticompetitive distortions of the level playing field in the EU and examines its compatibility with the European common market. Subsequently the PSR system, and auctioning under the EU ETS amendment is analysed under the same criteria.

---

86. The Dutch NOx emission trading system entered into force on the 1 Jun. 2005. See Staatsblad van het Koninkrijk der Nederlanden (2005a) and Staatsblad van het Koninkrijk der Nederlanden (2005b), Art. 1. See also Van Tol, I. & Oldenziel, H., (2002).

The analysis in Chapter 7 leaves the realm of Competition law and takes an auction theoretic focus. It examines what the present auction design challenges are that need to be addressed by the Commission in its implementing regulation due by 30 June 2010.

Chapter 8 summarizes the main findings of the present emissions trading system and places the gained insights into a broader perspective. In doing so it also analyses the challenges to allocation and Competition law presented by the amendment to Directive 2003/87/EC with regard to auctioning.

Chapter 2

Economic Foundations

1   INTRODUCTION

Economists are concerned with theoretical models that help to understand reality. Since reality is far too complex to be captured by a single model, simplifying assumptions are used in order to facilitate the theoretical analysis. As in other scientific fields, also economists have developed their own terminology and forms of expression that may not be immediately understood by other scholars. Since this book is intended for a wider public, this chapter intends to introduce the reader to the relevant economic concepts that are applied throughout the text in an easy and comprehensive fashion without striving for completion.

The chapter first presents a general equilibrium model and Pareto efficiency. Subsequently a short overview of the economic intuition of emission abatement and initial allocation is presented. This is followed by an introduction to industrial economics which gives the theoretical grounding for the discussions relating to competition and Competition law in Chapters 5 and 6. Giving a historical overview, this section will review the Standard-Conduct-Performance paradigm, note the existence of the Chicago school of Industrial Economics and review some empirics regarding the general link of industry structure, barriers to entry and profitability of enterprises. Thereafter economic concepts of monopolies, abuse, cartels and mergers and acquisition will be introduced.

For didactical reasons this chapter does, however, not address all economic fields that are used for the analysis. Since auction theory is complex in nature its treatment has been included in later chapters rather than in this one.

2   INTRODUCTION TO SOCIAL WELFARE, EFFICIENCY
               AND THE RELEVANCE OF INITIAL ALLOCATION

This section of the chapter explains the relevant concepts of Pareto efficiency,
allocative efficiency and social welfare. In order to foster a sound yet general
understanding of the importance of initial allocations of EU ETS emissions allowances, one can consider a general equilibrium model. General equilibrium theory is
a branch of microeconomics that seeks to explain production, consumption and
prices in the whole economy and starts by examining the behaviour of market
participants. These are thus models that – unlike partial equilibrium models
which analyse how individual markets function in isolation – are concerned
with the economy in its entirety. They are used if the linkages between individual
markets are so strong that they will affect other markets. Modern general equilibrium models are very complex and require substantial amounts of computation and
are therefore usually employing computers.

Leon Walras (1834–1910), the founding father of the general equilibrium
theory, proposed a series of models that took better account of the real world by
allowing for exchanges of commodities (consumption goods and several factors of
production) between agents (landowners, labourers, capitalists and entrepreneurs).[1] Walras proposed a dynamic process of price adaptations towards a
price that would be equating market supply and demand and thus lead towards
a general equilibrium. This is known as Walras's *tâtonnement* process and has been
tried to be described in literature as a fictitious auctioneer that calls out new relative
price ratios that allow market participants to compare their offerings and demand.
In this process trade between the market participants will only take place when
supply equals demand at equilibrium prices.

Since Walras the adjustment process of demand and supply in a general equilibrium context has been subject to research.[2] More universally known – also
among laymen – is the cobweb model of the market price adjustment of a partial
equilibrium market. Under specific assumptions[3] supply and demand adjust
towards an equilibrium. Within the framework of the cobweb model farmers
adjusted their agricultural production based on prices of the previous year.
Following a demand shock that triggered a rise in prices, and hence an increase
in the quantity supplied by farmers in the subsequent period, alternating periods of
excess demand are followed by periods of excess supply. Provided that demand is

---

1. For an account of Walras's contribution to modern economics and a description of the models'
    working, see Maks, H. & van Daal, J., (2002). This paragraph is based on this paper.
2. While the literature is largely beyond the scope of the text the interested reader is referred to
    Schinkel, M. – P., (2001), Ch. 3.
3. See for an elaborate discussion regarding the assumptions and workings see Kaldor, N., (1934).
    Assumptions underlying the cobweb model include that supply and demand are constant over
    time and not influenced by the transactions on the market and that adjustment of prices to
    different supply levels is immediate. Farmers are assumed not to be realizing their ability to
    influence prices.

elastic relative to supply,[4] quantities demanded and supplied move ever closer together as the price level approaches the market's equilibrium. Repercussions of changes in the spending power of the market participants (consumers and farmers) on other non-agricultural product markets are, however, not accounted for. While the cobweb model is thus inadequately reflecting reality, it is useful to depict how forces of demand and supply eventually adjust – though it goes without saying that not each and every economy has to be able to have a (single) equilibrium[5] nor will an equilibrium be always able to be attained.

Also in the geometrical model described below, reality is crudely simplified. While it considers more then one market, the range of employed variables is severely restricted and the means of exchanges are based on barter. Exchanges do however, not only take place in equilibrium, that is, at equilibrium relative prices, but already when market participants are benefiting from exchange in disequilibrium situations.[6] It is however not determined how market participants find beneficial trading opportunities. For the moment it is assumed that the 'contracts' concluded between the market participants are binding, implying that the outcomes are stable.[7] This model serves the object and purpose of explaining allocative efficiency, Pareto optimality, social welfare and to examine the relevance of initial allocations in the context of the EU ETS. The model described below resembles the standard microeconomic textbook example of a general equilibrium model that is attributed to the Edgeworth's Mathematical Psychics written in 1881 (1845–1926).[8]

Due to the object and purpose of this research, the theoretical complexity of the presented model is limited and restricted to a verbal description – the choice of not using formulas was deliberate.[9] It reflects an economy which is characterized by a production side and a consumption side. Each will be described in turn. Subsequently a third criterion, efficiency in the production mix, establishing the full equilibrium between the two sides will be explained too. Subsequently the model's implications when factors of production are allocated to producers are described in order to outline the relevance of initial allocations in the presence of market imperfections and a linkage to the EU ETS is made.

Production side:
In our small economy, the production side consists of two firms (firm F and firm S) that can employ a given amount of capital (K) and labour (L) as production factors to produce two different goods, food (F) and shelter (S). It is assumed that the

4. See Kaldor, N., (1934), 135.
5. The possibility of unstable or locally stable equilibria was already noted by Walras, see Maks, H. & van Daal, J., (2002), 9.
6. See Uzawa, H., (1962), 219.
7. For an elaboration see Uzawa, H., (1962), 219. An underlying assumption is that the process has a positive solution and starts with a positive initial endowment.
8. This presentation is partly based on Cullis, J. & Jones, P., (1998), Ch. 1.
9. The interested reader be referred to a standard microeconomic text book, as for example Katz, M. L. & Rosen, H. S., (1998), Frank, R., (1997), Ch. 16 but also Mamuth, H. A., (1992), Ch. 3.

Figure 1 Edgeworth Production Box

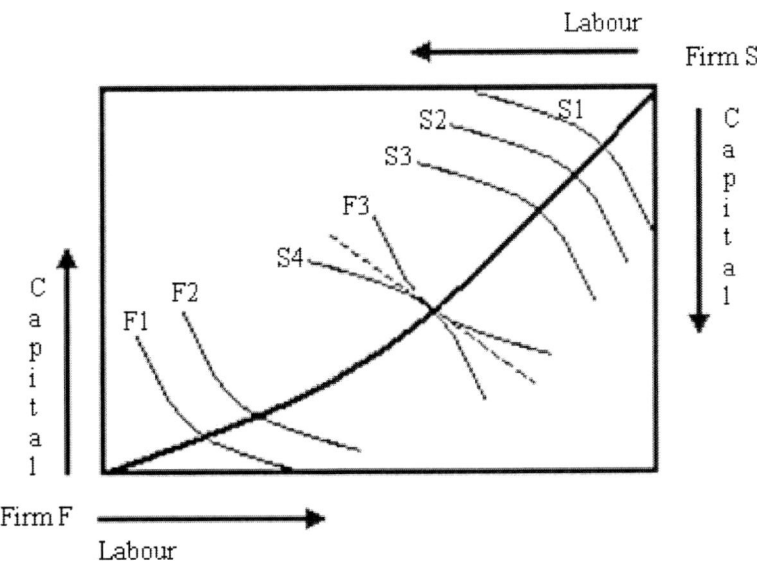

firms' production function, their ability to turn inputs into outputs, are constant or yielding decreasing returns to scale. A so called 'Edgeworth production box' can be constructed to understand the requirements for efficiency in production. Since the length and the height of this box represents the total quantity of factor inputs available in this economy, each point within it depicts a possible allocation of production inputs to firm F (producing food) and firm S (producing shelter). The total amount of inputs used by the particular firm is read from its origin; these origins are situated on opposite corners of the box. The origin of each firm is also the starting point of the firm's isoquant map. An isoquant map is a continuous mapping of isoquant curves[10] which represent the varying combinations of inputs that can be used to produce a given output. Here isoquants are assumed to have a convex shape. The further isoquant curves are shifted outward, that is, away from the respective origin, the more factors of production are available for production of the good and hence the more output will be generated. This is indicated by the increasing quantity of food and shelter that denotes the isoquants that are furthest away from their origin, as depicted in Figure 1.

The closer the food and shelter isoquants approach each other, the more profit opportunities can be realized by the firms and the smaller the differences in the firms' subjective valuation for factor inputs. If the isoquants are tangent to each

---

10.   Besides completeness, other underlying assumptions include transitivity and 'more-is-better'. Transitivity means that if A is preferred to B, and B is preferred to C, that necessarily A is preferred to C as well. The 'more-is-better' assumption is a simplifying assumption that more of a desired good is always preferred.

other, the production process of neither firm can be improved without deteriorating the production process of the other firm. In this case the two isoquant curves have identical slopes and both firms have an equal valuation for input factors. Economists say that here the marginal product of a factor of production (called the marginal product of labour (MPL) and the marginal product of capital (MPK)) is equal to its price (prices of the production factors are wage (w) for labour and capital rent (r) for capital). Furthermore, at such a point of tangency of the isoquant curves the marginal rates of technical substitution (MRTS), defined as the rate at which one input can be exchanged for another without altering the total level of output that is, the slopes of the isoquant curves, are equal.

Hence one can technically summarize the equilibrium on the production side as follows:

$$
\begin{aligned}
\text{MRTS}^F{}_{KL} = \text{MRTS}^S{}_{KL} &= (-)\text{MPL}^F/\text{MPK}^F \\
&= (-)\text{MPL}^S/\text{MPK}^S = (-)w/r
\end{aligned}
\tag{1}
$$

Thus the production side is in equilibrium if no further mutually beneficial exchanges of factors of production are possible. This is called Pareto optimality. Such Pareto optimal situations can be brought about through the interaction of individual market participants as described above, or through the involvement of a planner or an auctioning mechanism. An equilibrium produced by competitive markets will exhaust all possible gains from trade.[11] Two further aspects concerning Pareto optimality are of crucial importance. Firstly, allocative efficiency is also attained at a Pareto optimal point because the producers' valuation is in line with the allocation of resources. Those who value a production input most do have it. Secondly, in the presence of perfect competition on all markets, throughout all time periods and for all possible situations of the economy, the general equilibrium will give rise to a Pareto optimal allocation.

Before starting to discuss the equilibrium on the consumer side, an additional point has to be noticed. As stated above, the isoquant mapping for both products, food and shelter, is continuous, there are in fact many points of tangency. These can be summarized in the so called contract curve, depicted by the thick black line linking both origins in Figure 1, which represents all points of efficient allocations.

Consumption side:
Analogous to the above explained production side, a consumption side with two consumers, seeking to consume all products that are produced in this economy, is created. For simplicity any form of savings is assumed away. Here also an Edgeworth box is depicted (see Figure 2). This time, however, not with producers but with two consumers (A and B). The length and the height of the Edgeworth consumption box represents the total quantity of goods (food and shelter) available in

---

11. This is also called the first fundamental theorem of welfare economics. See Frank, R., (1997), 564.

Figure 2 Edgeworth Consumption Box

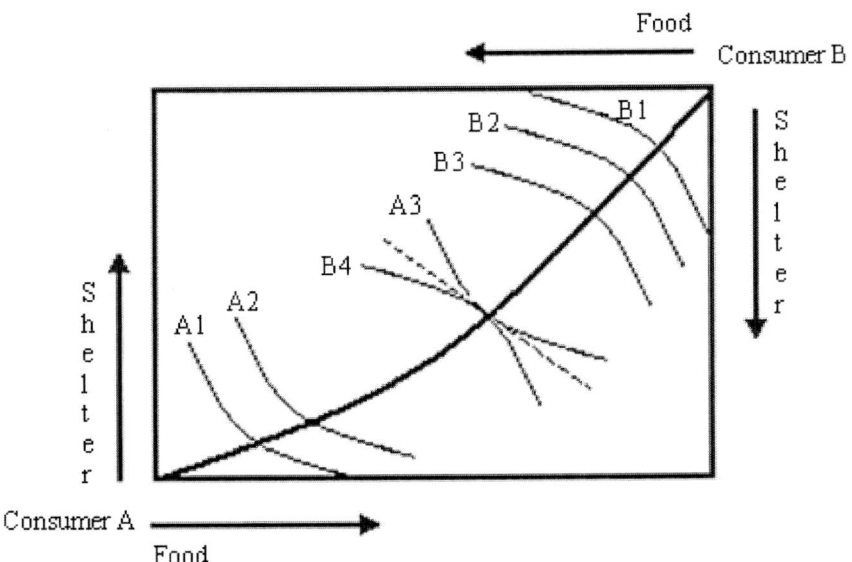

this economy. Each point within it depicts a possible allocation of goods to consumer A and consumer B. As was the case on the production side, the total amount of goods consumed is read from the origins situated at opposite corners of the Edgeworth consumption box. The origins are also the starting points of consumer A's and consumer B's utility map,[12] respectively. Utility curves, also called indifference curves, represent the varying combinations of consumption bundles that can be used to attain a particular level of consumer satisfaction. Here it is assumed that the utility curves have a convex shape. The further the utility curves are shifted outward, that is, away from their origin, the more goods are available to the consumer.

The closer the indifference curves of consumer A and consumer B approach each other, the smaller the differences in the consumers' subjective valuation for goods. If the indifference curves are tangent to each other, the utility of neither consumer can be improved without decreasing the utility of the other consumer. In this case the two indifference curves have identical slopes and both consumers have an equal valuation for the goods. Economists say that at such a point of tangency the marginal rates of substitution (MRS), defined as the rate at which a consumer is willing to exchange food for shelter, are equal for both consumers. Since the marginal rate of substitution represents the willingness to exchange one good for another it can also be described by a relative price ratio (P_F / P_S), the price of one good in terms of another. This also implies that the marginal utilities (MU) that the consumers derive from food and shelter are the same.

---

12.   Here again, underlying assumptions include completeness, transitivity and 'more-is-better'.

One can technically summarize the equilibrium on the consumer side as follows:

$$MU^A{}_F / MU^A{}_S = MU^B{}_F / MU^B{}_S = MRS^A{}_{FS}$$
$$= MRS^B{}_{FS} = (-)P_F / P_S \qquad (2)$$

When both the MRS of consumer A and consumer B are equal, no further mutually beneficial exchanges can occur because there are no differences in utility that can give rise to (further) exchanges. This is called Pareto optimality. Such Pareto optimal situations can be brought about through the interaction of individual market participants as described above, or through the involvement of a planner or an auctioning mechanism. Two further aspects concerning Pareto optimality are of crucial importance. Firstly, allocative efficiency is also attained at a Pareto optimal point because the consumers' valuation is in line with the allocation of goods. Those who value a good most do attain it. Secondly, in the presence of perfect competition on all markets, throughout all time periods and for all possible situations of the economy, the general equilibrium will give rise to a Pareto optimal allocation.

Since the utility curve mapping for both consumers is continuous there are many points of tangency. These can be summarized in the so called contract curve which represents all points of efficient consumption.

Efficiency in the production mix:
We have seen that both the production and the consumer side will be in equilibrium if no further mutually beneficial exchanges are possible anymore, that is, when the allocation of scarce resources is Pareto optimal. Yet, even if both sides are in equilibrium, the economy could still perform poorly when contrasted to satisfying the needs of its members.[13] Therefore a third criterion, namely efficiency in the production mix, is required. To attain such an efficient production mix, both sides need to be linked together. This is achieved via the so called production possibility frontier (PPF).

Every given point on the contract curve of the Edgeworth production box represents a particular production mix that can be produced within the economy. The quantities can be transposed into a PPF that represents the set of all possible output combinations that can be produced with the given factor endowments and given technology. The slope of each point of the PPF is called the marginal rate of transformation (MRT) and is defined as the rate at which one product can be exchanged for another one. At any particular point of the MRT the slope is equal to the ratio of the marginal cost of food (MC_F) and the marginal cost of shelter (MC_S) and consequently also equal to the relative price ratio of food and shelter.

$$MRT_{FS} = MC_F / MC_S = (-)P_F / P_S \qquad (3)$$

---

13. Frank, R., (1997), 567.

Figure 3 Production Possibility Frontier

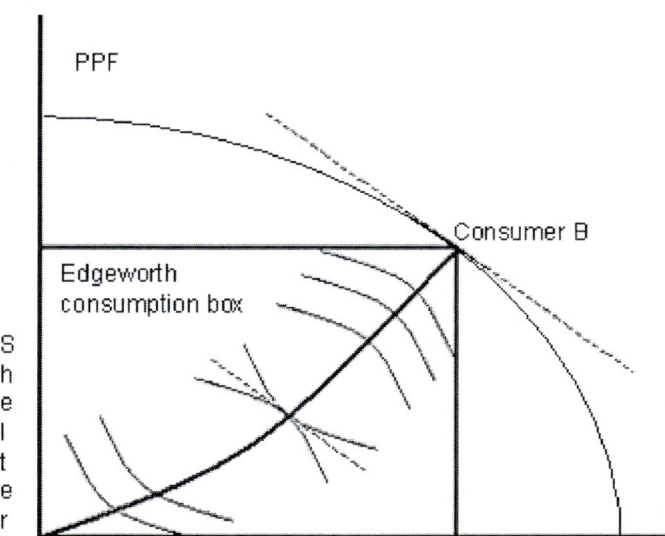

An economy can only be in equilibrium if the MRT is equal to the MRS. This is depicted by the two dotted lines in Figure 3.

$$MU^A{}_F / MU^A{}_S = MU^B{}_F / MU^B{}_S = MRS^A{}_{FS}$$
$$= MRS^B{}_{FS} = (-)P_F / P_S = MRT_{FS} \tag{4}$$

If for example the MRS for a good was larger than the MRT, this would imply that a consumer would be willing to pay more for a product than it costs to produce it. Hence there would be excess demand in the economy and the production mix would not be optimal.

As presented above, in a competitive general equilibrium model, both the production- and consumption side will be in equilibrium and give rise to the optimum product mix. Yet it should be noted that a society that has reached its competitive general equilibrium is not necessarily a good society nor necessarily a happy one. Pareto optimality and hence allocative efficiency is logically separable from distribution (equity). Whether society is better off under a situation in which a more equitable distribution prevails depends on its preferences. Since the economic analysis of societal preferences is quite complex and beyond the scope of the text it is assumed that society is best off if allocative efficiency is maximized that is, if an undistorted production point on the contract curve is reached.

The above finding of a competitive general equilibrium is, however, based on a series of assumptions including a given technology level and given initial endowments. In spite of the beauty of theoretical models, a model's explanatory powers have to be contrasted to the real world. The model described above does not specify the equilibrium adjustment process in detail but assumes that barter exchanges of initial endowments take place when ever one market participant can be made better off without the other one being made worse off. While it is one element to assume that individuals will only trade if they stand to gain from it, it requires much stricter assumptions for such exchanges to actually take place. The occurrence of such trade requires a very strong degree of information dissemination particularly if one allows for a larger number of market participants.[14] In addition such information and exchange processes would have to be costless (no transaction costs) in order to allow the rapid attainment of a market equilibrium. Both assumptions may be questioned.

In the model presented above, the set of feasible market equilibria (on the product market) is determined through the initial factor endowments and the existing production technology (the ability of firms to turn factor inputs into outputs is given by the isoquant curves). The trading path is confined to all those allocations that constitute a Pareto improvement *vis-à-vis* the initial endowment. Within that range the profit maximizing firms will observe the prevailing relative price ratio that equates market demand and supply and an equilibrium is established on the contract curve. An assessment of the adjustment process on the production market if an initial endowment is not supported by the initially prevailing relative price would require more detailed assumptions. Elements to be considered could include consumers' reactions to substantial quantity changes of the goods available, the nature of the produced goods (complements or substitutes) and the accuracy of producers to predict market demand (and failure to realize their ability to impact prices). Depending on these assumptions adjustment processes that in the above model were implicitly assumed to be immediate would require careful modelling and possibly extend over several trading periods. Such situations are thus not easily described by the simple model presented above.

Despite the above shortcomings this simple model allows to gain theoretical insights to better understand factors that distort a general equilibrium. These include, amongst others, taxes, monopolies and externalities. Each will be discussed in turn.

A tax levied on food on the production side for example has an effect on the relative price ratio. It makes food relatively more expensive since producers are able to keep only a part of the price while the remainder they have to transfer to the government. Producers basing their production decisions on net prices will have an incentive to produce more shelter and less food. Since consumers still base their consumption decision on gross prices, and if their preferences remain unchanged, there will be excess demand for food. Thus, a tax drives a wedge between price ratios of consumers and producers with the natural result that the product mix is suboptimal even though both the consumer and producer side can operate efficiently. Efficiency on the consumption side is restored through a change in the relative price ratio making food relatively more expensive than shelter.

Unlike companies operating in a perfectly competitive environment, monopolists recognize the effect their pricing behaviour will have on their profits. They will therefore not set prices equal to their marginal (production) costs but rather in such a

---

14. For a review of this issue the interested reader is referred to Madden, (1978).

way as to maximize their revenue.[15] The effect is analogous to the one with taxes. Here as well, too little of the monopolist's good is being produced and consumers are willing to buy more of a good than is being produced. Eventually a change in the relative price ratio will bring the economy into equilibrium. The same is true for cartels operating on the production market that collude to restrict output in order to reap additional profits.

Economists describe costs or benefits affecting other persons than those taking action as externalities. Pollution is an example for a negative externality while the beautiful flowers in a local fore garden is an example for positive externality. The economic aspects of externalities are treated in more depth in connection to the European Emissions Trading System discussed below. Therefore the author will limit himself here to highlight their implications for the operation of a simple general equilibrium model. Since the prices operators on the production side take into consideration when taking their production decisions do not reflect all costs or benefits to society at large, there is either too much or too little being produced. Consequently, also here there is a wedge between what consumers demand at a given price and what they are able to buy. Hence, the production mix is suboptimal.

After having introduced the above, it is appropriate to relate the preceding treatment to the EU ETS and consider the relevance of initial allocation. Initial allocation of emission allowances describes the insertion of endowments onto the production side. Emission allowances can thus be described as a factor of production required to produce output.

Initial allocation will be of relevance if it distorts competition on the merits. If initial allocations under the EU ETS allows some producers to recognize the effect their pricing behaviour will have on their profits they may find it attractive to charge above perfectly competitive prices. This could, for example be attempted through cartelization, as has been discussed above.

If an initial allocation would favour one producer more than another, substantial re-distribution between operators would be required. Depending on the assumptions about the credit rationing of operators and financing costs this could undermine competition on the merits. While the mere re-distribution would not be problematic, the change in production prices between operators could lead to differences in production costs and hence to changes in the quantities operators will be able to profitably sell on the market. Differences in goods supply and demand will impact the price ratio.

The price ratio could also be affected if one allows for different demand elasticities on the consumer side. If the consumers were for example dependent on a certain product (a 'necessity', inelastic demand) and could not freely adjust their demand as they are able to do for the other good (a 'luxury' good, elastic

---

15.   While a firm under perfect competition will set price equal to marginal costs, a monopolist will recognize the selling of any additional unit of production requires a reduction in the price that can be charged per unit. The monopolist will thus seek to equate marginal production costs to marginal revenue. For a more elaborate treatment see any microeconomic textbook such as Mamuth, H. A., (1992) for example.

demand), the producer of the 'necessity' good is able to pass on a relatively larger share of the production cost increases to consumers then the producer of a 'luxury' good. Such pricing behaviour could impact the relative price ratio of the goods and changes in consumer demand.

If initial allocation systems trigger adaptation processes, it is worth while to consider if possible allocation systems differ in the degree they impact the economy. This becomes particularly important if one assumes that the time required for adaptation is depended upon the magnitude of the distortion. Eventually a change in the relative price ratio will bring the economy into the general equilibrium. Yet in reality adjustment processes – even on emission trading markets that are organized through (energy) exchange houses – require time and allocations based on administrative decisions rather than the market mechanisms can be far off from optimal allocation points.

Looking at the market price of EU ETS emission allowances for the first trading period (2005–2007) it can be observed that the market prices were strongly positive and increasing over a period of sixteen months and only started to decline after publication of governmental reports indicated the presence of a very strong over allocation of emission allowances.[16] The market failed to realize this despite a strong over allocation of probably 160 million ton CO_2 (amounting to an over allocation of around 8%).[17] The market price decreased from above EUR 30 to well below 10 cents per emission allowance unit. Another insightful element is that the information asymmetry from which bureaucrats are suffering at both national and European level was indeed so strong that they were unable to detect such flaws and thereby put the environmental effectiveness of the first trading period into doubt.

In light of the above, one may therefore be inclined to conclude that an allocation mechanism that introduces an initial endowment in the neighbourhood of a later equilibrium and therefore requires relatively less working on the market, could have appealing properties. In such a case a review of existing allocation mechanisms in light of such insights may be well founded. Such an analysis is conducted in Chapter 3 within a partial equilibrium framework.

3   BASIC ECONOMIC INTUITION FOR ABATEMENT AND EMISSION TRADING

The preceding section has reviewed a general equilibrium model and explained the effects of market distortions. This part of the chapter presents on an intuitive level the rationale for emission abatement, emission trading and the importance of initial allocation. For this purpose, a partial market is considered. As allocations of EU Emission allowances are granted to operators and hence the producers of goods and services, the

---

16. For a review of the current EU ETS system see Weishaar, S., (2007d).
17. Data based upon Table 1 of IP/07/1094 of 13/07/2007. The data is not conclusive because it is based upon data from those twenty-two Member States that do have an electronic emission allowance registry.

consumers of products are not of focal interest to understand the effects of allocation in this book. Hence, in the remainder of the chapter a partial market analysis is used as a basis for explaining economic concepts. It is neither claimed that a partial market analysis substitutes a general equilibrium model nor that obtained insights can be directly transposed. Nevertheless it is possible to make inferences from both models and gain a better understanding of the underlying economic concepts.

3.1   WHY ABATEMENT?

Economic theory predicts that if a good is under priced, to much of it will be used. This proposition becomes particularly important if its 'excessive' use reduces the standard of living of other market participants. CO_2 did not have a market price and entrepreneurs did not take into account the 'negative externalities', that is, the negative effects they inflicted upon the environment. This 'market failure' can be overcome by internalizing the negative effects, that is, by bringing the good into the market price mechanism. The basic intuition behind this is that the price of the good private parties pay should be inflated to adequately reflect social costs in order to create incentives to use less of the good. This can be achieved by levying adequate taxes for each level of usage of the under priced good.[18] Figure 4 provides a graphical representation of this. The introduction of a so called 'Pigou Tax', which equals the vertical distance between a firm's marginal production costs (Private Marginal Costs) and the total costs inflicted upon society (Social Marginal Costs), leads to a price increase of the product from P1 to P2. Because consumers demand less at a higher price, the demanded quantity decreases from Q1 to Q2 and brings about a full reduction of the loss to society.

As opposed to the gain from the internalization of these externalities, there are also losses to individual groups of society. Part of the consumer surplus and pro- ducer surplus is redistributed to the State in the form of a tax transfer. While this does not constitute a loss to society, it constitutes a loss to consumers and producers that could negatively impact political support for the introduction of such a tax. There are, however, further effects that need to be considered. First of all because producers' production decisions are shifted upward to the social marginal cost curve, they are loosing part of their former producer surplus, as indicated by the triangle entitled 'loss in producer surplus' in Figure 4. Due to the reduction in overall production a loss to society is created that is not redistributed. This is indicated by the triangle entitled 'welfare loss'. The overall benefit to the society stemming from internalization of the negative external effects depends on the assumptions underling the marginal cost curves and the demand curves. In the case that triangles entitled 'loss in producer surplus' and 'welfare loss' are in sum smaller than the costs of society stemming from the externalities, indicated by the triangle entitled 'Costs to society to be internalized', the internalization will be beneficial. It should therefore be noticed that the losses to producers and consumers

---

18. This referred to as the 'Pigou tax'. See Pigou, A.C., (1949), Ch. 8.

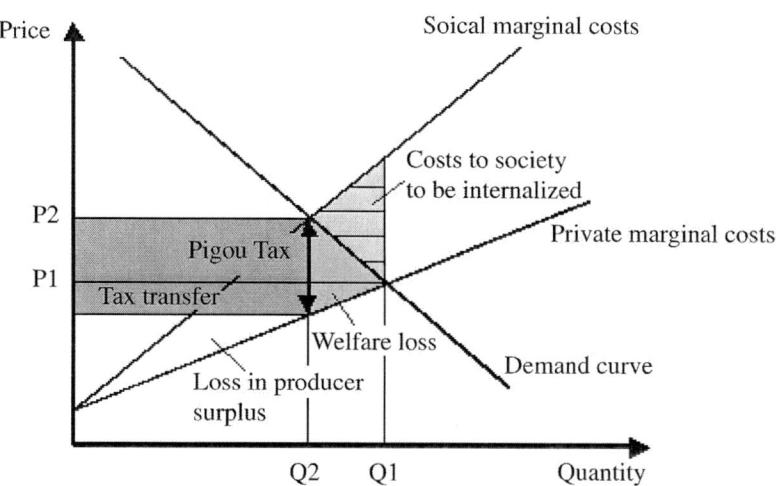

Figure 4 Internalization of Negative External Effects

Source: Based on Pigou (1949), Chapter 8.

could also outweigh the gain to society from the internalization. This however, ignores the possibility of other long run effects that negative externalities – such as for example environmental pollution – may give rise to.

Besides market-based instruments to internalize negative external effects, there are other means to change behaviour of private parties. Private law instruments such as Tort law liability schemes or Public law regulations, which can be sanctioned by Administrative and Criminal law, can be used.[19] The instruments employed in the European CO_2 Emissions Trading System are a combination of Public law rules, administrative sanctions and the price mechanism.

With regard to the introduction of CO_2 emission allowances, a contraction of the Gross Domestic Product (GDP) is expected. Despite the reduction in overall economic output, or rather, precisely due to this contraction, society is better off. The absolute loss to society that stems from the excessive use of scarce resources is reduced to a socially desirable level. Since the GDP only takes into account economically quantifiable data, it does not take into account the destruction of the environment and is thus not a viable measurement instrument for social wealth.

3.2   WHY EMISSION TRADING?

After having reviewed the basic intuition why abatement benefits society, we now examine how emission trading can reduce CO_2 abatement cost to society.[20] The

19. See Faure, M. & Skogh, G., (2003), particularly Chs 14 and 16.
20. Since allocation mechanisms and not emission trading systems are the focal point of interest of this chapter, the author will restrain himself to merely depict the underlying intuition of such trading systems.

Figure 5 Abatement Costs

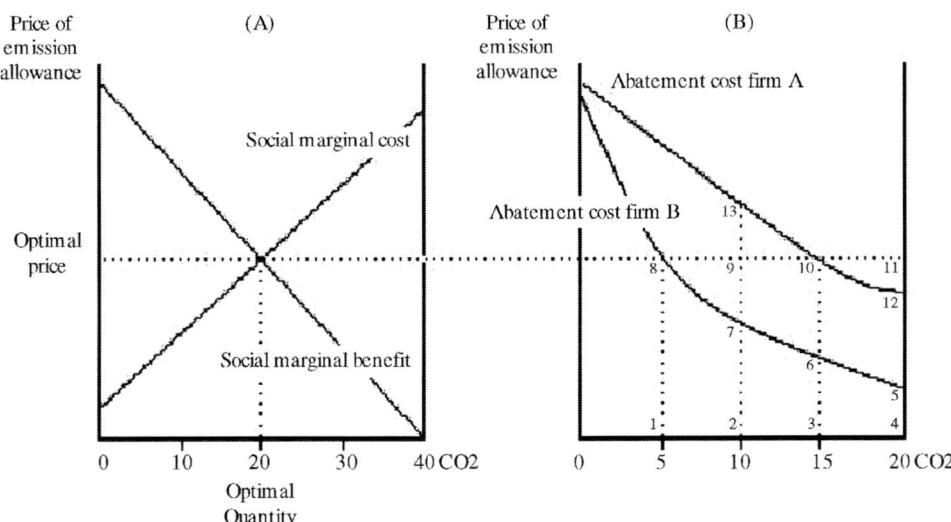

Source: Own representation.

optimal quantity of CO_2 emission should be reduced until Social Marginal Benefits from CO_2 emissions equal Social Marginal Costs from emissions. In Figure 5(A), society's welfare is maximized when emissions are restricted to 20 units of CO_2.

In a 'command and control' setting, society could order firms to reduce their emissions by an equal amount. Figure 5(B) shows CO_2 abatement cost structures of firms A and B which produce 20 units of CO_2 each. Under a command and control system, abatement costs for 10 units of CO_2 of firm A are given by the area 2, 4, 12, 13 and of firm B by the area 2, 4, 5, 7.

An Emission Trading System allows firms to freely exchange CO_2 emission allowances.[21] Firms are thereby enabled to determine themselves which firm will in fact abate emissions. Because it is cheaper for firm A to buy emission allowances from firm B, rather than to invest in abatement technology, firm A has an incentive to buy allowances. If, as in the previous example, allowances are restricted to 20 units, firms will review their marginal abatement costs and engage in trade. Firm B will reduce its emission by 15 units (costs are given by area 1, 4, 5, 8) while firm A will only reduce its emission by 5 units (costs are given by area 3, 4, 12, 10). Such an Emission Trading System generates the same abatement result at a lower cost than a command and control system. This can be seen in the fact that area 2, 3, 10, 13, representing the costs of firm A abating 5 additional units, is bigger than area 1, 2, 7, 8, which represents the costs of firm B abating 5 more units. Thus the overall

---

21. This finding has long been proposed by scholars. For a similar explanation see Tietenberg, T., (1994), 222 ff.

cost savings of the CO_2 emission reduction is the difference between firm A's cost to abate CO_2 units 10 to 15, (area 2, 3, 10, 13) and firm B's cost to abate CO_2 units 5 to 10 (area 1, 2, 7, 8).

According to the European Commission's own assessment[22] the EU's cost of climate policy of the Emissions Trading System will be between EUR 2.9 and EUR 3.7 billion. In the absence of a trading system, the environmental costs would amount to EUR 6.8 billion.

3.3 WHAT IS THE PROBLEM OF INITIAL ALLOCATION?

From a socio-economic perspective, there are four important considerations with respect to initial allocation. These regard price determination, allocative efficiency and hence social welfare and, in accordance with the overall objective of the Emission Trading System – which is the focus of this book –, environmental considerations. These issues will be discussed below.

3.3.1 Why Do We Need Prices?

The economic problem at hand is that an allocation mechanism has to allocate CO_2 emission allowances to firms in such a way as to distort markets in the least possible way. From a social welfare point of view, the operator placing the highest value on an emission allowance should be able to use it. Due to the fact that in reality, true CO_2 abatement cost structures of the individual installations are only known to operators themselves, the true value they place on CO_2 emission allowances is unknown to the allocating entity. In a market equilibrium framework this means that the allocating entity does not know the equilibrium point of the market.

Quite naturally firms would like to buy emission allowances for less than they are worth to them and are therefore reluctant to reveal their true value to the allocating entity. Market prices are able to overcome such problems of information asymmetry where one party knows more than another. In accordance with the law of demand and supply, prices reveal individuals' preferences. In particular, market prices fulfil two important tasks. Firstly, prices serve to redirect existing supplies of a product to those users that value the good most, this is also referred to as the 'rationing function of prices'.[23] Secondly, prices redirect resources from less productive uses to more productive ones. This is called the 'allocative function of prices'.[24]

With regard to the initial allocation of CO_2 emission allowances, however, the problem is that there is no such market price. In order to ensure both the rationing as well as the allocative functions of the price, there are two possibilities. Either the

---

22. European Commission, (2004a), 6.
23. See Frank, R., (1997), 43.
24. *Ibid.*

employed allocation mechanism leads to the revelation of 'true' values of allowances[25] or a well functioning trading system can reduce inefficiencies stemming from the initial allocation of emission allowances. In the parlance of an equilibrium model, one could say that either of the approaches would ensure an efficient outcome on the contract curve on the production side.

The revelation of true values in the perfectly competitive environment of an equilibrium model would not only lead to a final allocation of emission allowances but also to the exhaustion of all possible mutually beneficial exchanges from trade, that is, allocative efficiency. In reality there is of course no perfect information so that even if true preferences were known, trade between operators would still occur to adjust to their production to changes in demand.

An efficient trading scheme, by contrast, would allow all market participants to trade emission allowances at minimal costs. In the presence of a trading system with sufficiently low transaction costs, an efficient resource allocation will result irrespective of how allocation mechanisms distribute emission allowances.[26] Thus if an operator values an allowance higher than another operator that has been awarded a permit, both would engage in trade.[27] Hence, the absence of a price does not have to create severe economic obstacles to the selection of an initial allocation mechanism from an allocative efficiency point of view, as long as emission allowances can be exchanged with sufficiently low transaction costs. If, however, markets are not adjusting quickly, allocative efficiency on the production side may only be attained over time. A suboptimal production mix would result that temporary bares the attainment of an equilibrium. Distortions of competition stemming from suboptimal allocation can also distort relative prices and prolong market adjustments.

To summarize, the relevant criteria to be assessed are the revelation of operators' valuations of emission allowances (price determination), whether those market participants who value the CO_2 emission allowance most will be able to attain it (allocative efficiency) as well as adverse effects on the efficiency of the Emission Trading System. In principle, if either an efficient price determination mechanism or an efficient allocation of emission allowances can be ensured, the absence of a price will not per se constitute a problem from an allocative efficiency point of view. Yet an allocation mechanism that leads to a true revelation of preferences will be able to attain a market equilibrium more quickly.

3.3.2   Allocation

The preceding section has shown that CO_2 emission allowance allocation mechanisms can be equally desirable from an allocative efficiency point of view in the long run, but they do not necessarily lead to the same product mix in the short run. When an allowance with no intrinsic value is introduced into the market price

---

25. Auctions can lead to such desirable revelation of private values.
26. Coase, R. H., (1960).
27. Provided of course, trading partners are sufficiently informed.

mechanism, real value is created[28] due to market participants' willingness to attribute value to such an allowance. Windfall profits accrue to the issuing entity if emission permits are sold to the market participants. If allowances are allocated free of charge or below their true market value, windfall profits accrue to the recipients. The discussion who should be awarded a windfall profit can be led on a welfare economic level or on a social choice level.

With respect to the welfare dimension of this question, there may be a strong rationale to allocate wealth to those groups of society who invest most wisely[29] or to those who create the strongest impulses for economic growth.[30] Costs and benefits should be compared and resources should be allocated in such a way as to maximize social welfare.

From a social choice point of view, however, there may be strong normative grounds to prefer a different allocation. Society may refer to normative principles such as the 'polluter-pays' principle in order to determine how emission allowances should be distributed.[31] Besides such general principles, society could decide that consumers, producers or taxpayers should benefit the most from an allocation mechanism. Or society could, for example, decide that CO_2 emission allowances should be allocated to small and medium-sized enterprises in order to reward them for their labour-intensive production.[32] In such a case large operators, that are not awarded allowances, will have to acquire them. The monetary transfer constitutes a pecuniary, that is, a redistributive, effect[33] to the extent that it does not have a bearing on the overall social welfare. To the extent that such measures do affect the size of the overall economy[34] or distort competition on the merits and lead to a different product mix, they create undesirable effects from an economic point of view.

If particular firms may be put in a position of comparative disadvantage because their competitors benefit from a subsidy in the form of CO_2 emission allowances which effectively reduces their production costs, or if their production

---

28. This is analogous to the concept of 'seigniorage' when a government exploits the monopoly power of the central bank to create money as a means of raising real resources.
29. Here industrial policy issues come to mind.
30. A 'multiplier effect' denotes a phenomenon whereby some initial increase or decrease in the rate of spending will bring about a more than proportional change in national income.
31. For an early statement of the strong acceptance of the 'polluter-pays' principle see OECD (1974). In 1992 this principle was adopted at the UN Conference on Environment and Development and found its place among the soft law principles of environmental law mentioned in the Rio Declaration, see UN (1992), principle 16. The 'polluter-pays' principle is also inscribed in Art. 174(2) of the EC Treaty. Even though the 'polluter-pays' principle seems to be gaining stronger international recognition, its content remains abstract, often ambiguous and subject to divergent interpretations. For an assessment of the 'polluter-pays' principle and emission trading systems see Nash, J. R, (2000). The discussions regarding the 'polluter-pays' principle are beyond the scope of the research question and are therefore scarcely addressed.
32. It may be noted that many economists are critical of such arguments. Economists tend to emphasize overall social welfare and production efficiency and tend to prefer to leave such normative standpoints to politicians.
33. Cullis, J. & Jones, P., (1998), 133.
34. Through, for example, multiplier effects.

costs are unduly inflated, this may lead to strong market distortions and even adversely affect the Common Market. In this context State aid implications of initial allocation schemes under Article 87 EC Treaty and distortions of competition created by Member States falling within the ambit of Articles 81 and 82 are relevant and addressed in Chapters 4 and 5.

To summarize, the overall social welfare should be as large as possible, and mere redistributive effects of initial allocations can be 'rationalized' by social choice arguments. To the extent that market distortions caused by certain social choices give rise to competitive disadvantages and distort the market equilibrium, however, they should be subject to severe scrutiny.

3.3.3   Environmental Considerations

Even though scholars have been arguing that the initial allocation of emission allowances as such does not affect their use,[35] the above discussion has shown that initial allocations are important. If operators are maximizing their profits and the Emission Trading System works efficiently[36] reallocation of emission allowances will occur. In such circumstances market participants valuing an emission allowance the most will attain it, irrespective of its initial allocation, and give rise to allocative efficiency. From an environmental point of view, however, it does not matter who will eventually 'use' the emission allowances, that is, in which sector of the economy CO_2 is produced. It is the atmospheric CO_2 concentration dependent on the absolute amount of CO_2 emission and the timing of emissions which is of importance in terms of environmental effects.[37]

4   OVERVIEW OF INDUSTRIAL ECONOMICS

4.1   INTRODUCTION

After dealing with a equilibria the rationale behind emission abatement and the effects of initial allowance allocation, this section addresses the economic field dealing with competition: industrial economics. It serves the purpose to give a general introduction without striving for completeness.[38] This section of the chapter is subdivided into three interrelated parts. The first introduces the field of industrial economics. It will review the Structure-Conduct-Performance paradigm and presents some of the empirical findings regarding the link of industry structure,

---

35. Tietenberg, T., (2002), 3.
36. This again implies that transaction costs are sufficiently low so that the operator with the lowest abatement costs has every incentive to sell its CO_2 emission allowance on the market, rather than using it. Again this is an application of the Coase theorem.
37. Unlike other greenhouse gases CO_2 emissions do not give rise to a specific local effect. See STEM (2005), 15.
38. The interested reader be referred to other texts that give a more elaborate introduction such as Peepercorn, L. & Verouden V., (2007) but also Mamuth, H. A., (1992).

barriers to entry and profitability of enterprises. In the course of a historical over-view, the Chicago school will be introduced before reaching the new industrial economics approach. The second part of the section will depict the basic concept of social welfare costs from a simple static model of monopoly. Thereafter economic concepts of abuse, cartelization and mergers and acquisition and their implications for the general public are reviewed.

4.2   OVERVIEW

This part will give an overview of industrial economics as an integral element of the framework of this book. It will give a brief historical overview, dealing with both the Standard-Conduct-Performance paradigm and the Chicago school of thought and review relevant statistics.

The field of interest of industrial economics is the behaviour of firms on their specific markets. It regards the study of policies of firms towards their rivals and their customers and thus embraces prices, advertising, research and development. As both competitive and less competitive firms are examined, industrial economics is strongly related to microeconomics and in particular price theory.[39] While ele-mentary courses of micro economics focus on simple market structure models of perfect competition and monopolies, industrial economics has a strong focus on the myriad of structures between those two extremes,[40] and can thus be seen as being particularly interested in market structures as found in the real world.[41] As such it is also concerned with government policy regarding antitrust policy, regulation, and public ownership as well as with the determinants of firm behaviour, scale and scope of business organizations.

While antitrust legislation is intrinsically normative, industrial economics can be viewed as a positive[42] complement to determine which outcomes are superior in the presence of imperfect knowledge, product differentiation, transaction costs, ownership integration, research, development and innovation, and contractual rela-tions such as tying, resale price maintenance, franchizing, exclusive dealing and joint ventures.[43] Perfect competitive outcomes, the state where social welfare[44] is

---

39. Stigler, G. J., (1968).
40. That is if one would like to imagine a typology of market structures drawn along a linear continuum between the two poles of perfect competition, implying the maximum of consumer surplus, and a monopoly which can (but doesn't have to) yield the largest possible producer surplus.
41. Shughart II, W. F., (1990), 1.
42. For a clarification of the distinction of positive and normative approaches, the interested reader be referred to Friedman, M., (1953), 3–43.
43. Shughart II, W. F., (1990), 1.
44. Defined as the state when individuals are collectively as well off as they can possibly be that is, both allocative efficiency and productive efficiency are established. Though it shall be noted that this also entails some value statements, which are, however less arbitrary than other nor-mative criteria and are thus apt for economic analysis.

maximized,[45] are frequently taken as a benchmark to measure the desirability of market outcomes. Comparing welfare effects enables economists to give policy advice on, for example, antitrust legislation and regulation and to determine the effects of industry structure and business practices on social welfare.

At the root of industrial economics lies a methodological debate concerning the relationship between theoretical and empirical analysis, which even today gives rise to heated discussions.[46] Due to a growing dissatisfaction with the explanatory abilities of existing price theory, industrial economists of the 1930s were inspired by works of Chamberlain, who linked price theory tightly to enterprises and the real economy, that is, to concentration, product differentiation, and its legal side, to collusion, trade practices and barriers to entry.[47] Empirical analysis was put to practice in order to supplement economic theory.[48] Before Bain's first industry cross-section analysis of 1956,[49] case studies were common. In the late 1960s but also in the 1970s, larger cross-section analysis of industry data, often based on governmental census bureau data, appeared. The Structure-Conduct-Performance paradigm constituted the intellectual framework of these analyses.[50]

The Structure-Conduct-Performance paradigm[51] relies on casual observation of industries, institutional descriptions and metaphors linking[52] the industry's structural characteristics to firm's conduct and to economic performance. Structural determinants embrace the number and size distribution of sellers, the number and size distribution of buyers, product differentiation, entry conditions, cost structures, vertical integration, and conglomerates. The conduct dimension consists of pricing behaviour, product strategy, collusion, research, and product as well as process innovation, advertising, and legal tactics. Performance indicators are the firm's profitability, production, and allocative efficiency, dynamic efficiency (i.e., the rate of technological progress), employment, and equity. While the linkages among the three elements,[53] structure, conduct, and performance, are often viewed to 'flow in all directions',[54] industry performance is often viewed to be linked to its structural determinants for it is assumed that firm's conduct is causally related to industrial structure.

A particular focus of empirical research in the Structure-Conduct-Performance (S-C-P) context focused on the relationship between market structure, especially

45. See Debreu, G., (1959).
46. On earlier accounts see Morgan, T., (1988).
47. Bain, J., (1949), 130.
48. Mason's work, rejecting the development of theory for its own sake without proper empirical founding, is often regarded as being trend setting in this discipline. See Mason, E., (1939).
49. See Bain, J., (1956). Bain describes the market structure and the performance of large enterprises in twenty industries and analyses the relationship between industry idiosyncrasies and differences in performance.
50. Martin, S., (1993), 5.
51. Often referred to as the Harvard School of Industrial Economics.
52. See Bain, J., (1959), 36–38, 295–301, 310–315.
53. Martin, S., (1993), 8 speaks in this context of 'informal theoretical arguments' employed to link these elements.
54. Krouse, C., (1990), 416.

the degree of concentration, barriers to entry, and excessive prices and profits.[55] Despite considerable evidence for a weak positive relationship between profits and concentration and a somewhat stronger relationship between profits and entry barriers, not all analysis supports this finding. Studies conducted regarding small open economic settings,[56] did not find support for this relationship.[57]

Other critique of such positive correlation of concentration, barriers to entry and price, regards the use of accounting data, which is often used as an input in profitability analysis. According to Fisher and McGowan (1983),[58] accounting profit is not equal to economic profit and thus an imperfect basis for measurement. Analysing industry data of 1953–1957 Brozen (1971)[59] criticizes Bain's hypothesis concerning the long run equilibrium relationship[60] between concentration and profitability in Bain's sample (1936–1940).[61] Comparing both results, Brozen concluded that the convergence of both, above average profits of concentrated industries and the convergence of below average profits of unconcentrated industries, was attributable to a disequilibrium situation of Bain's sample.

As Martin (1993)[62] points out correctly, since both authors share in common the absence of independent tests for equilibrium situations of their data, either of them[63] could be mistaken. This example underlines the importance of independence tests of the equilibrium or disequilibrium nature of the used data. Another important critique, often associated with Demsetz, of the positive relationship between concentration and profitability is based on the argument that this relationship reflects the greater efficiency of large-scale operations[64] and not the market power of larger firms.

Peltzman (1977)[65] offers yet another explanation to the positive correlation of profits and concentration. Examining 165 four digit Standard Industry Classification (SIC) industries between 1947 and 1967, he finds that cost reductions were stronger in concentrated than in unconcentrated industries. Regressing changes in unit costs of concentrated markets and total revenues on industry price indexes he found that even though not all cost reduction was passed on to consumers, the net effect of concentration was to reduce prices substantially. Hence profits rise when concentration rises, not because prices increase, but because they fall more slowly

---

55. See Weiss, L. W., (1974).
56. See Pagoulatos, E. & Sorensen, R., (1976).
57. Mamuth, H. A., (1993), 311.
58. Fisher, F. M. & McGowan, J. J., (1983).
59. Brozen, Y., (1971), 351–369.
60. Brozen also questions Bains results on the basis of a sample bias in the industry selection and on the basis of data bias from firm selection, in essence entailing the 'efficiency critique' postulated by Demsetz a couple of years later.
61. See Brozen, Y., (1971), 352.
62. Martin, S., (1993), 462.
63. Or as a matter of fact, even both.
64. See the works of Demsetz (Demsetz, H., (1973); Demsetz, H., (1974); Demsetz, H., (1976)) but also Brozen, Y., (1971), 362, who arguably initiated the efficiency based critique.
65. See Peltzman, S., (1977).

than costs.[66] A generally reversed opinion regarding the positive relationship between concentration and prices is voiced by Shugart II (1990),[67] who claims that many cross-section analysis use biased samples, which are confounded by highly regulated industries.

Beginning with the studies of Comanor and Wilson (1967)[68] and Collins and Preston (1969),[69] industrial economists made use of extensive cross-section samples of industry level data and analysed particular determinants of firms' profitability. Comanor and Wilson established that advertising intensity has a stronger and more significant determinant of profitability than market concentration. The price-cost margin approach of Collins and Preston showed that barriers to entry were lower for producer good industries than for consumer good industries since product differentiation was expectedly different between both sectors. In addition to their finding that product differentiation is an important basis for competitive disadvantage, Collins and Preston showed that concentration raises price-cost margins only where small firms are at a competitive disadvantage.[70]

A third kind of econometric studies is based on firm level data. These allow the evaluation of changes of market share and its impact on market performance. Such studies generally conclude that market share has a stronger effect on the rate of return than market concentration.[71]

The Structure-Conduct-Performance framework developed out of dissatisfaction with the theoretical limitations and mere descriptiveness of price theory.[72] Based on informal theoretical arguments the relationship between market structure, conduct and performance was tested by innumerous econometric analysis. Current research in industrial economics, theoretical and empirical is not based on the S-C-P framework but on formal models. Reasons to be identified are threefold. Firstly, despite the largely independent development of the S-C-P paradigm and formal microeconomic analysis of imperfect markets, the evolution of both fields was largely parallel. Secondly, as econometric modelling became more sophisticated, the structural dimension in the S-C-P framework was treated exogenously which made formal theoretical models, such as existed within the realm of formal microeconomic analysis of imperfect markets, a necessary fundament for such structural equations. Thirdly, the strong use of game theory in oligopoly models gave industrial economists a formal tool to analyse strategic interactions to explain market performance. Since oligopoly theory was now capable of answering the questions it was unable to deal with in the 1940s, it replaced

---

66. In critique of this, Scherer postulates that in many industries examined by Peltzman, rising concentration was caused by product innovations (shift along the average cost curve) rather than cost reduction innovation (shift in the average cost curve), see Scherer, F., (1979). Peltzman, S., (1979) returns that a cost reduction was a cost reduction.
67. Shughart II, W. F., (1990), 95.
68. Comanor, W. S. & Wilson, T. A., (1967).
69. Collins, N. R. & Preston, L. E., (1969).
70. See also Martin, S., (1994), 206 and Martin, S., (1993), 468.
71. See Martin, S., (1994), 212 ff.
72. This passage is based on Martin, S., (1993), 8.

the S-C-P paradigm. The early 1970s to early 1980s marked a golden age of theoretical development in industrial economics. While empirical industrial economics and theoretical industrial economics were mutually enriching, divides still exist. Theorists are criticized for producing elegant models that are not applicable to the real world while empirically oriented industrial economists are criticized for conducting analyses with unclear and ambiguous theoretical founding.

The industrial economics school that succeeded the S-C-P paradigm in terms of influence on antitrust legislation in America is called the Chicago school. Among the most prominent supporters rank industrial economists as Bork, Demsetz, Stigler and others and also jurists such as Posner. Chicago school adherents have diverging opinions on several issues but also share common ground. Schmidt and Rittaler (1990)[73] reduce the fundamental understanding of the Chicago school to three aspects. Firstly, the school views the market process as a free interaction of economic subjects, without any governmental interference, as a game of the 'survival of the fittest'. Secondly, the role of the government or other public influence has to be minimized and restricted to the mere setting of the legal framework. Thirdly, the school is often regarded as being liberal-conservative and in favour of big business while being opposed to unions. Quite in contrast to the S-C-P paradigm, which views imperfect competition as appropriate to analyse the real world, Chicago regards perfect competition as having more explanatory capabilities.[74] As the long run competitive equilibrium model to explain real world behaviour is accepted, explanations involving market power and imperfectly competitive markets are rejected.[75] Chicago is often criticized for being excessively theoretical[76] and limiting its own research to criticize studies of the Structure-Conduct-Performance relationship.[77]

While the Chicago school achieved great influence in American political and legal circles[78] from the mid 1970s onwards, the claim that their views dominate academic economics is questioned and certainly untrue for the international debate.[79] The school's appeal to the legal community and policy makers alike is viewed to rest upon it's internally coherent rationale and it's capability to provide satisfactory answers without committing decision makers to explicit value statements.[80]

---

73. See Rittaler, J. B. & Schmidt, I. L. O., (1990), xiii.
74. See Martin, S., (1994), 9.
75. See Reader, M., (1982), 15 following.
76. Hildebrand, D., (2002), 151–152, claims that the Chicago school is too abstract, overemphasising economic theory and market forces, and merely emphasising allocative and productive efficiency while disregarding dynamic efficiency.
77. See Martin, S., (1994), 11.
78. See Rittaler, J. B. & Schmidt, I. L. O., (1990), xi.
79. See Martin, S., (1993), 449, n. 9 and elsewhere.
80. See Sullivan, L. A., (1980), 'Antitrust Microeconomics, and Politics: Reflections on Some Recent Relationships', 68 *California Law Review* 1 ff., 9, via Rittaler, J.B. & Schmidt, I.L.O., (1990), xii.

Contemporary industrial economics is less ideology driven but heavily influenced by game theoretic models. These models often focus on imperfectly competitive markets and their equilibrium and constitute reformation of empirical research.[81] Game theoretic models of the new industrial economics approach analyse strategic behaviour of individuals and firms in conflict situations to generate a conceptual understanding of their interactions and interdependencies.[82]

4.3   Economic Models[83]

The second part of this section will review economic concepts of monopolies, abuse, cartels, and mergers and acquisition. The monopoly section reviews a simple static model of antitrust theory. Associated social welfare considerations will be introduced. The section on abuse examines strategies a dominant firm can follow to maintain its market position and to increase its profits. The social welfare implications of Cournot quantity setting and Bertrant price setting oligopolies are examined in the passage on cartels. Difficulties of cartel creation and cartel stability are treated as well. The passage on mergers and acquisition introduces social welfare considerations of horizontal, vertical, and conglomerate mergers.

4.3.1   A Simple Static Monopoly Model

A monopoly describes a market characterized by a single supplier[84] and blockaded market entry.[85] As the sole supplier, monopoly's output will have a bearing on the quantity demanded.[86] Hence the monopolist equates marginal revenue, the change in total revenue per unit change in the quantity demanded, to the marginal costs of production in order to maximize profits.[87] Assuming the presence of a linear demand function, constant marginal costs and the inability of the monopolists to engage in effective price discrimination, we can simplify reality to fit the model presented in Figure 6.[88]

Monopolistic output will be restricted to Qm while price (Pm) lies above the competitive level (Pc). Taking the perfect competitive output as a benchmark, a part of what was formerly consumer surplus now accrues to the monopolist. Whether this is socially desirable or not is inherently a normative question and not the realm of economics. What, however, is of fundamental interest to economics is the aggregated cost to society which is derived from those consumers who

---

81. See Martin, S., (1994), 12.
82. Bester, H., (2003), 4 ff.
83. This part is largely based on standard industrial economic text books as: Viscusi, K., Vernon, J. & Harrington, J., (1995), Martin, S., (1993), Martin, S., (1994) and Krouse, C., (1990).
84. In contrast to this, a market characterized by a single buyer is called a monopsony.
85. This effectively rules out potential competition.
86. This is of course not the case for perfectly competitive settings.
87. Frank, R., (1997), 385 ff.
88. See Martin, S., (1994), 28.

Figure 6 The Dead Weight Loss

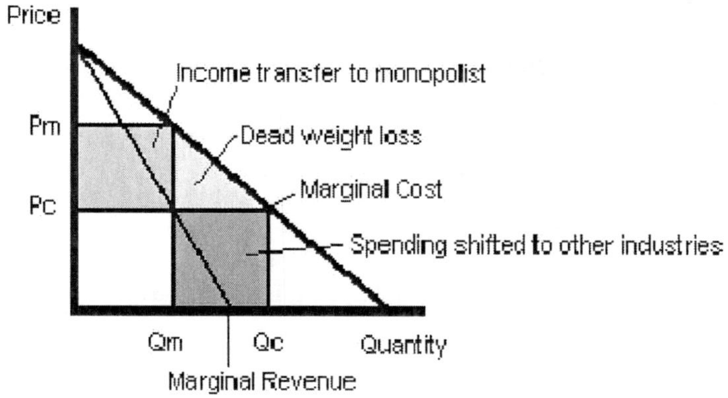

Source: Martin S. (1994) p. 28.

would have bought monopolist's produce at the perfect competitive price but are not able or willing to do so at the increased price level. This cost of unsatisfied wants, due to output restriction is a measurable and welfare reducing cost to society. It is called 'Dead Weight Loss' (DWL).

The first empirical study was conducted by Harberger in 1954.[89] Analyzing the DWL[90] for seventy-three industries, he estimated the negative welfare effects to be 0.1% of National Income for the United States. Critics of Harberger estimated normal rate of return and claimed that the use of average rates of return on capital would lead to the inclusion of monopoly rents in the mean estimate and hence to its overstatement. This in turn would lead to an underestimation of economic profit on sales and thus to an underestimation of the dead weight loss.[91] Empirical estimates of dead weight losses in the Harberger tradition depend crucially on the price elasticity of demand. The higher the price elasticity of demand, the higher the dead weight loss associated with increases in prices. The arbitrary selection of a price elasticity of demand of unity is thus questionable. Dead weight loss calculations without price elasticity of demand[92] estimate it to be as large as 50% of the monopoly profit and thus may constitute an upper bound for the loss to society.

Firm level data, as used by Cowling and Mueller (1978)[93] is expected to circumvent the negative effect that unproductive firms have on the industry

89. Hay, D. & Morris D., (1991), 582.
90. Harberger used an equation to approximate the DWL depending on the rate of return (r), revenue (Price of the monopoly (Pm) and Quantity of the monopoly (Qm) and price elasticity of demand (ε_{QP}) assumed to be equal to unity. Martin, S., (1994), 33 ff.
91. See Martin, S., (1994), 35.
92. See Martin, S., (1994), 34.
93. See Cowling, K. & Mueller, D., (1978), 731.

mean, which is used in Harberger like studies. Using firm level data of 734 US firms during the period of 1963–1966 and 103 UK firms during the period of 1968–1969, they find dead weight losses of 0.4% of corporate products. Calculating two different dead weight losses, one using Harberger's estimate with a price elasticity of demand of unity and a second one by assuming that the DWL is 50% of the monopolistic proceeds, they find that the estimates without any reference to price elasticity are about ten times as large as those in the Harberger's tradition.

Other possible sources which are believed to increase the social welfare costs of monopolies are derived from advertising, rent seeking,[94] and x-inefficiency.[95] While advertising can also be viewed as containing positive social benefits, potentially mitigating the adverse effect, rent seeking is commonly viewed as increasing the dead weight loss by a multiple[96] of the expected monopolistic profit. A third cost of monopolies is generated by the absence of competitive pressure, which is assumed to lead to all sorts of allocative inefficiencies. Scherer (1980)[97] believes that the x-inefficiency could be as large as 1% of the national product. Overall it appears that dead weight losses could be around 2% of national income.[98]

4.3.2   Abuse

Abuse is a term used to describe the action of a dominant firm of taking advantage of its ability to influence market structure.[99] A dominant firm is a firm large enough to recognize that its price increase will drive customers from the market. In contrast to monopolies, dominant firms have to take into account the reaction of other (potential) market participants. The incumbent firm's ability to set prices above the competitive level depends on the difficulty of market entry, that is, sunk costs, its assumed future production behaviour,[100] and the size of the market.[101]

In the presence of sunk costs, average cost curves of incumbents and the fringe firm differ. The incumbent can set market price low enough to make it impossible for an entrant to cover its average costs. A profit maximizing firm is thus effectively barred from market entry. This pricing strategy is called limit pricing. Whether an incumbent will choose to set limit prices, however, depends on the

---

94.   Rent seeking is to be understood as the utilisation of political processes in order to generate economic rents that could not have been generated in ordinary market transactions.
95.   X-inefficiency is to be understood as the wasting of production inputs in process of production.
96.   Depending on the amount of actors involved. See Cullis, J. & Jones, P., (1998), 293 ff. on rent seeking.
97.   See Scherer, F. M., (1980).
98.   See Mamuth, H. A., (1993), 318.
99.   This passage is predominantly based on Martin, S., (1994), Ch. 4 and Viscusi, K., Vernon, J. & Harrington, J., (1995), 164 ff.
100. A credible threat of an incumbent to expand output after market entry would not allow fringe firms to cover their average costs. Such practice is called predation. It would be interpreted as monopolization if it was proven that prices charged by the incumbent firm are below its average variable costs.
101. If the market is growing rapidly, the output the incumbent firm has to supply in order to keep new firms off the market rises.

fringe's expansion rate[102] and the net present value of future income.[103] Profit maximization will decide if a dominant firm will engage in limit pricing.

Other possible ways to maintain a dominant position include mergers, consent to cost increasing social welfare policies,[104] investment in excess capacity,[105] vertical integration,[106] product differentiation by research, and development or advertisement.[107] These are legally accepted means to maintain a dominant market position. They will allow a dominant firm to rise prices above competitive levels to the extent that these constitute a barrier to entry.

Examples of abuse include price discrimination,[108] tying contracts, predatory pricing and (vertical) contractual restraints as resale price maintenance, territorial or customer restrictions. Establishment of 'abuse of a dominant position' depends on the evaluation of market power and Court's interpretation of actions taken to constitute attempts of private monopolization.[109]

4.3.3 Cartels

Unlike monopolists, in real life many entrepreneurs recognize their interdependence with other market participants.[110] Models describing the behaviour of oligopolies generally can be distinguished into two categories. One regarding output setting models and the other one regarding price setting models. Industries which cannot flexibly adjust production schemes due to significant sunk costs, such as the car industry, are well described by the former category. Industries that set a price and sell as much as they can, like the insurance industry for example, are well described by the latter.[111]

If a firm considers its rivals' output levels when taking production decisions, it can assume that rivals will keep their output constant.[112] In the so called Cournot duopoly model each firm assumes that its rivals will maintain its current levels of

---

102. The fringe's expansion rate depends on technological factors and expected profits from expansion.
103. Decisive for the estimation of the net present value of future income is discount rate.
104. Agreement to strengthen worker security or introducing social welfare schemes will also increase the operation costs of fringe firms. This makes market entry more difficult.
105. Idle production facilities allow the expansion of production needed to lower prices if a fringe firm enters. The threat of excess capacity is, however, only credible if the investment is sunk and can not be sold to other industries after a fringe has entered.
106. Backward integration with the objective to coordinate supplies or forward integration to control the distribution channel will set fringe firms at a cost disadvantage and signal that the incumbent is determined to defend its market share.
107. Consumer loyalty will increase difficulty for new entries to compete with the incumbent.
108. This does not suggest of course, that all price discrimination is illegal.
109. For a good Law and Economics treatment of abuse under EC Competition law see Camesasca, P. D. N. & Van den Bergh, R. J., (2006).
110. See Viscusi, K., Vernon, J. & Harrington, J., (1995), 102.
111. *Ibid.*
112. The Cournot duopoly model assumes constant marginal costs and the absence of fixed costs. The duopoly model can be extended to a larger number of firms with analogues outcomes.

output.[113] Hence production decisions are based on the residual market demand. Since all entrepreneurs are assumed to have the same production and decision criteria, their production decision is symmetric and leads to a stable equilibrium in the long run.[114] The interdependence between market participants in a situation where output decisions are imperfectly coordinated causes the market price to be lower and total quantity produced to be higher than under a monopoly.

Extending the Cournot duopoly model to allow a larger number of firms, the long run outcome is driven down towards competitive equilibrium levels.[115] This is particularly true if firms are of equal size. If they are not of equal size[116] and the industry becomes more concentrated,[117] firms will be able to exert (some) market power and raise prizes. The ability of dominant firms to restrict output and raise prices depends on barriers to market entry and the violent nature of smaller competitors[118] to successfully challenge such attempts. But generally speaking, market concentration is associated with market power. While this is certainly exacerbating negative social welfare effects, it shall be noted that Cournot competitors are unable to maximize joint profits. This is due to their failure to take their negative effect of an additional unit produced on their competitors revenue into account. It is the recognition of this failure that gives rise to incentives to collude.

In contrast to Cournot, Bertrand assumed that firms would not compete as much on quantity produced but on prices. Hence Bertrand models emphasize product differentiation and price, rather than market concentration and quantity produced as being crucial determinants of market performance.[119] Producers are assumed to make pricing decisions and sell as much as they can.[120] If products are homogeneous a minimal price difference will lead to the capturing of the entire market. The outcome is thus necessarily the same as under perfect competition.[121]

The stronger products are differentiated, the less elastic is the demand curve around the industry's price average. For completely differentiated products,

---

113. Bester, H., (2003), 77.
114. Even if we assume an oligopoly with a dominant position and a fringe firm, recognition of interaction will lead to a stable equilibrium. This stable Cournot market equilibrium will of course only be stable if one firm decides to produce less than the level which would prevail under a perfectly competitive setting.
115. Hay, D. & Morris D., (1991), 72 ff.
116. Reasons for this may be a difference in costs. This would imply that the firm with the lower costs produces more than the higher cost firms. See Martin, S., (1994), 123.
117. Market concentration is often measured by the 'Herfindahl index'. What should be noticed however, is that even highly concentrated markets can lead to very competitive outcomes in the presence of fierce rivalry. This can be the case if production and development decisions can not be adjusted flexibly and involve considerable sunk costs.
118. Of particularly importance is not so much the act of challenging the dominant firm, but rather the belief of the dominant firm that rival firms will challenge it. The percentage change in competitors output that a firm expects in response to a 1% change in its own output is called conjectural variation. See Hay, D., Morris D., (1991), 62.
119. Bester, H., (2003), 95.
120. Viscusi, K., Vernon, J., & Harrington, J., (1995), 102.
121. Bester, H., (2003), 106.

demand for each product is close to the market demand curve, and the firms can realize similar profits as monopolists. However, the more similar the products, the closer the profits are to those of the competitive outcome.[122]

Which ever model of oligopolies one uses to describe a given industry, the inherent problem of failure to reach joint profit maximizing production outcomes creates incentives to collude. In the following part, determinants of collusion will be considered.[123]

Mere realization of interdependence and the recognition that the profits of every firm could be increased by restricting joint output is not enough to collude. There are several factors to be born in mind. Firm's control over price depends on market concentration, the degree of rivalry, and on product differentiation in the market. Furthermore, colluding parties have to agree on how they reach profit maximization.[124]

If negotiation is costly, reaching an agreement will be difficult. Costs will increase the more participants are in the market. Therefore, the higher an industry is concentrated, the easier it may be to reach an agreement.[125] Differentiated products, particularly in the presence of changing market demand,[126] make it difficult for firms to agree on the necessary means to reach joint profit maximization. Similarly, different cost structures[127] complicate the reaching of an agreement. This will be the case if joint profit maximization necessitates a reduction of production of firms with a higher cost structure. This is exacerbated if the influence of high cost firms on the cartel depends on its market share. Different inter-temporal preferences,[128] as indicated by different discount rates of present and future profits, will make it difficult to reach an agreement, since the desired outcomes for joint profit maximization differ. Furthermore, Competition laws generally forbid overt collusion. Costs of collusion and differences in risk averseness may make it difficult to reach agreements. Collusion can be reached by firms engaging in price leadership, a form of signalling, and by applying generally known pricing rules.[129]

Yet even if collusion is reached, the inherent problem of any such agreement is stability. As long as price exceeds marginal cost, firms have an incentive to increase output in order to increase profits. Since every firm has this incentive, cartels are inherently unstable. In an inter-temporal setting, cartel stability depends on the discount rate of future income. Discount rates must be low enough, that is, firms must value future profits sufficiently high to outweigh short run profits to be

---

122. This passage is based on Martin, S., (1994), 135.
123. See Hay, D. & Morris D., (1991), 75 ff.
124. See Hay, D. & Morris D., (1991), 75 ff.
125. High concentration indices suggest that there are fewer large market participants who need to agree. While this facilitates collusion, this does not imply that concentrated markets do not reach competitive outcomes.
126. Martin, S., (1994), 155.
127. Viscusi, K., Vernon, J. & Harrington, J., (1995), 119 ff.
128. See Martin, S., (1994), 155.
129. See Martin, S., (1994), Ch. 6.

realized from defecting the cartel agreement.[130] In such situations cartel agreements would be beneficial.

Maintaining an agreement on the profit maximizing strategy is easier if there are effective sanctioning mechanisms to enforce cooperation. Excess capacity may indeed lead to such a situation.[131] It lends credibility to the fear that defection of the cartel agreement reduces the expected future profits to competitive levels. Furthermore, vertical integration can allow suppliers to effectively undermine cartel agreements reached in the supplying industry. Vertical integrated suppliers[132] can sell a product at the agreed price to their downstream entity, but since the 'true cost' remains the production cost of the supplier, the downstream entity can undercut prices of its competitors. An increase in the market share and output of the vertically integrated entity reduces the demand of its competitors in the cartelized industry. Thus vertical integration has negative repercussions on the stability of the cartel agreement.

The effectiveness of such sanctioning mechanisms depends heavily on intertemporal preferences of the firms (the discount rate) and on the probability of detection. Concentrated industries find it easier to maintain agreements.[133] This is based on the fact that they are more likely to notice changes in industry output. Since they stand to lose more from a violation of the cartel agreement they have higher incentives to monitor compliance. Detection is easier if sales volatility and changes in customers are relatively low and if sales occur frequently. Detection will also be eased if government agencies publish the bidding results of offers received. Another way to facilitate detection are external organizations who accumulate information about the industry without being directly involved in production such as trade organizations. Firms may be reluctant to reveal sensitive information to it's rivals but may be willing to give information to a trade organization which monitors the market.

4.3.4   Merger and Acquisition

The final element to be discussed in this section is mergers and acquisition. Even though mergers and acquisition are not expressly discussed in this book the topic is briefly addressed because concentrations of firms can give rise to dominant positions that may be inclined to engage in abusive practices.[134]

From a market point of view, mergers and acquisition have the same effect.[135] They combine two formerly unrelated firms. Therefore in the following the term

---

130. The underlying assumption here is of course, that competitors do not engage in a second collusive agreement once they have been betrayed.
131. See Martin, S., (1994), 167.
132. *Ibid.*, 166.
133. This passage is based on Martin, S., (1994), Ch. 6.
134. For a well researched Law and Economics review the interested reader is referred to Camesasca, P. D., (2000).
135. At least when the acquisition is a total acquisition. If the acquisition covers only part of the shares outstanding, the anticompetitive effect can be expected to be smaller if joint action is not established.

'merger' shall be understood as meaning merger and acquisitions. Whether mergers are beneficial from a social point of view is much disputed. Some economic models predict that profits of firms participating in mergers see their joint profit fall. Other models predict that mergers lead to cost reductions[136] that may not be passed on to consumers.[137]

There are three types of mergers. Horizontal mergers describe mergers within one industry. Vertical mergers are mergers in related fields of trade. The third group of mergers is called conglomerate mergers. These are mergers of firms operating in unrelated industries. Each will be taken in turn.

In the presence of constant marginal cost curves and linear demand curves, horizontal mergers of quantity setting firms that become large enough to recognize that their increase in output has an adverse effect on marginal revenue will lead to a reduction of output. A reduction in the quantity produced creates an artificial scarcity and can trigger expansion of fringe firms.[138] The power to restrict output and increase prices depends on the barriers of entry and the ability of fringe firms to expand their output. The higher barriers of entry are and the lower fringe's ability to expand are, the higher are the rewards of a merger.[139]

If firms have downward sloping cost functions, mergers may be beneficial for the merging parties as well as for society. Benefits for society would be generated if the dead weight loss produced by raising prices would be offset by a reduction of production costs due to larger scale operations. Horizontal mergers may be beneficial for firms if an industry suffers from excess capacity and different cost structures.[140] A collusive agreement for a joint profit maximizing strategy is easier to realize if cost structures are less diverse. Horizontal mergers can also be beneficial if an industry is characterized by many equally efficient production plants. In such cases administrative fixed costs can be spread over more output. Young credit rationed firms may benefit from mergers which grant access to distribution channels and financial resources.[141]

Reasons for engaging in vertical integration via mergers and acquisition are based on efficiency gains and strategic considerations. Efficiency considerations will be discussed first. Vertical integration reduces transaction costs[142] of establishing a contract and allows more flexibility in long term contracts and the quality

---

136. While economic literature tries to capture these cost reduction in terms of increasing returns to scale, management literature describes advantages as synergy effects. These can be realized by streamlining superfluous operations. Differences in corporate culture are frequently identified as having a negative effect on the benefit of the mergers.
137. It shall be noted that from an economic point of view the question of distribution of consumer and producer surplus is not the decisive factor. What is of concern is the inefficient resource allocation leading to a dead weight loss which stems from high prices.
138. Viscusi, K., Vernon, J. & Harrington, J., (1995), 207.
139. Martin, S., (1994), 260 ff.
140. Martin, S., (1994), 266.
141. Martin, S., (1994), 267 ff.
142. Martin, S., (2002), 407.

demanded.[143] Vertical integration can reduce the risk premium charged by the supplier and secure loyalty. It makes the firm less vulnerable to pressure from suppliers. Lack of information can also serve as an incentive to integrate vertically. Firms selling to firms with different demand elasticities can price discriminate by integrating to high-elasticity segments of the final market and serve the low elasticity market at a profit maximizing higher price.[144]

Strategic reasons for vertical integration can give rise to concern from a social welfare perspective. If a producer supplies it's output to its vertically integrated division at a lower price than to rival firms in the integrated market, it has an incentive to use its output by itself. Similarly, a vertically integrated distributor may decide not to trade goods of rivals who operate in the same industries. This practice is known as foreclosure.[145] Foreclosure will make simultaneous market entry at both industries necessary. In the presence of transaction costs and a risk premium depending on the size of capital investment, foreclosure may lead to an increased barrier to entry.

The social welfare effects of conglomerate mergers depend on the relation between the relevant markets of the merging firms. Conglomerate mergers, which are of a product extending or market extending nature, can raise concerns. Examples are beverages and restaurants for the former and merger of companies in the same trade but in locally separated markets for the later. This may lead to reciprocity dealings,[146] be conducive to predatory pricing[147] and the reduction of potential competition. Potential competition will be reduced if a company acquires a firm in a market from which potential entrants might have entered the industry. Potential competition is also reduced if a firm acquires a company which is already well established on the market.[148]

Another danger related to conglomerate mergers is the threat of retaliation.[149] If a number of firms compete in several industries in which their market position varies, attacks in one industry can be retaliated in other industries. This is believed to contribute to the reaching and the maintenance of a collusive agreement. Besides the above mentioned strategic considerations, efficiency gains can also constitute an incentive for conglomerate mergers.

---

143. Martin, S., (2002), 406.
144. This passage is partly based on Martin, S., (1994), 270.
145. Martin, S., (2002), 404 ff.
146. Reciprocity dealings describe dealings which require business partners to buy certain products in order to engage in a contract in other fields of business. Viscusi, K., Vernon, J. & Harrington, J., (1995), 216 ff.
147. This would be the case if the firm engaged in predatory pricing can be cross-subsidized through other enterprises. Viscusi, K., Vernon, J. & Harrington, J., (1995), 217.
148. One particular example is cited by Viscusi, K., Vernon, J.& Harrington, J., (1995), 218: Federal Trade Commission v. Procter&Gamble Co. et al., 386 U.S. 568, (1967). Procter&Gamble's acquisition of Clorox, market leader in the American bleach market, reduced potential competition in the bleach market. Through this acquisition Procter&Gamble removed the strongest competitor it would face after its market entry.
149. Viscusi, K., Vernon, J. & Harrington, J., (1995), 216.

5   SUMMARY

This chapter has been laying the economic foundations for understanding the concepts that are being applied in the remainder of the book. Among the reviewed theories the effects of initial allowance allocation and industrial economics are of particular importance since they are frequently used. The treatment of emission trading and abatement is not expressly addressed any more but serves as important background information and fosters a better understanding of this text. For didactical reasons this chapter has not addressed auction theory that is complex in nature but will be treated in subsequent parts of this text. As has been the case with the economic concepts in this chapter, also auction theory will only be reviewed to the extent necessary to give an introduction to a wider academic public while not striving for completeness.

Chapter 3

Allocative Efficiency: A Static and Dynamic Perspective

1   INTRODUCTION[1]

Much has been written about the advantages and disadvantages of marketable permits and Emission Trading Systems. One element of crucial importance that has not yet sufficiently been addressed is the allocation of emission allowances.[2] It is not yet fully understood which allocation mechanisms are most desirable from an allocative efficiency and environmental point of view.

The issue of allocation has been addressed in the debate whether allowances should be auctioned or freely distributed (grandfathered). A good example reviewing the benefits of auctions over grandfathering schemes is presented by Cramton and Kerr (1999).[3] They find that auctions are superior because auction revenues can be recycled to reduce distortionary taxes (double dividend hypothesis[4]), provide incentives for innovation and avoid market distorting awards of politically contentious windfall profits. Despite their enthusiasm, the authors note that auctions may not be a first choice because vested interests are much in favour of free allocation. In contrast to this, Stavis (1997) presents a good discussion why grandfathering has been widely accepted.[5] The author cites greater political control and distributional impacts of free allocation as reasons why grandfathering systems are more readily accepted.

---

1. A version of this chapter will be published in Weishaar, S., (2007a) and Weishaar, S., (2007b).
2. Gayer, T., (2005), 1.
3. Cramton, P. & Kerr, S., (1999).
4. On the double dividend see for example De Mooij, R. A., (1999).
5. Stavis, R. N., (1997).

Some works have explicitly included market distortions in their analysis of initial allocation mechanisms. Parry et al. (1999) apply analytical and numerical general equilibrium models to analyse the efficiency impacts of revenue-recycling carbon taxes and CO_2 grandfathering allocation schemes in the presence of pre-existing distortionary labour taxes.[6] For such an environment they find that the tax interaction effect (stemming from higher output prices and falling real wages' impact on labour supply) considerably inflates the efficiency costs of CO_2 abatement policies, in particular for grandfathering allocations which does not generate government funds. They suggest that revenue-recycling could be a necessary condition for CO_2 emission abatement policies to enhance social welfare as long as the environmental benefits of CO_2 abatement are positive.

Frequently, comparisons between initial allocation mechanisms are examined within a closed economy setting. Some authors have, however, also examined implications of initial emission allowance allocation systems at international level. Helm (2003) for example recognizes the problem that there is no central authority charged with the power to determine initial allocation of tradable emission allowances on international level.[7] In his article the author compares endogenous choices of tradable and non-tradable emission allowances by countries which are participating in an Emission Trading System and finds that environmentally concerned countries tend to choose fewer allowances under a trading system. This positive effect may, however, be offset by incentives of less environmentally concerned participants to demand more tradable allowances. Maeda (2003) examines the implications of market power and initial allocation between participants of the Kyoto Protocol.[8] The author finds threshold levels that give rise to competition distorting market power. Literature building upon international trade theory will be reviewed in more detail below when an open economic model is introduced.

Academic articles addressing the relative standards as a basis for initial permit allocation are still scarce. Gielen, Koutstaal and Vollebergh (2002) compare emission trading with absolute and relative targets in a partial equilibrium model with an absolute emission cap.[9] They find that abatement efficiency is safeguarded by a relative standard, while operators benefit from scarcity rents as well as an output subsidy. The authors note that deadweight losses cannot be reduced since no funds are raised, emission constraints are more uncertain and monitoring costs are higher under a relative system. The high political acceptability of relative target systems is based upon firm's ability to expand production within certain limits without having to pay for additional emissions, less severe competitive pressure *vis-à-vis* third countries and the ease of combining relative target systems with existing regulation.

---

6. Parry, I., Williams III, R. & Goulder, L., (1999).
7. Helm, C., (2003).
8. Maeda, A., (2003).
9. Gielen, A. M., Koutstaal, P. R. & Vollebergh, H. R., (2002), 5 ff.

The present chapter adds to the existing literature by placing a coherent typology of emission allowance allocation mechanisms into an emission trading model and by analysing how various assignment mechanisms deal with issues such as price determination, allocative efficiency[10] and with environmental considerations. It thereby also joins the disciplines considering the allocation of initial allowances and international trade theory.

The most important multiunit auction systems, financial allocation mechanisms as well as free allocation mechanisms are compared with each other in order to analyse which allocation mechanisms are most desirable from an allocative efficiency and environmental point of view. It seeks to establish a ranking of allowance allocation mechanisms with respect to allocative efficiency and their environmental impact. This ranking is used in subsequent chapters to analyse how effectively the current EU ETS performs in light of allocative and environmental objectives.

The analysis presented in this chapter goes beyond the traditionally analysed auctioning and grandfathering allocation mechanisms and pays particular attention to relative standard allocation and specifically the new Performance Standard Rate Emission Trading System, which is being applied in the Dutch NOx Emission Trading System. It examines the differences between various allowance allocation mechanisms with regard to allocative efficiency and the environment. At first a perfectly competitive static close economy provides the framework of analysis. In a subsequent part it is examined how the findings change if one relaxes the strict theoretical assumptions and allows for a dynamic open economy setting, in which only one economy has introduced emission trading. Parallels are drawn to existing Emission Trading Systems such as the Dutch NOx or the European Emissions Trading System where convenient.

After placing allocation mechanisms in a theoretical framework of Emission Trading Systems (section 2), section 3 discusses allocative efficiency and environmental impacts of initial allowance allocation mechanisms within a static closed economic setting. After introducing the static closed economy model (section 3.1) various allocation mechanisms are reviewed (section 3.2). Because of the symmetry of findings, environmental considerations with respect to all reviewed allocation mechanisms are addressed jointly in section 3.3. Section 3.4 presents the findings of this section.

Thereafter the restrictive theoretical assumptions are relaxed and allocative efficiency is examined in a dynamic open economy setting in section 4. After an introduction (section 4.1) allocation mechanisms are examined with regard to their allocative efficiency (section 4.2). Subsequently environmental considerations are treated (section 4.3). Section 4.4 summarizes the main findings. Section five presents an overall conclusion of the chapter and a discussion of the findings.

---

10. Allocative efficiency can be defined as a condition in which all possible gains from exchange are realized. See Frank, R., (1997), 350. This implies that those market participants valuing a good most have been able to attain it.

2   EMISSION TRADING SYSTEMS

After having presented the general economic rationale behind CO_2 emission allowances and emission trading in chapter two, the conceptual framework on which the current chapter is based will be introduced. With regard to Emission Trading Systems, three different elements can be distinguished: these elements are the Quantity Setting, the Allowance Allocation Mechanism and the Trading System (see Figure 7).

On the first layer, 'Quantity Setting', society determines to what extent it wishes to reduce CO_2 emissions.[11] Implicit to the required cost benefit analysis to be executed by society is the determination of both the regional scope and the nature of the covered emitters. The scarcity of the emission allowances can be expressed in absolute terms, setting an absolute amount of emission allowances, or in relative[12] terms. Relative scarcity can be expressed as being dependent on prescribed industry standards or on a certain tonnage of CO_2 emission per unit of Gross Domestic Product (GDP). Independent of the basis on which scarcity is measured, an absolute amount of CO_2 emission permits has to be determined so that they can be allocated in the next stage. For systems that operate on the basis of relative standards in the absence of a binding or non-binding emission target, society also has to determine how much it wants to be emitting in the future but the emission allowances are not directly distributed to emitters. The determination of scarcity in future periods may be complicated by the setup of a relative standard system without a binding emission cap because emission reductions will be reduced if output rises. Timely adjustments may be required if emission exceeds the level desired by society. With regard to the European Emissions Trading System, Directive 2003/87/EC requires Member State governments to set absolute quantities of CO_2 emission allowances in their National Allocation Plans. The amendment to Directive 2003/87/EC establishes a more harmonized allocation plan that does not allow Member States the discretion to draft new National

---

11. To motivate the United States and major developing countries to effectively control their greenhouse gas emissions is a core challenge for the international climate 'regime' beyond the Kyoto Protocol. Studies addressing problems related to international emission allowance allocation include Lecocq, R. & Crassous, R., (2003) and Böringer, C. & Welsch, H., (2004). It should be noted that determining the socially desirable level of pollution is a non-trivial task in the presence of imperfect information about present and future benefits and damages of pollution and the problem of collective action. Some economic models have addressed these problems by treating emission as an endogenous variable. See for example Smith, A. & Yates, A., (2003).

12. One successful example of an allowance trading program which applies such relative criteria is the American lead trading program. The program provides incentives for the diffusion of cost-saving technology; for a brief review of the program see Kerr, S. & Newell, R., (2003), 320 ff. and UNEP, UNCTAD (2002), 9 and 23. Nash, J. R., (2000), 487 ff., in particular 488–289, emphasizes that the program was launched in conjunction with a planned phase-out of lead and that it was in effect operating under a cap. For an overview of Emissions Trading Systems see Zwingmann, K., (2007), 113 ff. but also Tietenberg, T., Grubb, M., Michaelowa, A., Swift, B. & Zhang, Z. X., (1999), 27 ff.

Figure 7 Emission Trading System

Emission Trading System			
Quantity Setting			
· Determines scarcity of emission allowances			
Absolute		Relative	
Allowance Allocation Mechanism			
· Brings allowances onto the market			
Auctions	Financial Allocation		Free Allocation
Auction models	Administrative criteria		Administrative criteria / Grandfathering / Relative standard base
Trading System			
· Cost-effective exchange of allowances which are already on the market · Prescribes the basis on which market participants have to submit permits			
Cap and Trade		No - Cap and Trade	
Absolute emission per Firm / Relative performance standard		Relative performance standard	
			PSR-System

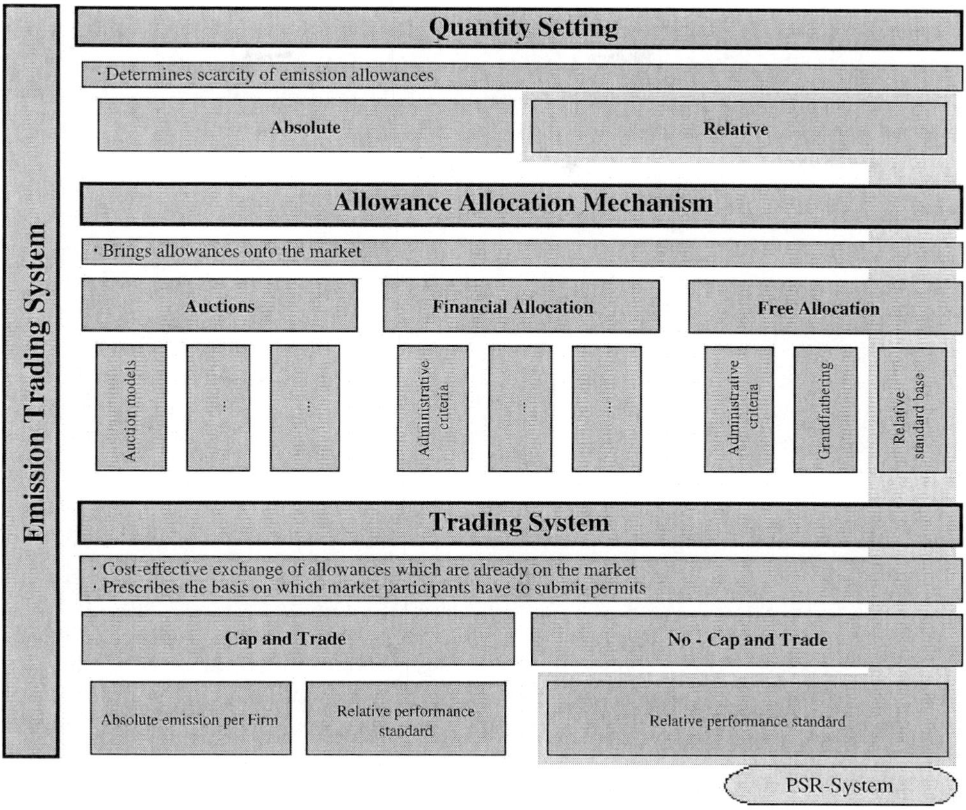

Source: Own representation.

Allocation Plans.[13] Allocation of emission allowances will be based upon fully harmonized implementing measures that will be adopted by the Commission by 31 December 2010 for the Transitional Community-wide rules for free allocation[14] and by 30 June 2010 for auctioning.[15]

An 'Allowance Allocation Mechanism' serves to initially allocate a predetermined amount of emission allowances. It therefore deals with the initial distribution of emission allowances among market participants but not with the operation of the emission trading market. Figure 7 shows a number of different generic formats of allocation mechanisms. All of them will be reviewed in this chapter. Auctions and financial allocation systems require market participants to pay for emission allowances. The difference between the two is that auctions allow market

---

13. See Art. 9, Directive 2009/29/EC.
14. See Art. 10a(1), Directive 2009/29/EC.
15. Article 10(4), Directive 2009/29/EC.

participants to determine the price they will pay in accordance with the prescribed auction rules while other financial administrative allocation mechanisms allow administrative bodies more discretion[16] in price determination and final distribution. In contrast to the two preceding mechanisms, free allocation mechanisms do not require payment. Administrative bodies can base their allocation decision on historic production data (grandfathering), on relative production standards or any other administrative criteria. In contrast to the direct distribution of allowances to individual operators, allocation under relative standards is directly dependent on the emission intensity of production. Operators complying with set emission standards receive the right to produce and may even be awarded credits they can sell on the market if they over-comply with the standards. Allocation under relative standards is thus more akin to a self selection process.

'Trading Systems' do not deal with issues of initial allocation. They treat the allowances that are on the market as given. Trading systems provide a cost-effective system of exchange and determine the criteria on which market participants are allowed to emit CO_2.

A (binding) 'cap' prescribes the maximum amount of CO_2 emission that a country is willing to emit. Under a 'cap and trade' system a maximum emission ceiling is installed. No emissions are permissible beyond that ceiling. Each firm has to submit at the end of a time period enough CO_2 allowances to comply with its prescribed obligations. These obligations can be based on the absolute emission per firm or a relative performance standard. In the latter case the standards have to be set in such a way that even in the presence of production variations the emission cap is observed. This could be achieved through frequent adjustments of the relative standards or the setting of security reserve. Also under a trading system that does not have a cap firms have to comply with their respective obligations to attain their emission targets or to yield emission allowances equal to their emissions at the end of a trading period. The difference between a cap and a non-cap trading system is that in the former the absolute amount of CO_2 emission is prescribed, while in the latter it is not. Yet also in the absence of a cap emission reduction goals can be set on environmental production standards.

One particular emission trading system merits special attention: the so-called 'Performance Standard Rate System', in short PSR. In this system emitting entities are free to produce but have to compare their actual emission level per unit of output with that prescribed by the government's benchmark. The emitting entity has to account for every emitted ton in excess of the benchmark and pay a punishment if it cannot attain CO_2 emission allowances via the Emission Trading System or benefit from past or future savings. One important specialty about the PSR trading system, which distinguishes it further from ordinary relative standard base benchmark allocation systems, is that the emission allowances are not created by the government and distributed to emitting entities. A legal act obliges emitting entities to meet particular emission targets, to have them verified

---

16. It should be noted that through the selection of auction rules administrative bodies are also able to influence prices.

by third parties and to report them to the government. While the government takes an active role in monitoring the compliance to the law, it is not participating as an actor. Government takes a *laissez faire* approach limiting itself to creating the rules to be followed by emitting entities. In this sense, the PSR system is operated by private entities who are able to sell their accredited savings to other market participants via a cost-effective trading system.[17]

It has been suggested for credit and trade systems that they were having high transaction costs because they required each credit generation to be approved by a regulatory authority.[18] The economic effect of different transaction costs associated with different allocation mechanisms would be to limit the degree of allocative efficiency by reducing the range over which mutually beneficial exchanges are possible.[19] Since the Dutch NOx system uses clearly defined benchmarks provided by law for monitoring and validation as well as centralized authorities,[20] for the purpose of this text it is assumed that such differences are negligible.[21]

As already indicated the field of interest of this chapter does not encompass the entire Emission Trading System but is restricted to allowance allocation mechanisms. Therefore it suffices to note that Emission Trading Systems have been subject to research[22] and that the European Emissions Trading System has been drawing from past experience.[23]

3   STATIC CLOSED ECONOMY

This part of the chapter consists of three parts. After introducing the concept of a static closed economy, the allocative efficiency of various initial allocation mechanisms within the framework of a closed static economy is examined. Subsequently the environmental impacts of these mechanisms are reviewed.

3.1   STATIC CLOSED ECONOMY MODEL

Because reality is very complex, models are used to gain a better understanding of intricate relationships between variables. Models construct an artificial framework of analysis based on assumptions how reality could look like. In this section,

---

17. This passages draws from European Commission (2003).
18. Tietenberg, T., Grubb M., Michaelowa A., Swift B. & Zhang Z.X. (1999), 26, 34 and 107.
19. See Harrision, D. & Radov, D., (2002), 52.
20. See Tweede Kamer der Staten-Generaal (2004), 22 ff.
21. For other allocation mechanisms it has been found that transaction costs differences are very limited. See Harrision, D. & Radov, D., (2002), Tables of Appendix B.
22. For a summary of market-based policies see Stavins, R. N., (2001) and Stavins, R. N., (2003). On the American experience with emissions trading see Pring, G., (2006) and for a review of emission trading systems see Nash, J. R., (2000), 487–496.
23. Kruger, J. & Pizer, W. A., (2004), 6.

a static closed economy model is introduced.[24] Unlike an open economy, a closed economy is not influenced by any form of international trade. Nothing can be exported or imported because there is only one economy within this framework. For the sake of analysis, this assumption simplifies the complex interactions within the economy to analyse the circular flow of national income. In this chapter we are, however, not interested in national income as such but rather in the allocative efficiency of CO_2 emission allocation mechanisms and its environmental impacts. Therefore only these issues are being analysed. In this part of the chapter the author has chosen a static economy which presents only a snap-shot but does not allow for strategic firm behaviour or changes in production technology. In the second part of the chapter these assumptions are relaxed to allow both dynamic firm reaction and trade.

3.2   COMPARISON OF ALLOCATION MECHANISMS
            AND ALLOCATIVE EFFICIENCY

Having introduced the static closed economy model and presented the importance and function of initial allocation mechanisms, this section will provide a comparison of allocation mechanisms. Criteria used for comparison are price determination, allocative efficiency and possible adverse effects of allocation mechanisms on trading systems. Of particular relevance for the establishment of allocative efficiency are the effective revelation of valuations for emission allowances and the absence of distortions of competition that is associated with the allocation of emission allowances. Only if it is known which operators participating in the EU ETS value an allowance the most can it be ensured that this operator will eventually hold an allowance after the initial allocation process. If this cannot be ensured, efficient trading mechanisms are needed to safeguard allocative efficiency. Though it should be noticed that the attainment of an equilibrium may be impaired through allocations that give rise to severe redistribution effects and distortions of competition. This section consists of two parts. The first part addresses auction mechanisms, the second examines other financial allocation mechanisms and free initial allocation mechanisms, including grandfathering and 'relative standard base' systems.

3.2.1   Auctions as an Allocation Mechanism

Auctions have become enormously popular and are being used in a large number of economic exchanges both in the public and private sector. Just two prominent examples are mobile phone licenses and the decentralization of electricity markets. They have also been used within the framework of emissions trading. Notable

---

24. The treatment in this chapter does not encompass detailed mathematical models in the strict economic usage of the concept but represents a logically reasoned discussion of effects that facilitate an understanding of the economic intuition.

examples of auctions that have been successfully employed to reduce emissions albeit to a varying degree include the Acid Rain Programme in the US[25] or the Virginia NOx auctioning system.[26] An auctioning system for greenhouse gas emissions that has been employed in the European context prior to the introduction of the EU ETS and successfully circumventing the deficiency of stakeholder support was the UK Emissions Trading Scheme. It was based on voluntary participation and essentially reversed the auction parties. In the UK Emissions Trading Scheme that ran from 2002 to 2006 the State offered incentive money to buy emission allowance reductions from undertakings.[27] Within the framework of the EU ETS auctioning has so far not played a decisive role. During the first trading phase an about 0.13% of allowances were auctioned – by many countries only within the framework of a new entrance reserve. This number increased to 0.5% of allowances during the second trading phase. Following the EU ETS amendment this will change substantially for the third trading phase. Post 2012 auctioning will be one of the most prominent allocation mechanisms. In order to explain the implications and capabilities of auction theory, here it is, however, assumed that 100% auctions is being employed. The possible complexities arising from multiple allocation mechanisms and an important secondary market are examined in later chapters.

There are four reasons rendering this allocation mechanism attractive. Firstly, an auction is designed to lead to self-revelation of the bidder's private values. In the presence of inherent information asymmetry, in which a potential seller is unable to determine the market value of a particular object, an auction mechanism can yield higher revenues than simply quoting a price or repeated negotiations with potential buyers. While this is very desirable from a theoretical point of view, it should be noted that bidders are generally reluctant to reveal their preferences because they fear that competitors could take advantage of it – protection of such information is crucial for firms. Secondly, auctions can be designed in such a way as to ensure allocative efficiency. It should be noted that efficiency here is to be understood as to award the bidder with the highest valuation for an object with the tender.[28] Thirdly, auctions legitimize transfers – of sales in the absence of market prices – which would otherwise be suspect. Prior knowledge of the auction rules provides bidders with a transparent framework of how their bids will be assessed while at the same time ensuring bidders that selling agents have clear and indiscrete tender selection criteria.[29] Fourthly, since no time consuming negotiation has to take place, auctions are fast allocation mechanisms. Though it should be noticed that

at last!

25. Evans & Peck (2007), 20.
26. *Ibid.*, 25.
27. See Defra, Enviros (2006), 28 following.
28. Implicitly assuming away the possibility of credit rationed bidders. See Milgrom, P., (2004), 57. Maximization of social welfare, defined as the maximisation of the sum of producer surplus and consumer surplus can be reached if side-payments (in the presence of budget balance constraints) are possible. In such cases Pareto optimal allocations are feasible in which one person is better off without someone else being worse off.
29. See Rothkopf, M. & Harstad, R., (1994), 368.

the development of an auction mechanism depends on the object being auctioned as well as its potential market, and can be a non-trivial, time consuming process. Similarly the time and expenses required to gather information and to design an optimal bidding strategy can be substantial.

This section of the chapter is subdivided into two parts. The first part introduces general auction formats. Drawing from that, the second part will present auction theory's contributions to CO_2 allocation mechanisms. A conclusion will summarize the main points.

3.2.1.1   *Auction Formats*

An auction can be understood as a set of rules, which translates information revealed by bidders by means of an allocation rule, and a payment rule into efficient outcomes. The challenge of auction theory is to develop auction rules which are tailored to the preferences of bidders in such a way as to provide Pareto optimal[30] allocations. Auctions do not only differ with regard to allocation and payment rules but also with respect to the amount of information they require bidders to reveal.

There are innumerous possibilities to design auctions. These models fall into several categories, or formats. A standard auction is an auction in which the highest bidder among potential buyers, or the lowest bidder among potential sellers wins. Since there is an almost perfect correspondence in results,[31] it is quite unimportant to distinguish between both forms.[32] Standard auctions are commonly distinguished into 'open' and 'closed' auctions. In open auctions bidders are aware of their competitors' bids while in closed ones they are not. Two examples of open auctions are the ascending price auction, also called the English auction, and the descending price auction, also known as the Dutch auction. Two examples of closed auctions are the second-price sealed-bid auction, frequently referred to as Vickrey auction,[33] and the first-price sealed-bid auction. The four standard auction types are presented in table 1. Since there are innumerous possibilities to design auctioning mechanisms, the main emphasis of the treatment of auctions in this book is limited to these standard forms.

---

30. Pareto optimality describes situations in which it is impossible to make one person better off without making at least someone else worse off. In the absence of side payments between bidders, i.e., when bidders are unable to compensate each other, Pareto efficient but suboptimal allocations can occur.
31. With the possible exception of the invalidity of reserve prices and treating zero as an implicit limit to acceptable bids. Despite the intuitive appeal of the later argument, Shubik, M., (1983), 39 ff. cites Herodotus reporting on Babylonian marriage markets which did include auctions starting at negative bidding values.
32. Rothkopf, M. & Harstad R., (1994), 366.
33. Named after Nobel laureate William Vickrey, who first presented this auction in his seminal paper on auctions. Vickrey, W., (1961).

Table 1 Standard Auction Types

Open auctions	Closed auctions
Ascending price auctions (English auction)	Second-price sealed-bid auction (Vickrey auction)
Descending price auction (Dutch auction)	First-price sealed-bid auction

In an open ascending price (English) auction, the price is raised by the auctioneer or by bidders themselves until only one bidder remains. At any particular point in time bidders know the level of the current best bid.[34] Such auctions are often used by auction houses like Sotheby's. In the open descending price (Dutch) auction, the price decreases continuously until one bidder accepts the current price. A well-known example where (sequential) open descending price (Dutch) auctions are used, is the flower auction in Aalsmeer (the Netherlands). In a closed sealed-bid auction bidders are only allowed to enter one bid, thus they are unable to react *ex post* to their rivals. In the closed first-price sealed-bid auction the highest bid wins, while in the closed second-price sealed-bid (Vickrey) auction, the highest bidder is only required to pay a price equal to the second highest bid.

After having reviewed the four standard auction types, the following section examines auction mechanisms which could be used to initially allocate CO_2 emission allowances.

3.2.1.2 Multiunit Auctions

Under Directive 2003/87/EC each tradable permit is defined as 1 ton of CO_2 equivalent over a designated period of time.[35] Later the trading may be extended to other greenhouse gases with an equivalent global-warming potential of 1 CO_2 ton.[36]

Because the benefit derived by operators from attaining 1 ton CO_2 equivalent is the same irrespective of the greenhouse gas they purchase, the good is clearly a substitute[37] and can thus be treated as a homogeneous that is, a similar good, from an auction design point of view.

Surprisingly little is known about efficiency properties of multiunit auctions. A lot of the conventional wisdom comes by analogy from single-unit auctions[38] but a sound understanding of how equilibria respond when assumptions about values

---

34. McAfee, P. & McMillan, J., (1987a), 702.
35. Defined in Art. 3(a) of Directive 2003/87/EC.
36. Defined in Art. 3(j) of Directive 2003/87/EC.
37. If obtaining one good makes the bidder willing to pay more for a second good, the goods are complements, if the bidder is willing to pay less, they are substitutes.
38. Ausubel, L. & Cramton, P., (2002), 1.

and information change, are not yet answered at any level of generality.[39] Since participants to an emission trading auction will generally acquire larger quantities of allowances, multiunit auctions are of core interest. As in the case for single-unit auctions, auction designers strive to reach two goals that may well be assumed to be close to the objectives of the government. They try to ensure an efficient outcome and to maximize government revenues. Even though these goals are closely related, here we merely focus on efficiency criteria as the relevant benchmark because revenue maximization focuses on redistribution of wealth rather than the maximization of consumer and producer surplus.

In this section, two general multiple auction methods that could be used to allocate CO_2 emission allowances are reviewed. Firstly, sequential auctions in which one allowance would be sold after the other and secondly, simultaneous auction models in which multiple CO_2 emission allowances are sold at the same time. The latter group is subdivided into open and closed auction formats. Because strategic firm behaviour such as for example, demand reduction undermines allocative efficiency, a short review is presented whenever appropriate.

Sequential auctions are easy to implement for auctioneers but are not very much favoured by bidders. One reason is the strategic complexity of bidding decisions and the likely price variation of sequential auctions of homogeneous goods.[40] Ashenfelter (1989) has termed this 'declining price anomaly'[41] and explained a falling price in subsequent auctions[42] in terms of bidders' necessity to acquire particular quantities, risk aversion[43] and uncertainty.[44] Ashenfelter's observation entails that on average bidders are risk averse (a necessary assumption we would have to make also for the EU ETS operators) and appears to be based on a more or less static supply and demand. A *ceteri paribus* assumption in this context may, however, not be realistic under the capped EU ETS. As firm output increases over time also their demand for allowances will increase. The effect of an increasing price may work against the effect of a declining price anomaly. Nevertheless, the empirical observation of such effects in auctions underlines the relevance of risk averseness as a relevant bidder characteristic to be taken into account by mechanism designers.

From an allocative efficiency point of view, auctions ensuring a single market price as for example, uniform-price auctions are preferred over sequential auctions because they mitigate the 'price risk' of paying too much for the same good.[45] In general, an auction format that cannot ensure that the bidder with the highest

---

39. See Börgers, T. & Van Damme, E., (2004), 43.
40. Inefficiencies from synergies and complementariness of goods appear to be smaller if goods are homogeneous.
41. Ashenfelter, O., (1989), 29 ff.
42. See Ashenfelter, O. & Graddy, K., (2002), 34–36 and Table 8 for a review of subsequent research.
43. Risk aversion implies that bidders dislike taking fair bids. McAfee, R. & Vincent, D., (1993) show that risk aversion can create declining prices.
44. See Neugebauer, T. & Pezanis-Christou, P., (2005).
45. Milgrom, P., (2004), 256.

valuation always receives the good cannot be allocatively efficient, and hence not be a viable first-choice option as a CO_2 emission allocation system. If such auction schemes were to be chosen, allocative efficiency would have to be established on an efficiently operating emission trading market.

After having reviewed sequential auctions, simultaneous multiunit auctions are considered. At first, simultaneous closed sealed-bid auction models are examined. Thereafter simultaneous open ascending-price auctions models are treated.

Due to their importance, three simultaneous closed sealed-bid auction models are reviewed here.[46] These are pay-as-you-bid, uniform-price auction and multiunit Vickrey auction. Each will be discussed in turn.

The pay-as-you-bid[47] is a closed sealed-bid auction in which bidders simultaneously submit demand schedules[48] for goods. Bidders win the quantity demanded at the clearing price and pay the particular price for each unit as indicated in their submitted demand schedule. In order not to pay unnecessarily high amounts for emission allowances, bidders have to estimate the market-clearing price and bid slightly above it. This exposes less informed bidders to the strategic risk of misjudging the clearing price and pay 'more' for identical goods. This increases the transaction costs to the parties of participating in a bid and may even deter potential bidders from participating, which in turn reduces the competitiveness of the entire market and hence its efficiency.

The uniform-price auction is a closed sealed-bid auction in which bidders simultaneously submit demand schedules for goods and pay the clearing price for every unit demanded at that particular price. In contrast to the pay-as-you-bid auction, this auction format has two advantages. Firstly, every bidder pays the market-clearing price which is equal to the overall marginal valuation. Secondly, in the absence of the danger of paying too much for the same good, less informed bidders are more inclined to participate in such auctions.

Both, the pay-as-you-bid auction and the uniform auction format can be expected to be inefficient in the presence of market power or collusion. In such cases they may give rise to inefficiency[49] inducing 'demand reduction' strategies.[50] In multiunit auctions dominant players recognize the interdependence of their bidding strategy and competitors' bidding behaviour. A strategy of self restricting the quantity demanded while bidding the minimum price to indicate interest in a number of units can generate large consumer surpluses. The inherent inefficiency stems from the fact that users with the highest value for a good do in fact prefer not to attain it; large bidders win too little and small bidders win too

46. See Krishna, V., (2002) and Ausubel, L. & Cramton, P., (2002).
47. Also called 'discriminatory auctions' or 'multiple price auctions'.
48. In this section all demand schedules are assumed to be downward sloping, i.e., more is being demanded if the price decreases.
49. Inefficiency is created by 'differential bid shading', i.e., when bidders with identical marginal values reduce their bids by different amounts so that awarding the bidder who values the item most is impossible. See Ausubel, L. & Cramton, P., (2002), 4.
50. For examples see Weber, R., (1997) and Ausubel, L. & Cramton, P., (2002).

much. Salmon (2003)[51] points out that if such behaviour is strictly unilateral, this does not amount to collusion. However, if it does involve strategic considerations exemplified by trigger strategies, such behaviour would amount to (tacit) collusion.[52]

The third sealed-bid multiunit auction format reviewed here does not suffer from demand reduction in private-value environments.[53] The closed second-price sealed-bid auction system (a multiunit Vickrey auction), is an auction in which bidders simultaneously submit demand schedules for goods. Bidders win the quantity demanded at the clearing price and pay an amount equal to the highest losing bid for each unit. Since sincere bidding is a dominant strategy, allocative efficiency distorting demand reduction will not occur.

In contrast to closed auction formats, in open multiunit auctions the price and the allocation of emission allowances are determined by open competition. In cases of considerable uncertainty with regard to future market developments and future technological developments, bidders' valuations may depend on information held by other bidders. The 'feedback' that bidders get in the process of bidding in ascending auction formats renders them more efficient than sealed-bid auction formats when it comes to solving complex allocation problems.

The open uniform-price ascending auction is the dynamic version of the closed sealed-bid uniform-price auction. For each slowly increasing price quoted by a fictitious auctioneer, bidders are allowed to observe and respond by quoting quantities they wish to purchase. Quantities demanded are added horizontally in order to determine the market demand. As long as demand exceeds supply, the price will be increased. Unlike in the closed sealed-bid uniform-price auction, in the open ascending-price auction bidders can be informed about other bidders' demanded quantities.

Open ascending-price auctions are easily implemented since bidders will only have to quote the quantity demanded and observe simple activity rules. As the simultaneous ascending (multiunit) auctions, this auction format is vulnerable to demand reduction and collusion. In cases where market power is limited, inefficiencies from standard ascending-price auctions are also expected to be low.

Kagel and Levin (2001) suggest that demand reduction can be stronger under the open ascending-price auction than under the closed sealed-bid uniform-price

---

51. Salmon, T., (2003), 5.
52. Distinguishing between tacit collusion and pure strategic firm behaviour is complicated if not impossible. In a multiunit auction both bidders could, for example, independently decide to pursue a 'demand reducing' strategy. The outcome would be identical to tacit and indeed outright collusion.
53. In private-value models bidders are assumed to have knowledge, which is strictly private to them such as for example the price they would be willing to pay in a tender. Common-value models in contrast, assume that the actual value is identical to all participants, but bidders do have diverging private information about this value. Unlike in private-value environments, in common-value environments Vickrey auctions do not always produce efficient equilibria. To the extent that emission allowances depend on the operators abatement costs, emission allowances are probably best thought of as depending on private values.

auction.[54] With respect to (tacit) collusion in multiunit open ascending-price auction environments, Ausubel and Schwartz (1999)[55] postulate the existence of a unique cooperative equilibrium if bidders are able to use backward induction.[56] If bidders fail to immediately reach a low price outcome, signalling[57] can be employed to 'negotiate' a mutually acceptable allocation. In a simultaneous ascending-price bid auction environment with a limited number of participants and known limit prices of fringe firms, Grimm, Riedel and Wolfstetter (2001)[58] cite a powerful example of how effectively signalling can be used to reach an almost immediate mutually acceptable strategic demand reduction.

While signalling and demand reduction certainly are strong points of critique of open multiunit ascending-price auctions, there are two factors which complicate effective collusion. Firstly, Brusco and Lopomo (1999)[59] show that the collusion becomes more difficult as the number of bidders relative to the number of items rises. Secondly, Brusco and Lopomo (1999)[60] show that considerable externalities or synergies across items negatively impact prospects of collusion.[61] The authors conclude that due to signalling,[62] collusion is possible, even in the presence of a high ratio of bidders to objects and under some complementarities in bidders' utility functions.

For its favourable efficiency properties the Ausubel auction deserves particular mentioning. Ausubel (2002)[63] proposed an efficient ascending auction design for homogeneous goods that eliminates incentives for demand reduction[64] and rewards the revelation of true values.[65] As in the ascending-price auction, quoted prices continuously increase until demand equals supply. The bidder demanding the highest quantity when the collective demand of all other bidders is one unit less than the total quantity supplied, will be rewarded one unit at the current price. Consequently the bidder with the highest valuation for a particular unit is able to secure it and pay the price indicated in the demand schedule. Notwithstanding the notable efficiency properties which the Ausubel auction demonstrates in both private-value and common-value environments, Manelli, Sefton and

---

54. See Kagel, J. H. & Levin, D., (2001).
55. See Ausubel, L. & Schwartz, J., (1999).
56. Bidders are assumed to imagine how the auction will be developing and to derive from this a mutually acceptable offer at the beginning of the auction.
57. By for example using the financially inconsequential digits of their bids, parties can signal their identity or indicate the market for which they are retaliating.
58. Grimm, V., Riedel, F. & Wolfstetter, E., (2001).
59. Brusco, S. & Lopomo, G., (1999).
60. Brusco, S. & Lopomo, G., (1999).
61. For an analysis of externalities in single-unit auctions, the interested reader is referred to Caillaud, B. & Jehiel, P., (1998).
62. See Cramton, P. & Schwartz, J., (2002) for an insightful analysis of FCC spectrum auctions.
63. Ausubel, L., (2002). Earlier versions of this paper date back till 1997.
64. For an experimental assessment see Kagel, J. H. & Levin, D., (2001).
65. Ausubel replicates the intuition of the Vickrey auction in a dynamic context.

Wilner (1999) find that this auction format generates incentives for bidders to engage in strategic overbidding to mislead competitors.[66]

The above can be summarized as follows. Sequential multiunit auctions are easy to implement but due to the complexity of bidding strategies they require and their inability to ensure allocative efficiency, they will not serve as an efficient means to allocate CO_2 emission allowances.

With respect to closed multiunit auctions, simultaneous sealed-bid uniform-price auctions are likely to be more allocatively efficient than pay-as-you-bid auction schemes because the costly strategic burden of misjudging the clearing price is not placed upon firms. Even in the presence of demand reduction, small firms would benefit from it and consequently have every incentive to participate in the auction which in turn would mitigate effectiveness of demand reduction strategies. Auction rules in both auction formats are easy to understand for bidders but the strategic complexity of placing bids is non-trivial. If one, however, would only consider efficiency from an auction theoretic point of view, the efficiency ranking of pay-as-you bid auctions and uniform-price auctions is ambiguous.[67]

In the presence of market power on emission allowance markets, the multiunit Vickrey auction is clearly more efficient since it does not give rise to demand reduction. In the absence of market power, a uniform-price auction is similarly efficient and due to the simplicity of the auction rules and price uniformity at first sight it may generate even more desirable results.[68] Since it is difficult for firms to develop good bidding strategies, multiunit Vickrey auctions, in which true bidding is a dominant strategy, may be socially more desirable. This will be the case if – and only if – firms could be convinced that the information they reveal is kept secret at all times.

With regard to open multiunit auction formats, it has been shown that open ascending auction systems are allocatively efficient if there is no market power. In those cases where there is market power on the emission trading market, the Ausubel auction may be a viable option.

One can therefore conclude that auction mechanisms can be used effectively for the allocation of CO_2 emission allowances. Auctions can solve the problem of price determination by allowing bidders to 'reveal' (at least part of) their preferences. If the number of bidders is sufficiently large and value revelation is appropriately accounted for auctions are capable of approaching and potentially even reaching the market equilibrium while their dependency on the market mechanism may be limited and mainly regard adjustments to changes in emission allowance demand. The express underlying assumption is of course that auctions are the main allocation mechanism – which will only be increasingly the case during the third trading period.[69] Auctions can, however, also give rise to strategic behaviour of bidders. Allocative inefficiencies stemming from for example, demand reduction

---

66. Manelli, A. M., Sefton, M. & Wilner, B. S., (1999), 3.
67. Ausubel, L. & Cramton, P., (2002), 16 ff.
68. Cramton, P. & Kerr, S., (1999), 6.
69. The superior allocation properties of auctions may not materialse, however, if bidders mistakenly use the secondary market price as a guidance.

could, nevertheless be corrected *ex post*, if large bidders are able to acquire emission allowances through efficient trading systems, so that the market equilibrium does not have to be impaired for long.

3.2.2 Administrative Allocation Mechanisms

After having reviewed how auctions can be used as an emission allowance allocation system, we turn to two different mechanisms. Both have in common the relative importance of administrative bodies. First, financial administrative allocation mechanisms will be briefly reviewed and afterwards free administrative allocation mechanisms will be introduced.

3.2.2.1 *Financial Administrative Allocation Mechanisms*

In the preceding section auction mechanisms have been reviewed. In contrast to these fully market-based instruments, other financial emission allowance allocation mechanisms can be considered. Such instruments are more akin to 'command and control' mechanisms to the extent that they attribute an active role to bureaucrats.

All or part of the emission allowances could be sold. Either on the basis that society wants to recoup expenses for the establishment of an Emission Trading System or because it would like to grant industries preferential access to emission allowances at a price which lies below a competitive market price.[70]

From an allocative efficiency point of view, it is certain that bureaucrats will perform less efficiently than purely market-based financial allocation mechanisms. Firstly, because civil servants are likely to be less well-informed about the particular needs of an enterprise than entrepreneurs themselves. And secondly, because firms do have an incentive to misstate their true valuation in order to attain more allowances.[71] Thus due to bureaucrats' inability to determine the optimal price and allocation correctly, any interference in the allocation mechanism will lead to less efficient allowance allocation. In other words it is unlikely that bureaucrats are able to allocate emission allowances in such a way as to reach the equilibrium of the market. In the parlance of the Edgeworth box, it is unlikely that the initial endowments will be distributed to firms so that they will directly be on their contract curve. In the presence of an efficient trading system, however, inefficient initial allocation of emission allowances can be remedied by the market[72] and an efficient allocation is safeguarded over time.

---

70. With regard to such schemes, State aid considerations under Art. 87 of the EC Treaty are relevant.
71. Such an incentive will be present as long as the expected rewards outweigh costs placed upon violators. Becker, G. S., (1968), 169–217, shows that disutility from trespassing legislation can be modelled as being a function of the detection probability, the level of punishment and the risk averseness of the violator.
72. This is again subject to the assumption that transaction costs are sufficiently low and that markets operate efficiently.

While administrative financial allocation mechanisms appear little appealing from an allocative efficiency point of view, it should be noted that society would only be worse off if the resulting allocation would not be reflecting its preferences. Even if there are normative grounds (such as equity considerations) which would justify an administrative influence in financial allocation mechanisms, it should be examined whether the positive effects cannot be created in a less distortive manner.

3.2.2.2   *Free Administrative Allocation Mechanisms*

A free administrative allocation mechanism does not require payments for emission allowances. Based on administrative criteria, allowances are distributed to market participants. There are numerous possibilities of how to distribute emission allowances. They may differ as much with regard to the social group they set at a comparative advantage as to the criteria they apply as a basis for allocation. Society may distribute allowances to virtually anybody. Firms may be rewarded for their labour intensiveness of production or families for having many children. Promising economic sectors, low-income families, art galleries, etc., are other examples. It is obvious that in free administrative allocation mechanisms normative distribution justifications based on society's preferences or industrial policy considerations prevail over allocative efficiency considerations.

In the presence of an efficient trading market those market participants who value an emission allowance most are able to attain it. Provided, of course, that there is no significant leakage of allowances because some recipients prefer to save rather than to use or sell them. Furthermore the liquidity of the emission trading market may be undermined if transaction costs for initial recipients may be high. This would unduly increase the financial burden which is placed upon those market participants which value emission allowances most and constitute a real waste to society. If society would like to support particular groups, it should consider using more direct means such as tax cuts or direct monetary transfers accruing from auction proceeds.

There are, however, two incompatible free allocation mechanisms which merit particular attention. These are the so-called grandfathering and the relative standard base allocation system. Each will be discussed in turn.

3.2.2.2.1   Grandfathering
Bureaucrats can allocate emission allowances for free on the basis of historic data. There are three bases which can be used to allocate allowances. Firstly, input-based (used historic energy input); secondly, output-based (e.g., kilowatt-hours of electricity production); or, thirdly, emission-based (direct or indirect that is, total emission from emitting facilities).[73] Furthermore, the base period for historic data has to be determined. As can be seen from these choices, allocating emission allowances is a non-trivial matter that is likely to generate a high administrative

---

73.   Harrision, D. & Radov, D., (2002), 60 ff.

burden and provoke both strong opposition from disadvantaged parties and welfare reducing lobbying.[74]

As other free allocation mechanisms, grandfathering systems cannot be expected to generate an allocatively efficient initial distribution of emission allowances. Since grandfathering systems will allocate allowances to operators directly and therefore to a societal group which naturally has a high valuation for CO_2 emissions, it is obvious that grandfathering mechanisms can be allocatively superior to other free allocation mechanisms.[75] Yet since allocation is based on historic emissions and is not based on abatement cost structures of firms, they will always be less efficient than auction models, even in the presence of perfect information.

3.2.2.2.2 Relative Standard Base Mechanisms

As discussed above, there are many possible criteria on which emission allowance allocation can be based. One can also envisage a free allocation on the basis of industry specific CO_2 emission standards. Such standards can be quoted, for instance, in CO_2 ton per production unit or in terms of CO_2 efficiency per amount of GDP produced. Such relative standard base allocation systems form part of the Dutch National Allocation Plan of the European Emissions Trading System, the Dutch NOx Emission Trading System and are also underlying the PSR Emission Trading System presented in Figure 7.[76]

The first Dutch National Allocation Plan[77] (Dutch NAP) allowed for a sector-specific correction factor which takes into account the relative energy efficiency of installations. Government intended to encourage operators to further improve their already notable technology and to attain world best practice standards by means of the so-called benchmarking covenants schemes.[78] Besides the relative energy efficiency, historic emission data (base period 2001–2002)[79] and (projected) sector

---

74. Cullis, J. & Jones, P., (1998), 93 ff.
75. How grandfathering schemes compare to relative standard base allocation systems (discussed below) has to be determined on a case by case basis.
76. The particularities of the PSR system will be discussed below.
77. The Netherlands, (2004a).
78. The Netherlands, (2004a), 17 and 20. The Dutch government has decided to reward energy-efficiency performance with the allocation of allowances and to prevent infliction of punishments on first movers. More than 80% of total industrial energy use is now covered by covenants. In exchange for compliance with the covenant plans, Government has promised not to impose further national measures on participating businesses which are directed towards CO_2 reductions. It is thus bound to abstain from taking additional national initiatives but is able to follow European interventions. See also The Netherlands, (2004b). It is noticeable that the Commission criticizes the second Dutch NAP on the grounds that the energy efficiency factor cannot be justified by reference it to criterion 7 of Annex III of the Directive (early action) and that it refers to recent and future actions rather than early actions. Accommodation of early actions cannot justify allocation beyond expected needs. See Commission Decision (2007), 12 and The Netherlands (2006), 16 and 41 ff.
79. The same is true for the second Dutch national allocation plan but here the base period has been updated to 2001–2005. For a review of the present allocation methodology see The Netherlands, (2006), 37 and Ch. 3.

growth (2003–2006)[80] as well as an overall correction factor ensuring compliance with the emission ceiling, are variables in the allocation formula. Thus, the Dutch NAP can be viewed to combine elements of the grandfathering allocation with the relative standard base allocation mechanism. It should be noted however, that the Dutch NAP has specific quantities allocated for the industry which – depending on society's preferences – may be altered with regard to other CO_2 emitting sectors or the tax payer.

While the first Dutch NAP represents a hybrid form of allocation mechanisms with a strong focus on historical emissions, the mechanism applied in the Dutch NOx trading system is generic in nature. The Netherlands have introduced a relative standard base system,[81] and, more precisely, a PSR system[82] for emitters of NOx gases. There are four reasons cited by the Dutch government for opting for relative standards.[83] Firstly, the high emission reductions, secondly, the strong differences in NOx efficiency between operators in the same industry, thirdly, compatibility of the relative standard with existing legislation, fourthly, that the relative standard does not impede growth of firms since they are not required to buy additional emission allowances when their production increases.[84] With regard to the third reason 'compatibility with existing legislation' it should be noted that the Dutch NOx system's flexibility is limited by the existence of Council Directive 96/61/EC of 24 September 1996 concerning integrated pollution prevention and control (IPPC Directive) that prescribes the use of Best Available Techniques. While both systems are formally compatible with each other,[85] national measures can only go beyond community measures and not fall short of them: national PSR benchmarks will have to be compatible Best Available Techniques levels for the installations or more stringent but are not allowed to be weaker.[86]

Whether such a 'provision' system[87] is allocatively efficient is, however, subject to the same reservations as the above-mentioned free allocation mechanisms. Because emission allowances are created free of charge, politicians will have to estimate abatement cost structures and industry abatement potential in order to ensure a welfare maximizing allocation. This is equally difficult to attain as under a grandfathering system but similarly not problematic if the trading system functions effectively and market adjustment is timely. In this case an allocative efficient solution will result if transaction costs are zero. Yet as was the case under

---

80. See The Netherlands, (2004a), 25.
81. Tweede Kamer der Staten-Generaal, (2004), 4 ff. See also Staatsblad van het Koninkrijk der Nederlanden (2005a) and Staatsblad van het Koninkrijk der Nederlanden (2005b).
82. VROM, (2003), 17.
83. Tweede Kamer der Staten-Generaal, (2004), 4 ff.
84. Tweede Kamer der Staten-Generaal, (2004), 5 ff.
85. See Van Tol, I. & Oldenziel, H., (2006), 211 ff. and Gestel, R. A. J., van Backes, Ch. W. & Teuben, R., (2002).
86. See Case number 200502867/1, *MOB and Waddenvereniging v. het college van gedeputeerde staten van Groningen*, of 15 Feb. 2006.
87. As explained in section 2 of this chapter, emission allowances under the PSR system are created by an administrative act based on relative standard bases for specific industries but are not transferred from the government to enterprises. Therefore they are 'provided' but not allocated.

grandfathering, the likelihood that a PSR system will by itself give rise to a market equilibrium is low.

With regard to grandfathering, relative performance base allocation mechanisms have the important normative advantage that 'early movers' will automatically be rewarded for having invested in abatement technology. Furthermore, in the presence of upward sloping abatement cost curves 'early movers' will have an above average valuation for CO_2 allowances if compared to firms in a grandfathering system. In such environments relative performance allocation mechanisms are allocatively superior to grandfathering systems.

3.3. ENVIRONMENTAL ASPECTS

A major objective of Emission Trading Systems is the creation of incentives for operators to apply environmentally friendly means of production. With respect to the European Emissions Trading System, it is clear that a cap-and-trade system is intended,[88] which in effect strives to restrict the absolute amount of CO_2 emissions to prescribed levels. Operators who are not able to surrender CO_2 allowances for all their emissions on every 30th of April are subject to an excess emission penalty and to the obligation to surrender them in the following year.[89]

In a trading system, emission reduction is achieved by those operators who's marginal abatement cost is less than, or equal to, the emission allowance market price. It is thus the scarcity of the CO_2 emission allowances that inflates the market price of CO_2 allowances and that eventually determines which firms will be selling and which firms will be buying allowances on the market. From an environmental point of view, it is irrelevant which firms emit CO_2 but decisive how the concentration of CO_2 in the atmosphere changes. CO_2 concentration and hence environmental performance is dependent on the absolute number of emission allowances and the timing of emissions brought into circulation and independent on their particular market price. Therefore it can be concluded that all initial allocation mechanisms operating under an absolute annual emission cap are logically separable from environmental effectiveness considerations. The absolute amount of environmental allowances is exogenously determined and is thus not influenced by such allocation mechanisms. This dichotomy can be seen in Figure 7, which clearly distinguishes between the Quantity Setting decision and Allowance Allocation Mechanisms.

With respect to the operation of abatement, there are some subtle differences between the initial allocation mechanisms which do not, however, give rise to

---

88. Articles 3(1) and 3(2), in conjunction with Annex B of the Kyoto Protocol, impose a specified absolute emission target upon Member States. Even though 'the cap' is not explicitly introduced in Directive 2003/87/EC, it can be inferred from Art. 11(1) requiring Member States to decide upon the total quantity of allowances they will allocate during the three-year period to the operators of each installation. Although it is not explicitly stated in the Directive, it is a general assumption that Member States impose specific absolute emission caps on industries.

89. Articles 16(3) and 16(4) of Directive 2003/87/EC.

different environmental effects. In auction systems, market operators will monitor allowance prices and decide on the spot if they buy CO_2 allowances or if they commit themselves to install abatement technology. Similarly with regard to financial administrative allocation mechanisms, the decision to buy or sell emission allowances is determined by the market price. Similarly, with regard to free allocation mechanisms such as grandfathering, firms will decide on the basis of the market price and their CO_2 abatement cost structure to buy or sell emission allowances and to install abatement technology. In allocation mechanisms based on relative industry standards, firms are awarded or provided with allocation permits on the basis of their relative CO_2 production efficiency but the decision to use them or to sell them does also depend on the market price.

Allocation mechanisms applying a 'relative standards base' do merit further attention. Such allocation mechanisms do form part of the PSR Emission Trading system which expressly operates in the absence of a cap.

In a PSR system, allowances are created on the basis of relative industry standards and the government requires firms to submit additional emission allowances if they do not fulfil the efficiency standards as prescribed by law per produced unit. This places environmentally minded 'early movers' at a comparative advantage with respect to those operators which have not been willing to invest in abatement technology. The PSR System combines both, a 'relative standard base' benchmark system that provides emission credits by reference to the actual emissions of the operator, with a trading system that operates in the absence of a direct cap for national industry and is therefore referred to as a PSR System.[90]

The Dutch NOx trading system, which only extends to industrial operators, does not have a direct emission cap. The overall amount of Dutch national NOx emission is, however, constrained by the Dutch ratification of the Gothenburg Protocol.[91] The Netherlands have agreed to restrict themselves to a maximum NOx emission of 266 kilotons in 2010, and in accordance with the EC Directive on national emission ceilings for certain atmospheric pollutants (NEC Directive)[92] committed themselves to reduce this amount even further to 260 kilotons. In accordance with society's preferences emission-reduction measures can be allocated to the various emitters concerned; however, a nation-wide emissions trading system which includes every NOx polluter, has not been established by law.[93] In addition it shall be noticed that the legislators discretion to design the NOx system is constrained by the strict standards of the IPPC Directive.[94] The NOx system's

90. VROM, (2003), 17.
91. The Gothenburg Protocol was signed in 1999 and entered into force on 17 May 2005.
92. Directive 2001/81/EC.
93. The transportation sector, receiving about 60% of the emissions allocations, is expected to encounter severe difficulties in meeting its emission target. Government will hold a security reserve to ensure compliance with Directive 2001/81/EC on national emission ceilings for certain atmospheric pollutants and strives to maintain the sectoral distributions. VROM, (2003), 14 and 29.
94. See Case number 200502867/1, *MOB and Waddenvereniging vs. het college van gedeputeerde staten van Groningen*, of 15 Feb. 2006.

efficiency would be impaired if another regulatory standard interferes with the benchmarks it sets and surpasses them in terms of stringency.

If policy makers were deciding to sharpen the benchmark in order not to permit an expansion of industrial emissions under the NOx Emission Trading System when industrial output increases, one may speak of a de facto emission cap. This highlights the essential positionality[95] of environmental protection standards. Thereby an environmentally beneficial 'race for top environmental standards' is created. Whenever the Dutch government determines that the overall use of emission allowances of a particular industry is sufficiently close to the maximum amount allocated to it, it would have to sharpen environmental-friendly production standards to create incentives for firms to invest in abatement technology. Pre-evaluation of the NOx emission trading system shows, however, that the PSR benchmark of a maximum of 40 gram NOx per giga joule is not stringent enough to meet the envisaged objective of 55 kilotons of NOx emissions for 2010.[96] Industry is predicted to surpass this amount by 12 kilotons unless benchmarks would be significantly reduced.[97]

Since it has been suggested that benchmarks would not be lowered beyond 40 gram NOx per giga joule and that the risks of not complying with the NOx emission cap of the NEC Directive would be for the government to bear,[98] it can be questioned whether the Dutch NOx trading system's adjustment mechanism is effectively designed[99] and whether a 'security reserve' of emission credits held by the government and used in case it fails to realize its commitments is a way forward.

The above discussion has shown that allocation mechanisms cannot influence the environmental effectiveness of Emission Trading Systems because allocation mechanisms are charged with the task to bring an externally determined and prescribed amount of emission allowances onto the market. Reservations, however, do arise when allocation mechanisms are combined with 'relative Performance Standard Rate' trading systems which are not subject to any form of cap.

Here the absolute amount of emission allowances in circulation and hence the overall amount of emissions can vary with respect to the total output of the economy. Whether a lowering of the social marginal costs of CO_2 emissions in Europe can tip the balance towards a conscious and normative preference of additional economic output is not only subject to societal preferences but also a question of both legal and economic considerations.

---

95. Frank, R., (1997), 169 defines a 'positional good' as 'a good whose value depends strongly on how it compares with similar goods consumed by others; . . .' and also referred to it as a 'status good'.
96. Van der Kolk Advies (2006), 2.
97. PSRs as low as 31 gram NOx per GJ have been mentioned. See Van der Kolk Advies (2006), 5 and 83.
98. Van der Kolk Advies (2006), 2.
99. Tietenberg, T., Grubb, M., Michaelowa, A., Swift, B. & Zhang, Z. X., (1999), 26, are critical about credit and trade schemes and assert that virtually no system has been associated with major economic or environmental gains.

Whether a relative PSR can work in parallel with a cap and trade system, its potential environmental and legal aspects are an interesting subject to further research.

3.4   SUMMARY

This part of the chapter has outlined the importance of initial allocations in an Emission Trading System. The comparison of various allocation mechanisms has shown that auctions are best able to solve the problems of price determination and can thus give rise to an allocation that allows the timely attainment of a market equilibrium or gives rise to a market equilibrium itself. Yet if auctions give rise to strategic behaviour of bidders efficient transaction mechanisms are needed to ensure allocative efficiency and the market equilibrium. The final selection of an auction mechanism is, however, dependent upon a magnitude of factors that include, market structure, the nature of bidders, the size of the market and information possessed by the bidders. Determination of an optimal mechanism design in the absence of such information is futile.

Due to the inherent information asymmetry bureaucrats cannot be expected to allocate emission allowances to those market participants that value them the most and consequently cannot ensure an allocative efficient distribution of emission allowances. In the presence of efficiently operating trading markets, however, this does not pose a problem because allocative efficiency can be ensured via market exchanges on such markets. In all administrative allocation mechanisms normative distribution considerations prevail over allocative efficiency. Allocative efficiency can, however, be reached in the presence of efficient secondary markets. Whether a society is better off under such systems strongly depends on its preferences.

With regard to free allocation mechanisms, it should be noticed that 'relative standard base' allocation mechanisms directly reward environmentally pro-active producers, while a grandfathering system does not have the same advantage. In the presence of increasing abatement costs 'early movers' will have higher abatement cost structures and hence also a higher valuation of CO_2 allowances. Consequently the relative standard base allocation mechanisms are allocatively superior to grandfathering systems in such environments.

From an environmental perspective, the discussion in a static closed economy has shown that the impact on the environment is dependent on the atmospheric CO_2 concentration and hence the amount and timing of emissions. The amount of allocated emissions is external to allocation mechanisms. Therefore all mechanisms reviewed can perform equally well. It should be noted, however, that this finding only holds as long as additional CO_2 emissions generated by economic growth do not offset positive environmental effects stemming from relative standards. In the latter case, environmental standards would have to be strengthened in order to attain identical emission targets.

4   DYNAMIC OPEN ECONOMIC SETTING

This part of the chapter consists of three interrelated parts. The first will give an overview of the relevant trade theory literature and the assumptions of an open dynamic economy. The second part examines how allocative efficiency in an open dynamic economy differs from a static closed economy. The third addresses environmental considerations. This last part is further divided into two sections. The first reviews variables that influences firm's production decision and hence the environment. The second analyses the environmental impact of the various initial allocation schemes.

4.1   THE DYNAMIC OPEN ECONOMY

After having examined allocative efficiency and environmental considerations in a static closed environment, some of the underlying assumptions are relaxed. We stop assuming that the world consists of only one economy but allow for international exchanges. An open economy allows for international trade through both import and exports. The economy is here assumed to be dynamic in the sense that it allows for the analysis of firm's strategic reactions to initial allocation mechanisms. By this it is meant that firms are thought to be able to relocate. Since effects of research and development are assumed not to take effect in the short run, they are not considered.

    Before elaborating upon the specific assumptions made for the open economic setting[100] discussed below, a general introduction into international trade theory is in order. The theory of international trade is built upon the ideas of the founding fathers of economic science: on the idea of absolute advantages and comparative advantages. The former was developed by Adam Smith (1776) who proposed that the degree of labour specialization would be dependent upon the size of the market.[101] His disciple David Riccardo (1821) theorized that relative cost differences give rise to international trade.[102] The idea of comparative advantages has been formalized in the Heckscher-Ohlin theorem of factor endowments.[103] In contemporary trade theory both comparative advantages and economies of scale are recognized as important determinants of trade. Most of the environment-and-trade literature is build on the comparative advantage model.[104]

---

100. The treatment in this chapter does not encompass detailed mathematical models in the strict economic usage of the concept but represents a logically reasoned discussion of effects that facilitate an understanding of the economic intuition.
101. Smith, A., (1937).
102. For a general overview see Krugman, P. & Obstfeld M., (1997).
103. For a general overview see Krugman, P. & Obstfeld M., (1997).
104. Kuik, O., (2005), 28.

Starting with Markusen (1975)[105] and Pethig (1976)[106] environmental externalities were discussed in trade literature. Extensive overviews of the empirical and theoretic results of the literature have been made elsewhere and will not be reproduced here.[107] An important finding of this literature is that environmental externalities negatively impact gains from trade and that the first best solution is to internalize negative externalities rather than to restrict trade.

Since the focal point of interest is the allocative efficiency and its environmental impacts of initial allocation mechanisms, Joint Implementation Mechanisms,[108] Clean Development Mechanisms,[109] economic growth and demand issues will not be discussed here. The economy allows for strategic firm reactions to the implementation of an Emissions Trading System such as the European ETS. Firms can react to the introduction of such systems by reducing output, investing in abatement technology and by shifting production abroad. This is assumed to be executed by shifting production abroad, even though it should be noticed that also abatement can be 'exported'. In particular with regard to the EU ETS, the linkage Directive grants Member States discretion[110] to allow operators to use Certified Emission Reductions (CERs)[111] and Emission Reduction Units (ERUs)[112] from project activities in the Community scheme up to a percentage of the allocation of allowances to each installation.[113] It is noticeable that this can already take effect in the first EU ETS trading period (2005–2007) with regard to CERs and from 2008 also with regard to ERUs and that the setting of percentages is only mandatory

---

105. Markusen, J. R., (1975).
106. Pethig, R., (1976).
107. The interested reader is referred to Rauscher, M., (1997).
108. The Joint Implementation (JI) mechanism is established by the Kyoto Protocol and allows parties listed in Annex I of the Kyoto Protocol to receive emissions reduction units for co-financing projects that reduce net emissions in an Annex I country. Article 30(3) of Directive 2003/87/EC contains the possibility to incorporate JI mechanisms into the European Emissions Trading System. The linking between the EU ETS and flexibility mechanisms is governed by Directive 2004/101/EC. See De Cendra de Larragán, J., (2006).
109. The Clean Development Mechanism (CDM) is established by the Kyoto Protocol (definition in Art. 12) and allows Parties listed in Annex I of the Kyoto Protocol to finance emission-reduction projects in countries which are not listed Annex I. Annex I parties are awarded certified emission reductions (CERs) for doing so. The goals of the CDM are two-fold: firstly to assist non-Annex I parties in pursuing sustainable development policies and in contributing to the ultimate objective of the convention and secondly to assist Annex I parties in meeting their emission targets. Article 30(3) of Directive 2003/87/EC contains the possibility to incorporate CDM mechanisms into the European Emissions Trading System. For an economics review of CDMs and JIs see Brander, L., (2003). The linking between the EU ETS and flexibility mechanisms is governed by Directive 2004/101/EC. See De Cendra de Larragán, J., (2006).
110. Member States' discretion is limited by the criteria in Annex III of Directive 2003/87/EC (as amended) and by requirements that the total use of ERUs and CERs must be consistent with relevant supplementarity obligations under the Kyoto Protocol.
111. From CDM projects.
112. From Joint implementation projects.
113. Article 11(a)(1), Directive 2003/87/EC as amended by Directive 2004/101/EC.

form 2008 onwards.[114] Excessive use of these instruments is intended to be foreclosed by the supplementarity requirement of the Marrakech accords.[115]

In order to better understand the implications of initial allocation mechanisms the possibility of 'exporting' abatement efforts via CDMs is assumed to be limited so that abatement takes place within the EU. It furthermore assumes that foreign environmental requirements are negligible or non-existent (polluter havens[116]) and that the only relevant variables influencing firms' relocation decisions are transport costs and profit opportunities abroad.

4.2   COMPARISON OF ALLOCATION MECHANISMS
           AND ALLOCATIVE EFFICIENCY

Having introduced the dynamic open economy setting, the next section will provide a comparison of initial allocation mechanisms. In particular, the differences to the static closed economy setting are emphasized. This section consists of two parts. The first addresses auctions while the second deals with financial and free initial allocation mechanisms, including grandfathering and 'relative standard base' systems.

4.2.1   Auctions as an Allocation Mechanism

The findings made in a dynamic open economy are very similar to those made in the static closed economy discussed earlier. There are, however a number of additions which need to be addressed. At first, collusion is treated. Thereafter, declining price anomaly will be reviewed.

As presented in the static economy, in-auction collusion is of particular importance in multiunit auctions.[117] Such issues as demand reduction and signalling have already been addressed in this chapter. Since the same negative effects hold true for a dynamic environment, they will not be repeated here. There is, however, an important addition: in a dynamic setting one does not only take into account in-auction collusion but also collusion occurring in a multi-sequential auction setting. Repetitive interaction of bidders provides the opportunity to retaliate against non-cooperating cartel members in later auctions.[118] In the presence of market power, an increased expected profit from collusion will distort the emission

---

114.   See Recital 5, Directive 2004/101/EC. This is subject to the requirement that the amount a Member State wishes to allocate has been specified in its National Allocation Plan, see Art. 30(3) of Directive 2003/87/EC.
115.   Supplementarity is defined as ' . . . the use of the mechanisms shall be supplemental to domestic action and that domestic action shall thus constitute a significant element of the effort made by each Party included in Annex I to meet its quantified emission limitation and reduction commitments under Article 3, paragraph 1'. See Marrakech Accords, (2001), 51.
116.   On polluter havens see Antweiler W., Copeland, B. R. & Taylor, M. S., (2001).
117.   Salmon, T., (2003), 10.
118.   See Martin, S., (1994), 163.

allowance market and impact allocative efficiency. Such form of collusion is beyond the reach of auction design theory and is dealt with in a Competition law context.

Similarly, the problem of 'declining price anomaly' which occurred only in a sequential auction framework in the static setting, is exacerbated in the dynamic economic setting. Here the temporal scope of simultaneous multiunit auctions is cast into a number of subsequent auctions over time. Hence bidders are forced to predict the evolution of market prices. A higher degree of uncertainty[119] is associated with demand and supply estimates in the more distant future, and expected to exacerbate the strategic complexities firms will have to deal with. The degree of uncertainty firms exposed to in both sequential and simultaneous multiunit auctions is expected to be higher in a dynamic environment than in a static one. High uncertainty may induce risk averse bidders to overbid even more in early auctions. Consequently strong declining price anomaly impedes allocative efficiency.

If Competition law would work suboptimally and fail to contain collusion occurring outside auctions, *ceteris paribus*, allocative efficiency would be worse than in a closed static environment. Similarly a higher degree of uncertainty associated with larger prediction errors of forecasts is expected to exacerbate declining price anomaly and hence negatively affect allocative efficiency and the prospects to reach the same market equilibrium in comparison to the static setting. If allocative efficiency cannot be attained through auctions, efficient secondary markets on which bidders can cost-effectively exchange emission allowances are needed.

4.2.2   Administrative Allocation Mechanism

After having reviewed how the allocative implications of auctions in an open dynamic economy differ from a closed static economy, we now turn to a set of different allocation mechanisms. Subsequently, the changes of both financial administrative and free administrative allocation mechanisms with regard to the dynamic setting will be analysed. Allocative efficiency is the relevant criterion.

4.2.2.1   *Financial Administrative Allocation Mechanisms*

The allocative effects of a financial administrative allocation mechanism are dependent on its particularities, but some general aspects can be highlighted. As in the static setting, allocative efficiency is expected to be suboptimal and the market equilibrium may not directly be attained because bureaucrats will normally not be fully informed about the needs of operators nor are entrepreneurs willing to reveal their true preferences.

Due to societal considerations, a financial administrative allocation mechanism can be created to cross-subsidize specific operators. Through the granting of

---

119.   See Neugebauer, T. & Pezanis-Christou, P., (2005).

more favourable prices, operators can be awarded windfall profits.[120] Independent of initial allocation, profit-maximizing firms will take the opportunity costs[121] of the awarded emission allowances into account and rationally decide to use or to sell them.[122] Therefore the same finding as in the static setting prevails: efficient secondary markets have to ensure efficient allocation of emission allowances.

4.2.2.2 Free Administrative Allocation Mechanisms

Unlike auctions or financial allocation mechanisms, free administrative allocation mechanisms do not require payments for emission allowances. As in the static economy, in a dynamic one the normative distribution justifications routed in societal preferences or industrial policy prevail over allocative efficiency considerations. Allocative efficiency is subject to the same reservations concerning leakages and transaction costs as presented in the static environment and has to be ensured on secondary allowance markets. Below two particular forms of free administrative initial allocation mechanisms are being discussed: (a) grandfathering, and (b) relative standard base initial allocation systems.

4.2.2.2.1 Grandfathering

As already outlined in the static setting, grandfathering systems cannot be expected to generate an allocatively efficient initial distribution of emission allowances and may not lead to a timely attainment of a market equilibrium. As in the static environment, grandfathering in the dynamic environment can be allocatively superior to other free administrative allocation mechanisms because the former distributes allowances to a set of market operators who can be expected to have a high valuation for emission allowances. Given, however, that grandfathering is based on historic emissions and not on abatement cost structures, this allocation mechanism will tend to be less allocatively efficient than auction models.

4.2.2.2.2 Relative Standard Base Mechanisms

Similarly to other free allocation mechanisms, the relative standard base systems, including the PSR system, are not allocatively efficient and may not give quickly rise to a market equilibrium. This is because emission allowances are allocated for free, leaving politicians and bureaucrats to estimate abatement cost structures and industry abatement potential in order to ensure a welfare maximizing allocation. This is as difficult to achieve as under a grandfathering system and can only be

---

120. Windfall profits are unexpected profits that accrue to the recipient and are beyond his control. Such windfall profits may give rise to State aid considerations under Art. 87 EC Treaty. Windfall profits in this context are to be viewed as non-distortionary lump sum subsidies. See on this point Bohm, P., (1999), 11.
121. Opportunity costs are commonly understood as the cost of something measured in terms of the benefit forgone by not using it for a different purpose. For a straightforward introduction to the relationship of opportunity costs and free allocation of emission allowances see Woerdman, E., Couwenberg, O. & Nentjes, A., (2006).
122. On this point see also Woerdman, E., (2003), 109.

remedied via an effectively functioning Emission Trading System. As already indicated in the static economic setting, 'early movers' are rewarded for investing in abatement technology. If 'early movers' have higher abatement cost structures,[123] a relative standard base allocation system will be more allocatively efficient. The finding of the closed static economy is thus repeated.

This section has generated results analogous to those in the closed static economy setting. Free administrative allocation systems are not allocatively efficient because bureaucrats are unable to award those market participants who have the highest valuation with emission allowances and may not give rise to a market equilibrium allocation but depend upon a secondary market for its attainment. In the presence of upward sloping abatement cost curves, however, relative standard base allocation systems are allocatively superior to grandfathering systems.

4.3   ENVIRONMENTAL ASPECTS

This part is decomposed into two sections. The first reviews variables that influence firms' production decisions and hence the environment. Subsequently the environmental impact of the various initial allocation schemes is analysed.

4.3.1   Dynamic Interaction of Firms and the Environment

This part of the chapter describes the complex interaction of an open economy in which emission allowances are introduced. Here it is assumed that from an environmental point of view it is irrelevant where in the world CO_2 emissions are reduced. Though it should be observed that this may not be the case with regard to other greenhouse gases.[124] Unlike in a static setting, a dynamic setting allows for strategic firm behaviour and the adaptation to its business environment. The actual effects of environmental legislation may therefore differ significantly from its envisaged objective. This section takes the afore mentioned assumptions into account and takes note of the international trade literature addressing carbon leakages. Carbon leakages can be defined as the counterbalancing effect occurring in other countries than those in which carbon emission mitigating policies are undertaken.[125]

Trade theory has identified four channels of carbon leakage.[126] Leakages can occur through (1) trade in energy goods, (2) through interactions of government policies, (3) trade in factors of production and (4) trade in goods and services. Since the first two are not directly relevant for the strategic decision making of

---

123.   This will be the case in the presence of an increasing abatement cost curve.
124.   In cases where the quantity of emission is dependent on the emission locus, e.g., for health reasons or public welfare, a 'cap on a cap' in the form of regulatory interventions has been suggested to limit negative effects. See Pring, G., (2006), 192.
125.   See Kuik, O., (2005), 3.
126.   This passage is based on Kuik, O. (2005), 14 ff.

undertakings it suffices to note their existence without discussing them in detail. The first way to generate leakages assumes that policies to reduce CO_2 emissions will lead to a (global) price reduction of CO_2 intensive fossil fuels and thus to an increase in demand of such fuels in countries that have not engaged in emission reductions. The second channel of carbon leakage, governmental policies, assumes that there is an interdependence between national policies. If one country follows a climate change reducing policy this may create incentive or disincentive (e.g., free ridging[127]) effects in other countries.[128]

As for the other two carbon leakage channels, trade in goods and services and in factors of production, firm's strategic responses have to be considered in more detail. The classical Heckscher-Ohlin model predicts that the equalization of commodity prices and factor prices can be attained through trade in goods and services and/or trade in capital.[129] If the strict assumptions of this model are relaxed,[130] the equivalence between trade in goods and trade in factors of production may break down: they are then complementary rather than substitutes.[131] It appears therefore in order to examine the strategic reactions of firms to CO_2 reduction policies with regard to trade in factors of production and goods and services jointly while indicating their differences at a later stage.

Due to the liberalization of international investment flows and the advances in transportation and communication, capital has become more mobile than ever before. In the 1990s globalization intensified when firms engaged in stronger vertical disintegration of production processes through outsourcing and process fragmentation. The ratio of global trade in goods and commercial services to world GDP has increased strongly from 8% in the 1950s to 19.8% in 1990. In the few years till 2002 this ratio increased by another 10%.[132] An important motivation underlying globalization is the desire to benefit from international price differentials of production inputs.[133]

Rising environmental standards will increase production costs and lower the rate of return on domestic capital, leading to a reduction of the competitive edge of production sites and thus eventually to capital outflows. This in turn will impact private income and government's ability to finance its expenditure via taxes. At present the knowledge on the direction and magnitude of the interdependences between environmental cost internalization and international competitiveness is limited.[134] Therefore the following discussion is restricted to a theoretical review of relevant strategic implications of firm behaviour that is followed by a short discussion of relevant literature in this field.

---

127. See Carraro, C., (1999).
128. Copeland, B. R. & Taylor, M. S., (2003).
129. Mundell, R. A., (1957).
130. See Mundell, R. A., (1957), 326 for qualifications and Krugman, P. & Obstfeld, M., (1997) for a general introduction to the Heckscher-Ohlin Model.
131. On this point see Springer, K., (2000).
132. See Van den Bossche, P., (2005), 8.
133. Jones, R. W. & Kierzkowski, H., (2001) and Ethier, W. J., (2002) for detailed discussions.
134. Alanen, L., (2003), 1112.

As already presented in this chapter, the economic intuition behind emission trading is the internalization of negative externalities. Incentives to produce in a socially and environmentally desirable way are created via the market-price mechanism by increasing costs of production. In the following a firm's profit maximizing reaction to increases in production costs and its associated environmental implications are reviewed.

The obligation to submit allowances on a predetermined date in accordance with specified criteria can entail budgetary consequences for operators. Whether these consequences will be positive or negative depends on CO_2 abatement costs, the prevailing market price of emission allowances, operating costs and stranded costs.[135]

Budgetary impacts will be positive if the market price of emission allowances exceeds the cost of CO_2 abatement to such a degree that an operator is able to profitably sell emission allowances on the market. Positive effects of emission trading are enhanced by mitigating the impact on operating costs by saving on expensive CO_2 intensive energy consumption.[136] There will also be a positive (or less negative) budgetary effect if an operator receives part of the needed CO_2 emission allowances for free or below their true market value. This is generally referred to as windfall profit.

On average, however, negative budgetary implications are to be expected, given the environmental objective of reducing anthropogenic CO_2 emissions. Besides the direct costs of emission allowances, administrative and operative costs are non-negligible. In addition to these costs, firms will incur accounting losses if existing investments will be rendered unprofitable through increases in variable operating costs. These costs are referred to as stranded costs. Particularly energy and thus CO_2 intensive sectors will be subject to considerable increases in production costs and may not be able to withstand competitive pressure.[137]

The above discussion shows that an important determinant to induce behavioural change of operators is to increase their production costs. Whether the introduction of a regional Emission Trading System such as the European Emissions Trading System will indeed lead to the intended reduction of global CO_2 emissions, is a different issue. Profit maximizing operators can reduce CO_2 emissions within the economy by investing in abatement technology, or by reducing production output. This is precisely the same finding as was discovered in the static setting. Allowing for dynamic interaction, however, one should also note the possibility of firms to adapt strategically to costly environmental protection schemes. Firms have three additional ways to react to such external price shocks. They can

---

135. Existing investments which are rendered unprofitable due to the introduction of environmental regulations are commonly referred to as stranded costs.
136. Fossil energy sources are rich in hydrocarbons, which liberate CO_2 during combustion. CO_2 is thus the result of a chemical reaction (combustion) that involves the hydrocarbons of the fossil energy sources and the oxygen from the atmosphere.
137. For a recent study addressing this issue see Reinaud, J., (2005).

(1) outsource abatement, (2) they can outsource production through movements of factors of production or (3) substitute domestic goods that contain the costs for CO_2 emissions for imported goods that do not. Each will be treated in turn.

4.3.1.1 Outsourcing Abatement

Firstly, in an open dynamic economy such as the one of the European Union, firms can 'outsource' CO_2 abatement. This can be done via Joint Implementation[138] and Clean Development Mechanisms.[139] Both mechanisms allow firms to benefit from lower abatement costs abroad. This in turn mitigates the cost burden placed upon domestic firms and hence the pressure to rely upon other means to contain CO_2 emission. Critics argue that this may create significant leakage and thus is not desirable from an environmental point of view. It should, however, also be noted that a lowering in the abatement cost structure enables operators to produce within the domestic economy and avoid additional CO_2 emissions associated with a shifting of production abroad.[140] From a legal point of view it has to be observed that the EU is bound by the supplementarity[141] requirements committed under the Marrakech Accords[142] to attain a significant element of its greenhouse gas reduction domestically. State liability issues may merit further legal attention.[143]

---

138. The Joint Implementation (JI) mechanism is established by the Kyoto Protocol and allows parties listed in Annex B of the Kyoto Protocol to receive emissions reduction units for co-financing projects that reduce net emissions in an Annex B country. Article 30(3) of Directive 2003/87/EC contains the possibility to incorporate JI mechanisms into the European Emissions Trading System. The linking between the EU ETS and flexibility mechanisms is governed by Directive 2004/101/EC and will allow ERUs to be used as of 2008. See De Cendra de Larragán, J., (2006).
139. The Clean Development Mechanism (CDM) is established by the Kyoto Protocol (definition in Art. 12) and allows Parties listed in Annex B of the Kyoto Protocol to finance emission-reduction projects in countries which are not listed Annex B. Annex B parties are awarded certified emission reductions (CERs) for doing so. The goals of the CDM are two-fold: firstly to assist non-Annex B parties in pursuing sustainable development policies and in contributing to the ultimate objective of the convention and secondly to assist Annex B parties in meeting their emission targets. Article 30(3) of Directive 2003/87/EC contains the possibility to incorporate CDM mechanisms into the European Emissions Trading System. For an economics review of CDMs and JIs see Brander, L., (2003). The linking between the EU ETS and flexibility mechanisms is governed by Directive 2004/101/EC. See De Cendra de Larragán, J., (2006).
140. See discussion below.
141. Supplementarity is defined as '. . . the use of the mechanisms shall be supplemental to domestic action and that domestic action shall thus constitute a significant element of the effort made by each Party included in Annex I to meet its quantified emission limitation and reduction commitments under Article 3, paragraph 1'. See Marrakech Accords, (2001), 51.
142. See Marrakech Accords, (2001), 51 and 52.
143. Campins Eritja, M., (2006), 78, discusses State liability before the Kyoto Protocol parties.

*4.3.1.2   Outsourcing of Production through Movements
                of Factors of Production*

Secondly, in an open dynamic economy firms can outsource production. A restriction of CO_2 emissions in one region can lead to increases in production costs which trigger changes in relative prices and shifts in trade patterns which work to offset positive CO_2 reductions.[144] It is appropriate to note that here the existence of 'polluter havens'[145] and ineffective international trade law sanctions[146] are expressly assumed. In such cases a mere suspicion that higher production costs may reduce profitability may lead to a redirection of investments *ex ante*, that is, before the introduction of cost internalization measures. It will, however, certainly lead to a redirection of future investments.

The profitability and hence occurrence of such outsourcing and the redirection of trade flows depends on transportation costs and profits reaped on the foreign market.[147] If it is more profitable to produce in the absence of any environmental cost internalization measures abroad and to serve the foreign and domestic market at the same time by shipping goods to the EU, profit maximizing firms will do so in the long run. Particularly so if growth prospects are more favourable abroad and transportation costs are expected to be declining over time. Production relocation would entail an increase of CO_2 emissions abroad with potential negative effects for the environment. Not only would the apparent EU emission be understating its true CO_2 consumption, but *ceteris paribus* even lead to a rise in CO_2 emissions due to an increase in the average transportation distance of goods marketed in the EU. This negative impact on global CO_2 emissions would be exacerbated if foreign producers used more carbon intensive production techniques.

As already indicated above, in reality the strategic decision of a firm to relocate is related to a large number of factors and is not limited to production costs alone. Relocation of enterprises is subject to a body of research known as 'new economic geography',[148] yet given that the present model does not provide for other influencing factors the relevant literature is not discussed further. It suffices to note that in the presence of relevant factors other than those related to production

---

144. Such effects have long been addressed in multi-regional studies. See for example Pezzy, J., (1992), and Edmonds, J., Wise, M. & Barns, D. W., (1995), Burniaux, J. M. & Oliveira Martins, J., (2000), Munskgaard, J. & Pedersen, K., (2001).
145. On polluter havens see Antweiler, W., Copeland, B. R. & Taylor, M. S., (2001).
146. For work analysing the possibilities to establish climate change levies under international trade law the interested reader is referred to Grimeaud, D., (2003), Kuik, O., Tol, R. S. J. & Grimeaud, D., (2003), Green, A., (2005), Van Calster, G., (2006).
147. In reality there are of course other political and socio-economic factors taken into account before a production decision is taken. For the sake of simplicity, only transportation costs and profit opportunities on the foreign market are taken into account.
148. See in general Fujita, M., Krugman, P. & Venables, A., (1999). Elbers, C. & Withagen, C., (2004) examine labour mobility as a function of environmental concerns and Calmette, M. F. & Péchoux, I., (2004) analyze the effects of clustering upon the probability of environmental disasters.

costs, it has been shown that relocation may not occur.[149] It appears that differentials in labour productivity, taxation and environmental protection are not of primary significance for investment decisions of multi-national corporations,[150] and that capital is less mobile than commonly assumed.[151]

While empirical literature seems to depict ambiguous effects regarding the linkage between environmental regulation so that no firm conclusions can be drawn,[152] simulation studies do not suggest that Kyoto Protocol induced capital flight will be of major significance.[153] While there seems to be a general consensus about the limited contribution of capital mobility to carbon leakage in the short run, Burieaux (2001) expects that industry relocation may become the primary source of leakage in the long run.[154]

4.3.1.3   *Substitute Domestic Goods for Imported Goods*

Besides outsourcing abatement and shifting production sites abroad undertakings can also alleviate their cost burden stemming from environmental regulation by substituting domestic goods that are subject to a CO_2 levy by foreign goods that are not. This can be described as the goods and service trade channel for carbon leakage. It has been found that the ease of substitution of domestic products by foreign ones significantly contributes to carbon leakage.[155]

Besides ease of substitution and increasing capital flows that follow suit,[156] an additional factor impacting carbon leakage is the structure of the market on which a particular undertaking operates. As was the case in the simple oligopoly models introduced in chapter two of this text, also in international trading contexts competitors can recognize their mutual interdependence.[157] This is also of relevance in the context of carbon leakages.[158] If production costs of large operators increase due to CO_2 taxations, this may lead to a reduction of production while carbon intensive productions in other countries may increase. Depending on the particular legislation applicable to dominant firms, this could contribute to carbon leakage. In addition it can be shown that in a model of monopolistic competition[159] the number of profitably operating firms on the home market can decrease as production costs

---

149. Baldwin, R. & Krugman, P., (2000), Jeppesen, T. & Folmer, H., (2001). Environmental regulation may, however, significantly influence destination choices of companies that have taken the decision to relocate, see List, J., McHone, W. W. & Millimet, D. L., (2003).
150. Wang, Z. K. & Winters, L. A., (2001).
151. See Gordon, R. H. & Bovenberg, A. L., (1996) for an information asymmetry based approach.
152. See Jeppesen, T., List, J. & Folmer, H., (2002), 36.
153. Babiker, M. H., (2001), Burniaux, J. M., (2001), Paltsev, S. V., (2001).
154. Burniaux, J. M., (2001).
155. See Gielen, D. & Karbuz, S., (2003), Paltsev, S. V., (2001).
156. An increase in demand of foreign goods may trigger investments in new distribution systems etc. See Rauscher, M., (1997), Ch. 3.
157. Krugman, P. & Obstfeld, M., (1997), 132.
158. Babiker, M. H. & Jacoby, H. D., (1999).
159. See Krugman, P. & Obstfeld, M., (1997), 127 ff.

rise because of environmental measures.[160] In such cases consumers' demand for product variety may be saturated through increased imports of foreign produce. This would imply that market structure can have a baring upon the strategic reactions of firms in the presence of production cost increases due to environmental legislation and can thus also have an effect on the overall carbon leakage. It can therefore be concluded that also under the trade channel carbon leakage can occur that reduces the overall effectiveness of national or multi-national abatement schemes.

The above section can be summarized as follows. It has been shown that the introduction of an Emission Trading System can inflate production costs of firms. In an open dynamic environment firms are able to react strategically to price shocks. Such firm reactions may have negative effects upon the environment. While the procurement of CO_2 abatement abroad lowers the overall cost burden on enterprises and permits domestic firms to continue their operations, relocation of production, substitution of domestic produce and market structure may lead to carbon leakage. The decisive factor for relocation, the substitution of domestic produce and the generation of associated negative environmental effects is the existence of a cost differential between environmental cost internalization on one hand and transportation costs on the other. Profit opportunities in foreign markets are also taken into account and counterbalance high transportation costs. Positive predictions of the evolution of transportation costs and profits earned abroad and dim prospects of permit prices can tip the balance towards relocation decisions. In this context it bears mentioning that the EU ETS amendment for the third trading period encompasses the inclusion of special provisions for sectors exposed to severe competitive pressure.

4.3.2   Initial Allocation Mechanisms and the Environment

After having reviewed general firm behaviour that can be triggered by production cost increases due to the introduction of an Emission Trading System, the following section reviews qualitative environmental differences stemming from the selection of different initial allocation mechanisms. Systems considered include auctions, financial administrative allocation systems as well as free administrative allocation systems. Here particular attention is devoted to grandfathering and the PSR system.

4.3.2.1   *Auctions*

Unlike in the static closed economy setting, auction mechanisms used in an open economy are capable of impacting the environment. Due to the mere fact that in auctions CO_2 emission allowances are sold, auctions do generate the strongest distributional effect of all initial allocation mechanisms. The cost burden placed

---

160.   Gürtzen, N. & Rauscher, M., (2000).

upon firms will lead to a reduction in CO_2 due to investment in abatement technology and the reduction of output to a socially desirable level. From an environmental point of view the critical element is the change in the production decision. The reduction on the rate of return of domestic capital can trigger the relocation of domestic production plants and lead to a rise in CO_2 emissions. This will be the case if an increase in transportation distance of goods is associated with more CO_2 emissions and if production abroad is more CO_2 intensive.

One interesting opportunity provided by auction systems is that they generate government funds in a non-distortionary manner. These funds can be channelled back to operators. Investment shifts will not occur if the collected funds are redistributed in form of tax cuts in such a way as to fully compensate operators for the expenses they incur.[161]

If funds are distributed via tax cuts to other groups of society, an increase in disposable income will lead to an increase in demand and hence to more overall world emissions. This will particularly be the case if foreign direct investment stimulates economic growth and profitable business opportunities in the host country.

The threat of the occurrence of declining price anomaly and collusion will distort the equilibrium market price. The declining price anomaly can give rise to an overshooting effect in the sense that the market price will at the beginning lie above its true equilibrium. This could, however, lead to short run over investment in abatement technology. Such positive environmental effects, may be partially depleted during the price adjustment towards its true equilibrium. Collusion will not lead to an overstatement but to an understatement of bidder's true valuations. This leads to adverse environmental effects to the extent that the lower market price for emission allowances gives rise to suboptimal investments in abatement technology and increases market penetration of foreign competitors.

4.3.2.2   *Administrative Allocation Mechanisms*

After having reviewed the environmental impact of auction mechanisms, this section reviews the impact of administrative allocation mechanisms on firm behaviour and consequently, its impacts on CO_2 emissions. First financial administrative allocation mechanisms are being reviewed. Subsequently free administrative mechanisms, including grandfathering and relative standard base mechanisms are examined.

---

161. The substitution of distortionary taxes on labour or profit could be substituted by the levy of non-distortionary taxes on CO_2 emissions. Ballard C., Shoven, J. & Whalley, J., (1985) estimate that one dollar raised through distortionary taxes can cost society one dollar 30 cents. Thus society, including operators, can be made better off by levying taxes through auctions. Such overcompensation is also found by Bovenberg, A. & Goulder, L., (2000).

4.3.2.2.1   Financial Administrative Allocation Mechanisms
As already indicated, environmental effectiveness in a dynamic open economy
does not only depend on the absolute emission cap but also on how operators adapt
to changes in their business environment. Emission reduction in a trading system is
only achieved by inducing operators to take externalities into account. This is
achieved by inflating the price of an earlier under-priced good.

Even profit maximizing operators who have been awarded windfall profits
under a financial administrative allocation mechanism have every incentive to
reduce emission if their production costs with regard to CO_2 increase. This
leads to a reduction in output, an increase in CO_2 efficiency or to a transfer of
production abroad. If transportation costs are sufficiently low as to allow a firm to
produce abroad and import goods at a profit, a profit maximizing operator has
every incentive to do so. This is rational behaviour irrespective of any windfall
profits which are granted to particular operators or sectors. This may lead to a
furtherance of the wedge between CO_2 production and CO_2 consumption within
the EU.

4.3.2.2.2   Free Administrative Allocation Mechanisms
Similarly, even in the case where all emission allowances are distributed for free,
overall scarcity is determined by the quantity supplied. Since in all cases reviewed
the initially allocated amount is assumed to be the same, the degree of scarcity and
hence also the prevailing market price will be the same. Firm's strategic reaction to
the introduction of emission allowances is similarly dependent upon transportation
costs and expected profits from abroad. Since the actual costs incurred by operators
are lower than under auction and financial administrative allocation mechanisms,
ceteris paribus, less relocation will occur.

4.3.2.2.2.1   Grandfathering.   Free allocation based on historical input, output
or emission data gives rise to strategic firm behaviour. Overall CO_2 emission will
depend on the market price of emission allowances, abatement costs, transportation costs and expected profits abroad. If it is profitable firms will shift production sites abroad and prefer to import rather than to produce within the EU. This
entails negative environmental consequences. Grandfathering, however, differs
from other free allocation mechanisms because it gives rise to additional strategic
firm behaviour. Before the base year is being set, firms do have every incentive to
emit more CO_2 and to postpone emission-reduction programs in order to increase
their future allocation. Depending on the basis upon which allocation in future
trading periods is made, such strategic considerations may also lead to time shifting
effects. How such negative environmental effects compare to other allocation
mechanisms is subject to further research but it seems that grandfathering based
on actual historic emissions is less preferable than free allocation mechanisms
based on for example, market share which do not give rise to such self-defeating
incentives.

4.3.2.2.2.2   Relative standard base mechanisms.   As has been shown in the static economy setting, a relative standard base allocation system can be applied to attain a predetermined emission cap provided that the industry standards are adjusted in accordance to increases in production. The costs accruing to particular operators are expected to reflect their abatement possibilities. Therefore the threat that existing investments will be transformed into stranded costs appears to be lower. Similarly, if transportation costs and potentially higher monitoring costs[162] were indirectly taken into account by assessing an industry's competitive position, relocation will be limited and consequently, environmental effects will be limited. This will, however, only be the case if monitoring costs are sufficiently low and uncertainty about bureaucratic action to raise prevailing environmental standards to maintain the CO_2 cap are not such as to induce firms to relocate. Since PSR systems reward early movers, they penalize late movers. This creates the danger that if standards are set too high, late movers bearing the relatively higher costs will have stronger incentives to relocate. Whether relocation can be limited through the introduction of relative standards that can contain adverse global CO_2 effects depends not only on the standard setting institution's ability to determine abatement capabilities but also on the stringency of the selected emission cap. If the abatement target is so large as to exhaust all possibilities to burden domestic industry, relocation will necessarily take place.

In the absence of a fixed emission cap, it is argued that CO_2 emission will rise more than under an absolute emission cap. While this is a valid reservation under a closed economy setting if decision makers are inclined to follow social considerations and lobbyists not to increase relative standards, this is not so evident in a dynamic open economy. Here, similarly, the reservations regarding decision makers' determination to adhere to goals committed to under International treaties are valid. Independent of the decision maker's resolution to contain CO_2 emissions, feasibility of CO_2 abatement is determined by profit opportunities abroad and transportation costs. Firms that can operate more profitably in a country which does not internalize negative externalities and still sell products in the domestic economy, will do so. The only means for the domestic economy to push internalization of externalities beyond the level of transportation costs is to shift existing taxation away from current tax sources[163] or to levy an import tax[164] to take the differences in CO_2 intensity of production attributable to particular goods[165] into account.

---

162.   Gielen, A. M., Koutstaal, P. R. & Vollebergh, H. R., (2002), 11 assert that monitoring costs for particular industries under a relative standard base mechanism will be higher if no existing relative standards can be referred to.

163.   One could think about increasing the tax burden on CO_2 while reducing the taxation for labour by an equal amount so that it is rational for firms to adjust to the new taxation rules by reducing the amount of CO_2 emission without having incentives to relocate production facilities.

164.   Import taxes serve to protect producers' surplus at the expense of consumers' surplus. While the overall social welfare effect is ambiguous, it is assumed to be lower for small economies. See Krugman, P. & Obstfeld, M., (1997), 193 ff.

165.   An 'import tax' is the same as an import tariff, or customs duty. Tariffs on most products are bound in the schedules of the WTO Members. Unilateral increases in tariffs are generally a violation of WTO law (Art. II GATT 1947) and are permitted only in certain very restricted cases (e.g., anti-dumping, countervailing, safeguards).

4.4   SUMMARY

This part of the chapter has shown that allocative efficiency is more difficult to attain in an open economy. The attainment of allocative efficiency on the market is complicated in a dynamic setting by collusion occurring outside the auctions and can only be contained via the application of Competition law. Similarly a higher degree of uncertainty associated with larger prediction errors of forecasts is expected to exacerbate declining price anomaly and hence negatively affect allocative efficiency. If allocative efficiency cannot be attained through auctions, efficient and sufficiently liquid secondary markets are needed. As in the static closed economy setting, allocative efficiency cannot be attained via administrative distribution systems which emphasize normative distribution considerations. Efficiently operating secondary markets allow operators to realize the windfall profits they have been awarded via financial or free administrative allocation mechanisms. In the presence of upward sloping abatement cost curves, however, relative standard base allocation systems are allocatively superior to grandfathering systems. Thus the finding made with respect to the static setting has been confirmed.

In a dynamic setting firm's reactions to an increase in production prices has an important bearing on the environment. The decisive factor for international competitiveness and the decision to relocate production and the generation of associated negative environmental effects, that is, carbon leakage, is the existence of a cost differential between environmental cost internalization on one hand and transportation costs on the other. Profit opportunities in foreign markets are also taken into account and counterbalance high transportation costs. Positive predictions of the evolution of transportation costs and profits earned abroad and dim prospects of permit prices can tip the balance towards relocation decisions.

Distributional impacts on particular firms can be exacerbated or mitigated through the selection of an initial allocation mechanism. Auctions generate the strongest cost burden on particular firms and thus also the strongest incentives for firms to relocate if the funds that are levied in a non-distortionary manner are not redistributed to bidders. If other parts of society are awarded parts of the funds, it can be expected that an increase in disposable money will lead to increased demand and CO_2 emissions. While collusion will not generate adverse environmental effects, declining price anomaly may lead to a temporary overinvestment in abatement technology. The cost burden placed upon firms is less severe under a financial administrative allocation mechanism than under auctions since permits are assumed to be marketed below their true market price. Free administrative allocation systems place the lowest direct cost burden on enterprises and thus generate the least incentive for firms to engage in environmentally unfriendly production relocation. Yet it should be noticed that grandfathering systems can induce operators to postpone emission-reduction projects in order to be awarded more emission allowances. Such incentives are not present for relative standard base mechanisms. In addition stranded costs are expected to be limited. Under a cap and trade system – or in case of its absence if a given emission quantity is observed – and assuming that the set relative standards take into account abatement costs and industry's competitive position, and that monitoring costs and uncertainty costs are limited, relative standard base systems can perform in the most environmentally efficient manner.

In a dynamic open economy the upper boundary for the internalization of environmental externalities is dependent upon the international competitive position of domestic firms and not determined by domestic regulation prescribing absolute emission targets. In such cases the existence of a cap or its absence is not of practical relevance since this differentiation does not affect firm's production decisions. Hence both from an environmental and competitiveness point of view the most cost-effective system is most desirable. Depending on the particular mechanism design, this might be a relative standard base system which may be an element of a PSR Emission Trading System. In the alternative well working support schemes for exposed sectors – such as those that are envisaged for the post 2012 trading period – need to be provided for. Such schemes will, however, impact the relative prices between goods that are being produced within the EU and lead thus to a distortion of the economy.

5 CONCLUSION

This chapter has placed various initial CO_2 emission allocation mechanisms into a typology and analysed them within a static closed and a dynamic open economy setting according to their price determination and allocative efficiency and the environment in order to determine which allocation formats are most desirable from an allocative efficiency and environmental point of view. A sound understanding of how different allocation mechanisms compare with each other is essential to examine if allocative efficiency can be safeguarded under the EU ETS if one assumes that market adjustment processes are not leading to immediate and full adjustments.

With regard to the static closed economy the chapter shows that only auctions are capable to solve the problem of price determination and to attain allocative efficiency and a market equilibrium. This is, however, assuming that auctioning is employed at a very large scale so as to make any secondary market irrelevant. In the presence of collusion or declining price anomaly, however, the equilibrium outcome can differ from the perfect competitive one. Since in administrative allocation mechanisms normative distribution considerations prevail over allocative efficiency, they have to depend upon efficient secondary markets to attain allocative efficiency. Relative standard base mechanisms award 'early movers' and thus are allocatively superior to grandfathering schemes.[166] Relaxing the strict assumptions towards a dynamic open economy confirms the above findings with only one qualification. Due to an exacerbated risk of collusion and a potential declining price anomaly attributable to a larger degree of uncertainty, auctions perform less well in the dynamic closed economy than in a static closed economy. Therefore it

---

166. Here it is expressly assumed that the PSR standards are rightly defined.

can be concluded that only auctioning mechanism can ensure allocative efficiency and that the other allocation formats that can be applied under the EU ETS do rely on cost-effective exchange systems to safeguard allocative efficiency and that PSR systems are preferred to grandfathering allocation schemes.

With regard to environmental aspects, initial allocation mechanisms under a static closed economy are unable to affect the environment as long as a strict dichotomy between the quantity setting decision and the distribution issue of CO_2 prevails. This, however, is not the case for a dynamic open economy setting. The introduction of an emission allowance trading system increases production costs of operators. In the framework considered, transportation costs and profit opportunities abroad constitute the upper limit of the internalization of negative environmental externalities because operators are able to adapt to price increases in their production costs by for example, relocating production facilities abroad or by substituting domestic goods by imported ones if it is profitable for them to do so. Global CO_2 emission will rise due to relocation if the average transportation distance increases[167] or if production abroad is not subject to equivalent environmental regulations.

In the absence of adequate compensation auctions generate the highest burden on operators, lead to strong incentives to relocate and may thus perform worst from an environmental perspective. Its performance may only partially be mitigated by short-term positive environmental effects stemming from declining price anomaly. The relocation inducing burden under financial administrative systems is less severe than under auctions but not as low as under free administrative allocation systems. Due to the non-existence of time shifting incentives and low stranded costs, relative standard base systems are more environmentally friendly than grandfathering mechanisms. This will be the case in a cap and trade system as well as in the absence of a cap as long as the overall emission quantity is not increased, that standards are set in such a way as to take abatement costs and industry's competitive position into account and that monitoring costs and uncertainty costs are sufficiently low. Irrespective of the existence of an emission cap, the upper limit to the internalization of environmental costs is given by operators' ability to evade such costs. Therefore it can be concluded that indeed initial allocation mechanisms have a bearing on the environment and that the financial impact of allocation mechanisms may merit more attention than the discussion of capped trade and trade without a cap may suggest. From a social welfare point of view it may even be possible that the costs associated with the failure to attain allocative efficiency is outweighed by benefits stemming from the internalization of negative externalities. In such circumstances the selection of a cost-effective administrative allocation mechanism may be more desirable than the selection of auctions that attain the market equilibrium at the expense of potentially limited environmental benefits.[168]

---

167. Here the underlying assumption is of course that CO_2 emissions are positively correlated with transportation distance, implying that emissions increase as distance increases.
168. The magnitude of such effects is subject to further research.

In this context it bears particular mentioning that the EU ETS amendment contains express provisions to address exposed sectors and to mitigate the cost pressure steming from the introduction of auctions. The co-existence of auctioning and special policies directed towards exposed sectors in the period post 2012 may, however, not only have a positive impact upon the environmental effectiveness of the scheme but also distort relative prices between auctioning and exposed sectors. The negative effect of distortions of relative prices should be outweighed by the positive gains in terms of environmental effectiveness of the scheme to increase social welfare.

Chapter 4

Compatibility of Allocation Formats and Directive 2003/87/EC

1   INTRODUCTION

After having established a typology of emission trading systems in the previous chapter and after having identified the most desirable allocation mechanisms with regard to allocative efficiency, price determination and the environment in the framework of a static closed and a dynamic open economy, this chapter assumes a legal perspective. It examines if the main allocation systems discussed in the previous chapter, that is, auctions, grandfathering and PSR are compatible with the Directive establishing the EU ETS in its original version and its amendment that will have to be transposed by Member States by 31 December 2012. This chapter therefore constitutes the necessary methodological link between the general economic findings of chapter three and the EU ETS. Only if the allocation mechanisms presented in the previous section are compatible with the Directive can their assessment under EU Competition law be meaningful for policy makers.

    This chapter is subdivided into seven parts. After the introduction, the requirements established by the Directive regarding an allocation model are presented in section 2. Taking the gained insights into consideration, the subsequent sections examine if the previously discussed allocation mechanisms are compatible with the Directive. Section 3 examines auctions and section 4 addresses the grandfathering system. Subsequently the compatibility of the PSR system is examined (section 5). Thereafter the EU ETS amendment is treated. Here it is examined with which criteria the proposed allocation mechanisms need to comply and which effects they could have on allocative efficiency. A conclusion summarizes the main points and assesses the findings from a Law and Economics perspective to answer the question if allocative efficiency is safeguarded under the current form and the amendment.

2   ALLOCATION FORMATS UNDER
                DIRECTIVE 2003/87/EC

As has already been elaborated in the first chapter of this book, Directive
2003/87/EC establishes the EU ETS and obliges Member States to operate it.
This section of the chapter examines the legal requirements under the Directive
that serve as a legal framework for allocation mechanisms. The Directive orders
Member States to identify those undertakings falling within its scope, to grant them
greenhouse gas emission permits, to issue a proportion of the total quantity of
allowances allocated for a trading period every year by the 28th of February,[1]
and to establish effective monitoring and verification procedures.[2] Besides monitoring the functioning of the system[3] and reporting upon the application of the
Directive,[4] Member States are in particular charged with the allocation of emission
allowances.[5] Since allowance allocation is the core interest of this section of the
chapter, stipulations governing National Allocation Plans (NAPs) will be considered with regard to this task.

      Member States draw up NAPs in which they state the total quantity of allow-
ances they intend to allocate during a particular period and how they propose to
allocate them.[6] From this it can be understood that the EU ETS is designed in the
form of a cap and trade system.[7] Any allowance allocation mechanism employed
under the EU ETS has therefore to be compatible with an absolute emission cap.

      NAPs are subject to scrutiny by the Commission to the extent that they are
incompatible with Article 10 or Annex III.[8] Article 10 contains particular thresh-
olds concerning free allocation and distribution means that are not for free (auctioning). In the first three year period at least 95% of allowances must be allocated
for free, while in the subsequent five year period up to 90% of allowances are to be
allocated free of charge. Auctioning will, however, become the dominant form of
allocation under the EU ETS amendment. Prior to making its legislative proposal

---

1. Article 11(4), Directive 2003/87/EC.
2. Articles 4–6, Directive 2003/87/EC. Related legislation includes Commission Regulation (EC)
   No. 2216/2004 on national registries and the Commission Decision on the avoidance of double
   counting of Kyoto project mechanisms under the EU ETS, see Commission Decision 2006/780/
   EC, C(2006) 5362.
3. Articles 14 and 15, Directive 2003/87/EC.
4. Article 21, Directive 2003/87/EC.
5. Articles 9–11, Directive 2003/87/EC. Emission allowances are defined in Art. 3(a) as '. . . an
   allowance to emit one ton of carbon dioxide equivalent during a specified period, which shall be
   valid only for the purpose of meeting the requirements of this Directive and shall be transferable
   in accordance with the provisions of the Directive'.
6. Article 9, Directive 2003/87/EC.
7. Though it should be observed that the cap is softened by the possibility of allowing the use of
   CERs and ERUs. See Art. 30(3), Directive 2003/87/EC.
8. See Art. 9(3), Directive 2003/87/EC. In Case T-387/04 *EnBW Energie Baden-Würtemberg AG v.
   European Commission* [2007] n.y.r., para. 112–115, the Court distinguishes the preliminary
   control mechanisms of Art. 9(3), Directive 2003/87/EC from the formal approval taking procedure under Art. 88(3) EC Treaty. It thereby clarifies the legal standing of the Commissions
   Decisions on the National Allocation Plans.

in January 2008, the Commission was still expecting that unless the Directive were to be amended in the course of the review process of the EU ETS,[9] that the threshold of 10% auctioning for the trading period 2013–2017[10] would be applicable. Last but not least Article 11(1) and (2)[11] set deadlines until when each Member State has to decide upon the total quantity of allowances it will allocate and the allocation of those allowances to the operator of each installation.

Article 9 of the Directive requires that NAPs have to be based on objective and transparent criteria, including those stated in Annex III of the Directive.[12] The majority of these requirements come in the form of vague criteria enlisted in Annex III of the Directive and are further elaborated in the two Commission guidance notes[13] that assist Member States in the drawing up of their NAPs. These Annex III criteria require that the total quantity of allowances to be allocated by Member States is consistent with its obligation to limit emissions pursuant to Council Decision 2002/358/EC,[14] Council Decision 93/389/EEC[15] and the Kyoto Protocol and with a path towards achieving or over-achieving its obligations.[16] Besides these requirements allocations must be consistent with the (technological) potential of the activities to reduce emissions[17] and take into account new entrants,[18] clean technology and public comments.[19] NAPs are also required to specify the maximum amount of CERs and ERUs which operators can use under the EU ETS.[20] NAPs may also accommodate early actions, best available

---

9. See Art. 30(2)(c), Directive 2003/87/EC.
10. European Commission (2006) MEMO/06/452 at 7. It should be noticed that this is not stipulated in Art. 10 of the Directive 2003/87/EC but based on the aforementioned Commission Memo.
11. For the first trading period this the deadline was 31 Sep. 2004 and for the second trading period it is 31 Dec. 2007.
12. Article 9, Directive 2003/87/EC. It may also be observed that Art. 9(3) of Directive 2003/87/EC uses a review period during which it can react to submitted NAP that is longer than the period prescribed by Art. 4(5) of Council Regulation (EC) No. 659/1999 of March 1999 laying down detailed rules for the application of Art. 92 of the EC Treaty, OJ L 83/1 of 27 Mar. 1999. On this point see Zwingmann, K., (2007), 170 and Case T-387/04 *EnBW Energie Baden-Würtemberg AG v. European Commission* [2007] n.y.r., para. 112–115.
13. COM (2003) 830 final, and COM (2005) 703 final.
14. Decision concerning the approval of the Kyoto Protocol to the United Nations Framework Convention on Climate Change and the joint fulfilment of commitments there under.
15. Decision for a monitoring mechanism of Community CO_2 and other greenhouse gas emissions.
16. Annex III, Recitals 1 and 2 Directive 2003/87/EC. The requirement of overachievement only regards Decisions 2002/358/EC and the Kyoto Protocol.
17. Annex III, Recital 3 Directive 2003/87/EC. See also Decision 280/2004/EC concerning a mechanism for monitoring Community greenhouse gas emissions.
18. New entrant is defined in Art. 3(h) of Directive 2003/87/EC as 'any installation carrying out one or more of the activities indicated in Annex I, which has obtained a greenhouse gas emissions permit or an update of its greenhouse gas emissions permit because of a change in the nature or functioning or an extension of the installation, subsequent to the notification to the Commission of the national allocation plan'. It bears mentioning that the Art. 3(h) of the Amendment to Directive 2003/87/EC has been altering the definition of new entrants.
19. Annex III, Recitals 6, 8 and 9 Directive 2003/87/EC.
20. This amount is specified as a percentage of the allocation of allowances to each installation. The percentage must be consistent with the Member State's supplementarity obligations under the

technologies and contain information on the manner in which non-EU competition is taken into account.[21]

Besides these more general criteria that are relevant for NAPs, a number of Annex III criteria are of particular importance with regard to allocation mechanisms. Criteria 4 and 5 require that NAPs have to be consistent with other Community legislation, including Competition law provisions.[22] One particular element that is not expressly referred to in the Directive itself but that is enshrined in Article 174(2) EC Treaty is the polluter pays principle. Even though this principle remains largely beyond the scope of the text, it is briefly considered because allocation mechanisms differ considerably in light of this principle.[23]

Quite importantly, another criterion, criterion 10, stipulates that NAPs have to contain a list of the covered installations with the quantities of allowances intended to be allocated to each.[24] It therefore appears that the Directive envisages a direct link between the quantity of allowances to be allocated and each installation prior to the starting of each trading period.[25] While the Directive does not clarify the scope of the concept of 'intent', the Commission takes a narrow interpretation in order not to undermine the efficient functioning of the trading market by endangering the stability and predictability of the supply on the allowance market.[26] The Commission's rejection of any *ex post* allocation scheme[27] is exemplified by its decision regarding the German NAP[28] and appears to be following a strict interpretation of the wording of Article 11(1) and (2). The Court, however, clarified in its case C-347/04 of 7 November 2007 that the Commission erred in law in its assessment that the German *ex post* adjustment contravened Community law.[29] The Commission's view point thus strongly contrasted to the view of the Court and the position of thirteen Member States that did try to include *ex post* adjustment schemes into their second NAPs[30] despite the existing guidance note.[31] Given the strong position taken by the Commission to avoid *ex post* allocations they seemed to be an important criterion with which allocation formats needed to comply.

In light of the above it can therefore be stated that there are a number of relevant elements that a potential allocation mechanism under the current legislation had to fulfil in order to be viewed to be compatible with Directive 2003/87/EC. An allocation mechanism must be compatible with a cap and trade system, and not

---

Kyoto Protocol and decisions adopted pursuant to the UNFCCC or the Kyoto Protocol. See Annex III, Recital 12 Directive 2003/87/EC as amended by Directive 2004/101/EC.
21. Annex III, Recitals 7 and 11 Directive 2003/87/EC.
22. Annex III, Recitals 4 and 5 Directive 2003/87/EC.
23. See Nash, J. R, (2000).
24. Annex III, Recital 10 Directive 2003/87/EC.
25. See Art. 11(1) and (2) and Annex III, Recital 10 Directive 2003/87/EC.
26. COM (2005) 703 final, 16.
27. *Ibid.*
28. European Commission, (2004), Decision concerning the German NAP, C(2004) 2515/2 final, 07 Jul. 2004, 3 and Art. 1(b).
29. Case T-374/04 *Germany v. Commission* [2007] n.y.r., see also Weishaar, S., (2008).
30. IP/05/762 of 20/06/2005.
31. COM (2005) 703 final, 16.

embrace *ex post* adjustments. In the subsequent sections it will be examined how the three allocation mechanisms, auctions, grandfathering and PSR can be assessed under these elements.

3   AUCTIONS

As presented in chapter three, auctions are purely market based instruments that are capable of solving the problem of price determination faced by administrative allocation mechanisms. Auctions are mentioned in the Directive in Article 30(c) within the context of reviewing the possibilities for further harmonizing allocation mechanisms for the trading periods after 2012. Since up to 5% and respective 10% of allowances can be sold in one form or another, it is justified to examine how auctions can be assessed under the issues listed in the previous section.

With regard to the requirement that auctions should be compatible with a cap and trade system, it suffices to notice that auctions can be employed to bring a predetermined amount of emission allowances on to the market. If the rules of the remainder of the Emission Trading System are such as to limit the absolute amount of emission allowance, it can be concluded that auctions can be compatible with an absolute emissions cap.

With regard to the conformity of allocation mechanisms with Competition law, auctions do not distort competition on the merits or give rise to any unequal treatment as such. As has been shown in chapter three, there are mechanism designs that are more prone to collusion and even abusive practices than others but this does not suffice to condemn all auctions per se as distorting competition on the merits. It is also noticeable that because auctions require competitors to pay for their allowances, there is no element of State aid (Article 87 EC Treaty) that is potentially problematic under free allocation mechanisms. While this appears to hold true in general, this wholehearted embrace of auctions from a State aid perspective will have to be qualified when one derogates from full auctioning such as was examined previously. In the EU ETS amendment auctions are slowly becoming the central allocation method but the secondary market will retain a central role. While full access to auctions is also envisaged in the third trading phase, here the general nature of the auctions is more questionable since the electricity sector is specifically singled out. Hence the State aid assessment is not quite as straight forward as under earlier schemes.

Linked to the fact that auctions require payment for each and every unit of emission is the effect that they are also complying with broad notions of the polluter pays principle that go beyond the Pigovian notion[32] of cost internalization. Within the framework of Figure 4, it is important that polluters take the social costs of emissions into account. This is represented by the area between private marginal and social marginal costs. Nash (2000), however, notes that residual pollution emissions, that take place after an emitter has complied with administrative

---

32. See Pigou, A. C., (1949), Ch. 8.

standards, are costless and that this is inconsistent with the polluter pays principle.[33] Nash therefore appears to be requiring that each emission unit is being paid for and not that the costs to operators are increased such that the optimal amount of pollution is generated. Since in an auction operators have to pay for each emission allowances this criterion is satisfied and it can be concluded that auctions are consistent with such a broad conceptualization of the polluter pays principle.[34]

Last but not least it has to be examined if auctions are at odds with the criterion of *ex post* adjustments. Interpreting the wording of Article 11(1)–(2) and Annex III criterion 10 strictly it would appear that auctions that are executed after the deadline for NAPs has passed, are incompatible with the Directive. The basic rationale being that only after an auction has been concluded can it be known which operator will hold which allowances. Considering the Commission's guidance notes it is apparent that auctions are mainly suggested for new entrants. In its first guidance note for example the Commission allows periodic auctions for new entrants,[35] requires the statement of form but not of time when these auctions will be held[36] and clarifies that no restrictions regarding the scope of participants is permissible.[37] From the wording of the guidance note it appears that this limitation is only applicable to new entrants but not extending to incumbent undertakings. If incumbents were also able to participate in auctions intended for new entrants this would imply that an express link between allocations and existing installations cannot be safeguarded. Because in the first trading period auctions are mainly used for bringing not used new entry reserves onto the emissions trading market,[38] this issue has not been problematic. With regard to the second trading period it appears that the situation is different.

Also in its guidance note for the second trading period did the Commission address the issue of auctions. Here the Commission requires the statement of time and the intervals when auctions are to be held.[39] Examining Member States' second NAPs it is noticeable that the European Commission accepted the Belgian NAP even though it does not specify which persons are allowed to participate in auctions.[40] It therefore appears that auctions will be held to be compatible with the Directive, even if it is not ensured that they are executed in due time before the start of the trading period – as prescribed by Article 11(1) and (2) of the Directive – so as to ascertain which installations will be holding which allowances.

The above can be summarized as follows. Auctions can be used as an allocation mechanism under the Directive. Auctions can be compatible with a cap and

---

33. Nash, J. R, (2000), 508.
34. Nash, J. R, (2000), 506.
35. COM (2003) 830 final, para. 55.
36. COM (2003) 830 final, Annex, s. 1, 24.
37. COM (2003) 830 final, para. 57 states that restricting persons to participate in such auctions is not in conformity with internal market rules.
38. See NAPs for the period 2005–2007 of Austria (2004), Greece (2004), Portugal (2004), Slovak Republic (2004).
39. COM (2005) 703 final, VIII, Nap summary table, 30.
40. See Belgium (2006), 192 and Commission Decision (2007) regarding Belgium Nap.

trade emission trading system, are assessed favourably under Competition law aspects and are in line with the wide constructions of the polluter pays principle. Regarding the element of *ex post* adjustments conformity rests upon the question if Article 11(1) and (2) of the Directive are being observed or not. To the extent that incumbents are allowed to compete with new entrants at auctions, and a very strict interpretation of the Article were applied, this could appear to be questionable. In addition it should be observed that the usage of auctions is limited to 5% during the first and 10% during the second trading period.[41]

4   GRANDFATHERING

Grandfathering allocation systems use historical emissions as a basis to allocate emission allowances for free. As is the case with auctions, grandfathering systems are dependent upon the particular rules of the remainder of the Emission Trading System to be compatible with the requirement of an absolute emissions cap.

They differ from auctions with regard to the second requirement of Competition law. Design issues could bring a particular grandfathering system into conflict with the Directive. Distortions of competition can stem from the selection of particular historical base years or unequal allocations that have varying effects upon competitors. These aspects are extensively addressed in subsequent chapters and will therefore not be extended beyond what is necessary at this stage.

There is, however, an issue that is not contingent upon the design of a grandfathering system: conformity with the polluter pays principle. Even though operators will take the value of grandfathered emission allowances into account for their production decision and one could therefore argue that the negative externality is brought within the market price mechanism, grandfathering allocation schemes are unlikely to satisfy a broader conceptualization of the polluter pays principle. If this principle were taken to be constructed so broadly as to entail direct expenses of emitters for every unit of pollution that is being emitted,[42] then grandfathering systems do clearly fail to comply with this requirement of the Directive.[43]

Grandfathering systems do not encounter difficulties with the last requirement of *ex post* adjustment. On the basis of historical data it can be determined which installation is to be awarded which quantity of allowance.

In light of the above, it can therefore be concluded that depending upon the particular notion of the polluter pays principle that is being applied and depending upon the finding distortions of competition, grandfathering systems can be compatible with the Directive.

---

41. See Art. 10, Directive 2003/87/EC.
42. See Nash, J. R., (2000), 505.
43. For a diverting opinion see Woerdman, E. & Arcuri, A., (2006).

5   PSR

After examining the compatibility of grandfathering allocation systems with the Directive, this passage considers the PSR system. In light of the above, a PSR system would face difficulties regarding a number of particular design aspects. These regard absolute emission reduction targets, Competition law aspects, the polluter pays principle and *ex post* adjustments. Each will be treated in turn.

In light of the typology presented in chapter three, it is clear that a PSR system, although in principle operating in the absence of a strict emission cap, would have to be compatible with a cap and trade system in order to be permissible as an allocation system under the EU ETS. This could be achieved within the system's design by setting PSR standards in such a way as to ensure compliance with the absolute emission reduction targets. In order to be able to estimate total emissions over a period of time the emission of the total industrial output for all installations falling under a PSR has to be predicted. Given that the exact usage of emissions is complicated by the fact that the operation of a benchmark will have an effect on actual production, adequate forecasting is complicated. It is therefore expected that a PSR scheme requires frequent stock taking reviews in order to determine if the PSR is appropriately set. Once it is apparent that the PSR is adequately determined, the time periods between stock taking reviews can be increased. Because a cap and trade system requires that the maximum emission amount is observed, economic growth could lead to an unaccounted increase of emissions that would endanger compliance with the absolute targets. In such cases PSR standards would have to be increased in due time in order to ensure full compliance with the cap. The question of compliance of a PSR scheme with an absolute cap and trade system is thus dependent upon the strength, ability and institutional independence of the authority charged with standard setting to adequately set and adjust PSR benchmarks.

With regard to the second issue, Competition law, a detailed review of the PSR scheme merits particular attention. Since such a review is executed in the two subsequent chapters, here the reference is limited to the most obvious concern that can be brought against the system. The Directive emphasizes the importance of not violating State aid legislation[44] and it can be stated that if a PSR scheme is incompatible with Competition law provisions it cannot be used under the EU ETS Directive. The State aid dimension of the PSR scheme will be discussed below in great detail and will not be examined here. Yet it can already be observed that concerning the Dutch NOx system – an epitome of a PSR system – the Commission has taken the view that a PSR system constitutes State aid.[45] The Court of First Instance (CFI) overturned the Commission's decision (case T-233/04) on the grounds of selectivity but the Commission has appealed to the European Court of Justice (ECJ) – this case is still pending before the ECJ (C-279/08). The issue of State aid is treated below.

---

44. See for example Recital 23, Art. 11(3) and Annex III Recital 5 of Directive 2003/87/EC.
45. European Commission, (2003), Steunmaatregelen van de Staten N35/2003 – Nederland Systeem van verhandelbare emissierechten voor NOx, C(2003) 1761 fin, 24 Jun. 2003.

Also with regard to the third element that is being considered, the polluter pays principle, it appears that the PSR system encounters difficulties. Under such a system it is expected that operators will take (technological) measures to enhance their greenhouse gas efficiency of production. Since they thereby bear the costs for pollution one could therefore argue that the negative externality is brought within the market price mechanism, and that PSR systems satisfy the polluter pays principle. If the polluter pays principle is viewed to entail direct expenses of emitters for every unit of pollution that is being emitted, then it is clear that the fact that polluters do not pay for their residual emission in excess of governmental benchmarks, is not compatible with this legal principle.[46]

The final aspect of the legal issues considered here, *ex post* allocation, merits particular attention from a PSR allocation scheme perspective. In accordance with Article 9(3) of the Directive, the Commission is allowed to monitor the conformity of NAPs with among others, recital 10 of Annex III. This recital states that NAPs must contain a list of covered installations and the quantities of allowances the Member State intends to allocate to each. While the Directive does not clarify the broadness of the concept of 'intend', the Commission takes a narrow interpretation. This position could be derived from its guidance note for NAPs under the first allocation period[47] and was expressed in its decision regarding the German NAP,[48] is endorsed by the explicit wording contained in the guidance note addressing the second allocation period[49] and can therefore be viewed to reflect the strong position of the Commission on this issue. In its decision regarding the German NAP[50] the Commission stated with regard to plant closure and transfer rules that criterion 10 in Annex III of Directive 2003/87/EC required the quantity of allowances to be allocated to each installation to be stated *ex ante* in the NAP.[51] Similarly German provisions regarding allocation adjustments in light of unforeseen low capacity utilization of installations were not entertained.[52] In light of the wording of Article 11(1) which obliges Member States to state the total quantity of allowances to be allocated and the allocation of those allowances to the operator of each installation, the Commission regarded a narrow interpretation to be well founded and likely to be upheld by the Court.[53]

46. See Nash, J. R, (2000), 505 ff., in particular, 508. The interested reader is also referred to De Cendra de Larragán, J., (2005).
47. COM (2003) 830 final, para. 97 and 100.
48. European Commission, (2004), Decision concerning the German NAP, C(2004) 2515/2 final, 07 Jul. 2004, 3 and Art. 1(b).
49. COM (2005) 703 final, 4 and 16.
50. European Commission, (2004), Decision concerning the German NAP, C(2004) 2515/2 final, 07 Jul. 2004, 3 and Art. 1(b).
51. European Commission, (2004), Decision concerning the German NAP, C(2004) 2515/2 final, 07 Jul. 2004, 3 and Art. 1(b).
52. European Commission, (2004), Decision concerning the German NAP, C(2004) 2515/2 final, 07 Jul. 2004, 3 and Art. 1(c).
53. Private party proceedings against Commission Decisions concerning NAPs that are based on Art. 9(3) of Directive 2003/87/EC are complicated by the Court's differentiation between the preliminary control mechanisms of Art. 9(3) of the Directive and the formal approval taking

The Court did, however, not follow the line of argumentation of the Commission. In its teleological interpretation, in particular examining the sub-objective of safeguarding the internal market, and in light of the self-limiting effect of the Commission's guidance, the Court concluded that the Commission erred in law. It failed to prove that criterion 10 of Annex III [54] limited Member States' freedom of action as to the forms and methods of transposing the Directive into national law. Also with regard to criteria 5,[55] concerning equal treatment, the Court rejected all submissions by the Commission. The CFI held that the Commission erred in law since it failed to prove that *ex post* allocations favoured new entrants and that *ex post* adjustments created incentives for undertakings to overstate demand while being less closely supervised by competent authorities. Furthermore the Court ruled that the Commission failed to state sufficient reasons for its decision and thus infringed Article 253 EC Treaty. Consequently the Commission's decision was annulled.

Despite the clarity of this Court ruling, it can be postulated that the practical effect is rather limited. With regard to Germany and the first trading phase the *ex post* corrections had to be undertaken within a relatively short period of time and in the presence of very low emission allowance prices.[56] For the second trading phase the impact was also limited since the German national allocation law (ZUG 2012)[57] had already been enacted prior to the publication of the case.

In light of the above it can thus be summarized that a PSR system is an allocation mechanism that is based on relative standards, that the usage of emission allowances is dependent on the emission efficiency of production and the total quantity produced, and that it is not possible to establish a direct link between an installation and the total emission quantity to be allocated to this installation before the start of a trading period. It can therefore be concluded that PSR systems would violate a strict interpretation of the Directive – as seems to be favoured by the Commission – with regard to Criterion 10 of Annex III, Article 11(1) and (4) concerning *ex post* adjustments.

_____

procedure under Art. 88(3) EC Treaty, see Case T-387/04 *EnBW Energie Baden-Würtemberg AG v. European Commission* [2007] n.y.r., para. 112–115.

54. Criterion 10 of Annex III of Directive 2003/87/EC reads as follows: 'The plan shall contain a list of the installations covered by this Directive with the quantities of allowances intended to be allocated to each.'

55. Criterion 5 of Annex III of Directive 2003/87/EC reads as follows: 'The plan shall not discriminate between companies or sectors in such a way as to unduly favour certain undertakings or activities in accordance with the requirements of the Treaty, in particular Articles 87 and 88 thereof.'

56. Weishaar, S., (2008), 150.

57. Gesetz über den nationalen Zuteilungsplan für Treibhausgas Emissionsberechtigungen in der Zuteilungsperiode 2008 *bis* 2012 (Zuteilungsgesetz 2012–ZUG 2012) from 7 Aug. 2007, BGBl. I 2007, 1788.

6   THE EU ETS AMENDMENT

The amendment to Directive 2003/87/EC fundamentally changes the present allocation format away from grandfathering towards auctioning and harmonized benchmarking. It substitutes Member State discretion under the NAPs to design and choose allocation formats that are presently subject to the criteria listed in Annex III and Commission scrutiny by harmonized allocation formats. It is not yet clear how these harmonized allocation rules will look like for the various undertakings but it appears certain that the Commission will have an important influence on their design through the commitology procedure. The examination if allocative efficiency is safeguarded under the scheme is thus necessarily restricted until more information is released by the Commission.

With regard to auctioning, Article 10(4) of the amendment obliges the Commission to adopt by 30 June 2010 a Regulation on timing, administration and other aspects of auctioning in order to ensure that auctions are conducted in an open, transparent, harmonized and non-discriminatory manner. In particular the auction process must be predictable for companies, and grant companies full, fair and equitable access to the scheme. Here it is noteworthy that information availability, cost-efficiency and market access for small emitters are of particular concern to the legislator. How harmonized these auction mechanisms will be is not yet known but it may be possible that only one or a few mechanisms will be permitted by the Commission. Member States' ability to influence the Commission is, however, possible during the discussion process of the commitology procedure that is employed in the drafting of the auctioning regulation. Besides the possibility that a 'one size fits all' approach will be opted for that will prevent auctions from realizing their full potential in terms of allocative efficiency and revenue generation, it is also noteworthy that the secondary market will remain a defining feature of the emission allowances market. This conclusion is based upon the large but decreasing number of free allocation that is envisaged during the third trading phase. This in turn may tempt auctioneers to strongly take the price information presented by the secondary market into consideration when designing their bidding strategies. To the extent that this market is not in equilibrium, it may also obstruct or slow down the process of reaching market equilibrium outcomes at auctions.

Also with regard to benchmarking very limited discretion appears to be left to Member States. The Commission is charged with designing community-wide and fully harmonized implementation measures for allocating allowances by the 31December 2010[58] – again the drafting of these measures are done within the framework of a commitology procedure. These Community-wide *ex ante* benchmarks should give incentives for reductions in greenhouse gas emissions and energy efficient techniques by taking account of the most efficient techniques, substitutes, alternative production processes, high efficiency cogeneration, efficient energy recovery of waste gases, use of biomass and capture and storage of carbon dioxide, where such facilities are available, while not giving incentives to

---

58.   Article 10a(1), Directive 2009/29/EC.

increase emissions.[59] Besides these criteria, the EU ETS specifies that these *ex ante* benchmarks in individual sectors or sub-sectors have to take the average performance of the 10% most efficient installations in a Community sector or sub-sector between 2007 and 2008 as the defining principle and orders the Commission to consult relevant stakeholders.

Benchmarks will thus be set for specific sectors or sub-sectors on historic emission data. In principle this approach is comparable to grandfathering from an allocative efficiency point of view in the sense that it does as such not take into account technologic advances nor the different abatement cost structures of various sectors. In this context it is positive that not only a uniform cross-sectoral correction factor can be employed to hedge for such sectoral differences[60] but also that the Commission can draw upon the expertise of operators themselves. While the later may help to foster allocative efficiency it also bears risks: regulated undertakings may have every incentive to strive for a more favourable allocation and to seek to misrepresent their true situation. A secondary market can therefore be expected to be crucial to ensure allocative efficiency on the EU ETS market. Last but not least it may also be observed that a benchmark set on product basis may lead to an increased of free allocated emission allowances if sectoral growth should exceed expectations. In which ways the Commission will take account of such eventualities is not yet known.

7   CONCLUSION

This chapter has analysed the compatibility of auctioning, grandfathering and PSR allocation systems with Directive 2003/87/EC. It has been shown that auctions are compatible with the Directive. Grandfathering schemes do comply with the requirements of compatibility with a cap and trade system, Competition law and *ex post* allocation but encounter problems with regard to the polluter pays principle. It is not expected that the violation of this EC law principle is so grave as to lead to the incompatibility of this system.

The question of compatibility of the PSR system with the Directive is in particular dependent on a detailed Competition law analysis that is conducted in the two subsequent chapters, and the need of *ex post* allocations. The incompatibility with *ex post* allocations is in particular related to the interpretation the Commission gives to criterion 10 of Annex III and based upon the stipulations contained in Article 11(1) that both seek to oblige Member States to expressly identify the exact allocations granted to particular installations. Since the PSR system's design necessitates *ex post* allocations, this system is excluded from the scope of the Directive – at least if the Commission's view is taken as the

---

59.   See Art. 10a(1), Directive 2009/29/EC.
60.   See Art. 10a(5), Directive 2009/29/EC.

relevant benchmark. For which particular political considerations such provisions have been established is not known but it may be related to transparency reasons.[61]

Comparing the above legal findings with those obtained in the previous chapter, it has to be noticed that auctions are the most desirable allocation format from an allocative efficiency perspective and they are also allowed under the EU ETS Directive. A significant draw back regarding the likelihood that allocative efficiency will indeed be attained regards the scope of their application. Not more than 5% during the first and 10% of allowances during the second trading period can be auctioned before the beginning of the trading period or to new entrants.[62] It may therefore be questioned if this amount is sufficient to fully realize the superior allocative efficiency properties of this allocation mechanism as to quickly attain a market equilibrium if adjustment processes are time consuming. Its overall social welfare effect is dependent upon the degree to which the limited cost burden and the associated limited inducement of relocation is able to compensate society through internalization of negative external effects. The concern if auctioning can realize its full allocative efficiency potential still remains valid albeit to a much lesser degree under the proposed amendment for the third trading period where only by 2020 up to 70% of emission allowances will be auctioned.[63]

Another point of critique regards PSR systems. Even though they are allocatively superior to grandfathering systems, they are excluded from the scope of the Directive on the basis that they do contain *ex post* adjustments and thus are unable to provide for an express link between an allocation and an individual installation three months before the beginning of a trading period.

The proposed free allocation format contained in the EU ETS amendment is based upon an *ex ante* determined sectoral benchmark. *Ex post* adjustments of the benchmarks may, however, be necessary if output growth in a number of sectors would be such that the total amount of emission allowances to be allocated for free would be larger than predicted and endanger the attainment of the binding 30% criterion contained in Article 10(a)11.

Even though the EU ETS is unlikely to attain a timely market equilibrium based solely on initial allocation mechanisms, it is still important to examine to what extent the EU ETS is able to give rise to perfectly competitive outcomes. In order to analyse this aspect the following two chapters address anticompetitive distortions stemming from the European trading system since they work to undermine the market forces that ensure an optimal mix of production. State measures taken under the EU ETS can impact firms' propensity to collude and abuse while the granting of subsidies is tantamount to a comparative advantage that is rationalized on environmental grounds. The next chapter therefore examines if the core

---

61. Article 17, Directive 2003/87/EC would disclose different information if a PSR system were to be employed.
62. Incumbents can benefit from auctions if a set new entrants reserve has not been depleted. See NAPs for the period 2005–2007 of Austria (2004), Greece (2004), Portugal (2004), Slovak Republic (2004).
63. This figure does, however, not take into consideration emission allowances granted to exposed sectors.

provisions of Competition law could be employed to contain anticompetitive effects originating in State measures that give rise to collusion and abuse. Chapter six addresses State aid issues. In the remainder of the text grandfathering, PSR systems and auctions, in particular under the EU ETS amendment, are examined. Although excluded under the present Directive, PSR systems are considered because they have been shown to be allocatively superior to grandfathering systems and because they may constitute a feasible approach for negotiating sectoral agreements on an international level.

Chapter 5

EU Emissions Trading System and Articles 81 and 82

1   INTRODUCTION

Competition on the merits is an important condition for the attainment of allocative efficiency. Abusive or collusive actions conducted by firms impede the reaching of optimal allocations and create welfare losses. In order to address the question if allocative efficiency can be safeguarded under the EU ETS it has to be analysed if Competition law – the area of law specifically designed to address distortions of competition – is able to contain them.

This chapter examines how anticompetitive distortions originating from Member State action under National Allocation Plans in general and by application of the grandfathering allocation mechanism and the Performance Standard Rate (PSR) Emission Trading System in particular, are dealt with under European Competition rules. It also examines how the allocation systems in the EU ETS amendment are to be assessed from a Competition law perspective. An inter-disciplinary approach making use of both industrial economics concepts and European Competition law is chosen as the framework for analysis.

The chapter is structured into several interrelated parts. Firstly, current anti-competitive distortions created by Member State measures inducing cartelisation and abuse are reviewed (section 2). The second part reviews those parts of EC Competition law that can be employed to contain such State measures and their evolution (section 3.1). Within this section the joined application jurisprudence under Article 81 (cartelisation), and Article 82 (abuse) is examined (sections 3.2 and 3.3). Subsequently Article 86 (directed against public undertakings) is treated (section 3.4). An economic appraisal and summary emphasises the main findings of this section of the chapter (section 4). Subsequently it is examined what role the joint application jurisprudence may play in light of the EU ETS amendment.

2   ECONOMIC PROBLEM

This section reviews the effects on competition that can be created by measures taken by Member States within the framework of the current European Emissions Trading System and that distort the attainment of allocative efficiency on the production market. It is examined in which ways the National Allocation Plans, the grandfathering allocation method and the PSR system create incentives for undertakings to behave in a welfare reducing way. The two issues identified from an industrial economics point of view, are barriers to entry and entrenchment of market shares that lead to the accentuation of undertakings' interdependence. Each will be treated in turn.

2.1   BARRIERS TO ENTRY

'Barriers to entry' can be described as factors that impede undertakings from entering a particular market. Which factors do constitute such barriers has been subject to decades of heated academic debate between the proponents of different antitrust schools of thought. Without reviewing the economic merits of the debate, the author will limit himself to cite factors that have been identified to constitute barriers to entry. Harbord and Hoehn (1994)[1] have presented a typology of such barriers. They distinguish between (1) absolute cost advantages that are enjoyed by the incumbent but are not borne by the new entrant (patents, copyrights, exclusive contracts with input suppliers), (2) first-mover advantages (economies of scale, sunk costs, product differentiation, advertising, goodwill, capital requirements), (3) vertical foreclosure and exclusion (refusal to supply, tying, exclusive contracts, vertical integration), (4) predation and (5) entry impediments (licensing,[2] certification and product registration requirements). According to this typology, the barriers to entry created by the introduction of an Emission Trading System, all fall within the first category, regarding absolute cost advantages incumbents enjoy over new entrants.

Not affording new entrants equal treatment can set them at a direct financial disadvantage.[3] Such barriers of entry can be erected by National Allocation Plans in general or be closely related to a particular allocation format.

---

1. This passage is based upon Harbord D., Hoehn, T., (1994), 415 ff.
2. Even though emission allowances could be regarded as constituting a licence for emitting CO_2 within the European Emission Trading System, in this typology the concept of 'licensing' is associated with a largely costless but time consuming process and hence not reflecting the situation at hand.
3. It should be noticed, however, that not granting new entrants equal treatment through a comparable (free) allocation of allowances can give rise to environmental benefits to the extent that free allocations constitute a production subsidy. In the presence of such a subsidy undertakings may prefer environmentally unfriendly means of production over environmentally friendly options. The incentive compatibility argument is also underlying transfer rules that grant emission allowances

New entrants can be set at a cost disadvantage when national reserves for new entrants are depleted. This is in part attributable to the large degree of discretion Directive 2003/87/EC affords Member States with regard to the treatment of new entrants,[4] but also related to the ambiguity existing in the equally authentic language versions.[5] If in such situations new entrants will have to buy emission allowances on the market instead of receiving them for free as their incumbent competitors, they will suffer from – *ceteris paribus* – higher operating costs.[6] While such problems are inherent if grandfathering is used as a form of allocation, it should be noted that this is not the case under a PSR system. When emission allowances become scarce under this system, environmental benchmarks can only be increased uniformly because the system does not differentiate between new entrants and incumbent undertakings.

Negative discrimination of new entrants does not only arise from resource depletion and the possibility to have new entrants pay for their allowances but also when criteria to allocate emission allowances differ between incumbents and new entrants. Under the Dutch NAP for 2005–2007 for example, new entrants could never receive more allowances than needed while some specified incumbent top performers could benefit from up to 10% more allowances.[7] Thus whenever new entrants are set at a comparative disadvantage, barriers to entry are created.

The PSR Emission Trading System does not generate such impeding and hence undesirable effects. By way of construction of the system, new entrants are subject to the same government benchmarks and hence are allowed to emit as much as incumbent firms.

By design, new entrants under a PSR system do not suffer from unequal treatment. While new entrants cannot be disadvantaged with regard to potential scarcity and uncertain emission allowance costs once they are on the market, they remain subject to other comparative disadvantages as are existing under the grandfathering system, to the extent that the potential competitor is credit rationed,[8] investments in the market are sunk,[9] or the financial burden of borrowing money is positive. In such cases new entrants need – *ceteris paribus* – a higher level of profits to find it appealing to invest. To the extent that the particular market under

---

to (outdated) production plants that are being substituted for more environmentally friendly ones. See Åhman, M., Burtraw, D., Kruger, J., Zetterberg, L., (2007).

4. It only contains the obligation for Member States to take the need to provide access to allowances for new entrants into account. See Art. 11(3) and Annex III criteria 6 of Directive 2003/87/EC.

5. While the English version speaks of the obligation to take into account the need of new entrants, the Dutch version of the text speaks of the necessity to keep emission allowances available for new entrants.

6. Notable exceptions include the German and Polish NAP for 2005–2007, see Germany (2004), 37 and Poland (2004), 39.

7. See the Dutch NAP for 2005–2007, The Netherlands (2004a), 27.

8. This implies that a potential entrant is unable to attain sufficient funds to make the necessary investments to enter the market.

9. Sunk costs refer to irrecoverable costs once invested. They do not have a bearing on a firm's decision to exit a market but constitute a decisive factor for market entry.

consideration is not perfectly competitive, such barriers to entry give incumbent firms the possibility to raise market prices and reap positive economic profits.[10] A reduced threat of new entries does not only have positive effects on cartel stabilisation but is also liable to give rise to suboptimal use of scarce resources due to reduced competitive pressure. In the absence of empirical evidence, it is impossible to determine the importance of these effects. What however is certain is that Competition law is essential to contain any potential cartelisation and abuse that may originate in Emission Trading Systems.

As presented above, from an industrial economics point of view, barriers to entry are criticized because they facilitate cartelization and abuse of market power. If markets are characterized by insufficient competition between incumbents and above normal rate of returns, new undertakings will enter and bring the market back into equilibrium. Barriers to entry have the effect to impede such entrance and the market is unable to correct itself. This is particularly lamentable from a social welfare perspective since the incumbent's reduction of output and increase of prices does not only earn it positive economics profits but also creates a direct loss to society. The social costs consist of unsatisfied demand that could be but is not met (also called dead weight losses) and inefficient resource allocation.[11]

It should be noted however, that the existence of barriers to entry is a necessary but not a sufficient condition for this to occur. Even in those cases where strong barriers exist, competition among incumbents can be fierce and lead to perfectly competitive market outcomes. It is only in those cases where competition on the market fails to generate true equilibrium outcomes that the correcting mechanism of new entrants is needed. It is however clear that entry barriers give incumbents the possibility to increase prices to a certain degree before new entry or expansion of existing firms becomes profitable.[12] The higher the barriers to entry, the more attractive it will be for incumbent firms to collude and to increase prices. Not only do stronger incentives influence incumbents' propensity to collude, but at the same time cartel stability is enhanced by effectively precluding new entrants.

The preceding section has shown how barriers to entry serve to delineate markets. If an entry barrier is erected or strengthened on a market characterized by the existence of strong and large firms, particular problems are created.

Reduced threat of new entrance places strong firms in a situation where they can better exploit their size and develop towards acquiring a dominant position. It should be noted that from an economic point of view the creation of a dominant position per se is not a problem as long as it merely reflects a natural tendency on the market to create large enterprises[13] and occurs in the absence of any abusive

---

10. Positive economic profit is defined as profit above a normal rate of return which merely reflects the opportunity costs of capital. Positive economic profit is logically separable from accounting profit.

11. Frank, R., (1997), 412, defines this so called x-inefficiency as a condition in which a firm fails to obtain maximum output from a given combination of inputs.

12. Economists refer to this as limit pricing. See for example Martin, S., (1994), 70 ff.

13. Markets which are characterized by strong returns to scale will give rise to what is commonly referred to a natural monopoly.

practises. Yet all else being equal, entry barriers may not only increase the expected profits of serving a larger share of the market, particularly when there is the possibility to be able to raise prices without having to fear strong competition. While this is not tantamount to mean that firms will automatically engage in abusive behaviour such as predatory pricing, it should nevertheless be noted that the incentives to do so increase together with increased expected profits.

Grandfathering large firms relatively more emission allowances[14] sets them at a comparative advantage *vis-à-vis* small incumbent firms and enables them better to expand their market share. It should be noted that such problems do not exist under the PSR system because here all competitors receive allowances based on their performance.[15]

2.2   Entrenched Market Shares

Besides barriers to entry, another element which has a bearing on firms' propensity to collude is the entrenchment of market shares. Under a capped grandfathering system and given market demand, competitive pressure on less efficient producers is – *ceteris paribus* – reduced. The rationale behind this is that part of the comparative production cost advantage of more competitive producers is absorbed by the necessity to buy additional emission allowances. Undertakings loosing market share are able to withstand competitive pressure better due to an influx of capital, when they sell emission allowances they are grandfathered on a historical basis while actually producing less than before. Consequently, gaining market share at the expense of less efficient producers, is rendered more difficult since less efficient producers are able to cross-subsidize their products and undertakings are more likely to recognize their interdependence. This is expected to have repercussions on the firms' propensity to collude, particularly if market concentration is high and markets are prone to collusion. In contrast to grandfathering, compensation for losers of market shares is absent under a PSR system and thus the PSR system will not give rise to collusion.

This section of the chapter has shown that measures contained in NAPs and the application of emission allowance allocation formats can impact market structure and influence undertakings' propensity to collude and abuse. If undertakings engage in such practices, it will be to the detriment of society at large. On a theoretical level, PSR gives rise to less instances of anticompetitive concerns than grandfathering. To what extent this is indeed true is, however, subject to future empirical research.

---

14. In the absence of any empirical studies, anecdotal evidence from the Dutch sugar industry is indicative. Complaints about unequal treatment setting smaller operators at a disadvantage have been raised.
15. This finding is based on the express assumption that new entrants and smaller incumbent firms are not credit rationed.

3   LEGAL ANALYSIS

The previous section has shown that National Allocation Plans, and to a varying degree allocation mechanisms, can lead to distortions of competition that influence firms' propensity to collude and abuse. If changes in market structure do not only translate into increased incentives for undertakings to engage in illegal activity but actually lead to a real change in their conduct, Competition law enforcement has to be strengthened in order to contain such negative effects. The legal problem and social dilemma at hand, however, is that measures are taken by national legislators which create costs to society – e.g. the distortion of allocative efficiency on the production market – that are unnecessary, avoidable and against which legal remedies are weak.[16] *relevant?*

With regard to the EU ETS, government measures are the source of anticompetitive distortions. These distortions have, however, been subject to compensatory justification balancing acts under State aid investigations where the European Commission compared environmental benefits to anticompetitive distortions. Upon finding a positive societal effect, State measures were declared to be compatible with the common market.[17] Even though it is unlikely that the ECJ would call into question the substance of the Commission's Decision, a number of interesting academic questions arise.

Legislative interventions[18] by Member States are largely contained through the application of the four freedoms, while EC Competition law is geared to the containment of competitive distortions arising in particular from undue behaviour of firms. The legislator has introduced Article 86 EC Treaty to contain State measures affecting public undertakings and the granting of special or exclusive rights to undertakings by bringing them within the field of application of EC law in general and Competition law in particular. The scope of this Article was and is, however, too narrow to address all possible anticompetitive measures Member States can take. The resulting legal gap in the Treaty, placed some State measures inducing cartelisation and abuse beyond reach of the four freedoms and Competition law alike. The European Court of Justice ('ECJ') has recognized this shortcoming.

Based on Articles 10(2), 3(g), 81 and 82 EC Treaty, the Court developed in 1977 jurisprudence establishing that Member States are obliged to abstain from taking measures which could deprive Articles 81 and 82 of their effectiveness (*effet utile*). Through this the Court established jurisprudence that was intended to contain undue State interference with the objectives of the EC Treaty. In its case law the ECJ established clear parallels to Article 86(1) EC Treaty. This jurisprudence was applied

---

16. In Case T-28/07 Fels-Werke and Others vs. European Commission [2007], para. 61 ff., a German undertaking's challenge of the Commission's decision on the German NAP was declared inadmissible by the CFI because it was not individually concerned.
17. See Weishaar, S., (2005).
18. As contrasted to subject matters dealt with under State Aid legislation.

in a myriad of contexts including price regulations, social security provisions and maximum credit rates.[19]

This section of the chapter starts with a review of the evolutionary process of the joint application jurisprudence. Thereafter its current application in connection to cartelisation and subsequently abuse is presented. Both culminate in an analysis of the EU ETS. The fourth part of this section deals with Article 86 and examines whether it can be employed within the framework of the Emission Trading System.

3.1   DEVELOPMENT OF THE JURISPRUDENCE

This section presents the evolution of the joint application jurisprudence. It shows how the Court recognizes the gap between competition provisions directed towards undertakings (Articles 81 and 82) and Member States (Article 86) and how it searches for the right approach to contain anticompetitive State measures. It is noticeable how the Court first takes a restrictive approach, becomes proactive, before summarising its jurisprudence in the *van Eycke* (1988) case, and limiting the scope of application of the jurisprudence. Even though this passage also treats the joint application jurisprudence with regard to Article 82 EC Treaty, due to the fact that the most relevant developments have only been taking place recently, they are contained in the section addressing the current application of the jurisprudence under this Article.

The ECJ acknowledged the important link between Articles 3(g), 10(2) and 81 and 82 in *INNO* for the first time.[20] A Belgium code fixed retail prices on the basis of excise tax labels giving rise to a situation where price competition among retailers was effectively ruled out. Since the Belgium Court viewed the legislation as encouraging abuse by tobacco producers and importers, it referred – amongst others – a preliminary ruling question to the ECJ[21] asking whether such national provisions were compatible with Article 82 EC Treaty.

In its assessment the ECJ acknowledged that Article 82 is directed against undertakings but that at the same time Member States were under the duty not to maintain in force any measures that could deprive this provision of its

---

19. Since the joint application doctrine requires infringements of Art. 81 or 82 to be applicable it is not surprising that the Court did not apply it in the context of collective agreements between management and labour, in pursuit of social policy objectives such as the improvement of conditions of work and employment which do not fall within the ambit of either Article. See Joined cases C-115/97 to C-117/97 Brentjens' Handelsonderneming BV v. Stichting Bedrijf-spensioenfonds voor de Handel in Bouwmaterialen [1999] ECR I-06025, para. 66; Case 219/97 Maatschappij Drijvende Bokken BV v. Stichting Pensioenfonds voor de Vervoer- en Havenbe-drijven, [1999] ECR I-06121, para. 52.

20. Though it is noticeable that the Court recognized the importance of the *effet utile* of Community law in the context of Competition law already in Case 14/68, Walt Wilhelm v. Bundeskartel-lamt, [1969] ECR 00001, para. 6.

21. Case 13/77, N.V. GB-INNO-B.M. v. Association des détaillantes en tabac, [1977] ECR 2115, para. 24.

effectiveness,[22] as stated in Article 10(2).[23] The Court recites the prohibition to enact or maintain in force measures contrary to Competition law contained in Article 86(1) EC Treaty[24] but avoids to attribute responsibility to Member States by stating that '... likewise, Member States may not enact measures enabling private undertakings to escape from the constraints imposed by Articles 85 (now 81) to 94 (now 89) of the Treaty'.[25] By emphasizing that Member States may not shield undertakings as to undermine the *effet utile* of EC Competition law, the Court avoids to clarify whether national regulation encouraging cartelization or abuse constitutes a violation of the Treaty. This is particularly lamentable because it was the first time when the ECJ was directly asked whether a national measure encouraging an infringement by undertakings was to be condemned under Competition law rules or not.

Even though the Court did not state whether such national measures were illegal, the case is nevertheless noteworthy for three reasons. Firstly it linked Articles 3(g), 10(2) and 82 EC Treaty together by stating that Member States were under the obligation not to adopt or maintain in force any measures which could deprive Competition law of its effectiveness. Secondly, it stated that Courts had to consider Articles 3(g), 10, 30, 34 and 81 to 89 EC Treaty when assessing systems of resale price maintenance,[26] which constitutes a prominent field of government intervention. Last but not least, it should be noticed that the Belgian Court avoided to approach the case under Article 81 even though the national measure fixed directly resale prices, a measure which is expressly condemned by Article 81(1)(a) EC Treaty.

Also in subsequent case law, the Court did not take the opportunity to cement its approach and appeared to interpret the scope of the joint application jurisprudence more narrowly. In *van Tiggele*[27] for example, the ECJ merely addressed – in line with the preliminary ruling questions raised – free movement of goods and State aid issues and failed to comment on the curtailment of competition on the merits through the establishment of a compulsory minimum resale price maintenance scheme. It failed to take the opportunity to opine about the jurisprudence even though an assessment under Article 81 EC Treaty would have merited special attention.

---

22. Case 13/77, N.V. GB-INNO-B.M. v. Association des détaillantes en tabac, [1977] ECR 2115, para. 31.
23. Case 13/77, N.V. GB-INNO-B.M. v. Association des détaillantes en tabac, [1977] ECR 2115, para. 30.
24. Case 13/77, N.V. GB-INNO-B.M. v. Association des détaillantes en tabac, [1977] ECR 2115, para. 32.
25. Case 13/77, N.V. GB-INNO-B.M. v. Association des détaillantes en tabac, [1977] ECR 2115, para. 33.
26. Case 13/77, N.V. GB-INNO-B.M. v. Association des détaillantes en tabac, [1977] ECR 2115, para. 36.
27. Case 82/77, Opebaar Ministerie v. Van Tiggele, [1978] ECR 25.

Worse still, several years later, in the *Buys* case[28] the Court was asked to judge upon the legality of national rules to freeze product prices under joint application of Articles 10 and 81 EC Treaty. By emphasizing that the presence of an agreement between undertakings was a *conditio sine qua non* for the application of Article 81, the Court returned to its strict reading of the Treaty text.[29] It continued by stating that if a national measure were to violate the principle contained in Article 10 EC Treaty, not to jeopardize the objectives of the Treaty, the assessment of the compatibility of the measure would not depend on Article 81 but on the provisions governing the measures.[30]

In another preliminary ruling proceeding relating to the lawfulness of national provisions obliging foreign producers and importers of medicine to charge the same prices as they were charging in the country of production, the issue was resolved solely within the context of free movement of goods. The Court thereby deemed it superfluous to comment on the interrelationship between Articles 3(g), 10, 81 and 82.[31]

Expanding upon its former ruling in *Buys*, the Court emphasized in *Duphar*[32] that Articles 81 and 82 EC Treaty were applicable to undertakings and therefore not relevant to an assessment of the question whether national legislation is in conformity with Community law.[33]

In *Jan van de Haar*,[34] the facts of the case were very similar to the ones presented in *INNO* and the ECJ endorsed its former rulings regarding the joint application jurisprudence. Yet this time with regard to Article 81, it drew on the essential findings of *Inno* and *Duphar* case by concluding that Member States are under an obligation not to enact measures enabling private undertakings to escape the constraints imposed by Articles 81 while endorsing that competition rules are intended to govern the conduct of undertakings in the common market and that competition law rules are therefore not relevant to the question whether national legislation is compatible with Community law.[35]

In light of the above cited cases, the ECJ approach to the joint application of Articles 3(g), 10, 81 and 82 can be summarized as follows. Observing the precise wording of the Treaty, the Court requires the presence of an agreement for

28. Case 5/79, Procureur général v. Hans Buys and Han Pesch and Yves Dulieux and Denkavit France SARL, [1979] ECR 3203.
29. Case 5/79, Procureur général v. Hans Buys and Han Pesch and Yves Dulieux and Denkavit France SARL, [1979] ECR 3203., para. 30.
30. Case 5/79, Procureur général v. Hans Buys and Han Pesch and Yves Dulieux and Denkavit France SARL, [1979] ECR 3203., para. 30.
31. Case 181/82, Roussel Laboratoria BV and others v. État néerlandais, [1983] ECR 03849, para. 6 and 26.
32. Case 232/82, Duphar BV and others v. The Netherlands State, [1984] ECR 523.
33. Case 232/82, Duphar BV and others v. The Netherlands State, [1984] ECR 523, para. 30.
34. Joined Cases 177 and 178/82, Criminal proceedings against Jan van de Haar and Kaveka de Meern BV, [1984] ECR 1797.
35. Joined Cases 177 and 178/82, Criminal proceedings against Jan van de Haar and Kaveka de Meern BV, [1984] ECR 1797, para. 24.

Article 81 EC Treaty and secondly holds that Articles 81 and 82 are only applicable
to undertakings but not to Member States. Furthermore the Court stated that
national legislation jeopardizing the attainment of Treaty objectives, would be
held to be incompatible with Community law independent of Articles 81 or 82
EC Treaty. This maxim can however only be applied satisfactorily as long as – as
was the case in the aforementioned cases – other Treaty provisions regulating the
free movement of goods for example, are applicable. If read together with the
preceding findings, the last element presents a paradox. Member States are obliged
not to take measures that would protect undertakings from the full force of Competition law (*effet utile*) but claimants cannot rely on Articles 81 or 82. Consequently the legal grounds for condemnation of national legislation encouraging or
obliging cartelisation or other types of infringements would be found in the joint
application of Articles 3(g) and 10(2) EC Treaty independent of other Competition
law provisions. The tragedy of this finding is that both provisions are neither
directly effective nor are they capable to develop legal effect on their own.
Sadly the Court failed to clarify what this entails in substance. One can therefore
conclude that Member States' obligations under Competition law rules, were still
unclear.[36]

Only one year later, the Court decided to distance itself from the formerly
applied minimalist approach.[37] In *Leclerc*[38] – as was the case in *Inno, van Tiggele*
and *van de Haar* – the Court was faced with a retail price maintenance scheme
which allowed manufacturers and importers to set prices which would be endorsed
by law. Drawing on its previous judgement of *Wilhelm*[39] and *Inno*, the Court
stated: 'Whilst it is true that the rules on competition are concerned with the
conduct of undertakings and not with national legislation, Member States are
non the less obliged under the second paragraph of Article 5 (now 10) of the Treaty
not to detract, by means of national legislation, from the full and uniform application of Community law or from the effectiveness of its measures, even of a
legislative nature, which may render ineffective the competition rules applicable
to undertakings.'[40]

The Court concluded that a resale price maintenance scheme, obliging publishers (unilaterally) to fix compulsory resale prices, did not require agreements to
be concluded between publishers and retailers and continued to note that:
'...legislation, which renders corporate behaviour of the type prohibited by
Article 85(1) (now 81(1)) EC Treaty superfluous, by making the book publisher
or importer responsible for freely fixing binding retail prices, detracts from the

---

36. See Neergaard, U., (1998), 41.
37. Gyselen, L., (1989), 42.
38. Case 229/83, Association des Centres distributeurs Édouard Leclerc and other v. SARL 'Aublé
     vert' and others, [1985] ECR 1.
39. Case 14/68, Wilhelm v. Bunderskartellamt, [1969] ECR 00001.
40. Case 229/83, Association des Centres distributeurs Édouard Leclerc and other v. SARL 'Aublé
     vert' and others, [1985] ECR 1, para. 14.

effectiveness of Article 85 (now 81) and is therefore contrary to the second paragraph of Article 5 (now 10) of the Treaty'.[41] Consequently, Articles 3(g), 10(2) and 81 EC Treaty are applicable with regard to resale price maintenance schemes irrespective of whether they originate in cartels or in Member State delegations.[42]

Despite its findings, the Court, drawing upon the principle of conferral of power contained in Article 5 EC Treaty, notes that purely national measures regarding book trade are not yet subject to Community competition policy and consequently does not apply the jurisprudence.[43]

A few weeks after this judgement, the Court issued a second ruling dealing with retail price maintenance and the joint application jurisprudence. In *Cullet*[44] the ECJ clarified that in the case before it, the national legislation did not intend to compel suppliers and retailers to conclude agreements or to take any other action of the kind referred to in Article 81(1) EC Treaty,[45] but on the contrary entrusted the responsibility for fixing prices to public authorities.[46] Because in *Cullet* the State did not delegate the power to fix prices to private undertakings but to public authorities,[47] the Court concluded that it was unable to apply Article 81(1) to Member State measures. This is therefore to be regarded as a qualification to its ruling in *Leclerc*. Only if a State delegates price fixing power to undertakings and thereby makes it superfluous for them to collude, Member State regulation is viewed to undermine the effectiveness of Article 81(1) and consequently to be held unlawful. If a Member State, however, establishes a retail price maintenance scheme by itself, it is not deemed to be incompatible with the common market. As noted correctly by van Gyselen (1989), such delineation is artificial because the regulatory choice will depend to a large degree on the particularities of the relevant product market.[48]

In the *Asjes* case, the Court questioned whether the joined application jurisprudence prohibits Member States from applying national provisions laying down compulsory approval procedures and enforcement systems for tariffs that resulted from an agreement, decision by associations of undertakings or a concerted practice

---

41. Case 229/83, Association des Centres distributeurs Édouard Leclerc and other v. SARL 'Au blé vert' and others, [1985] ECR 1, para. 15.
42. Gyselen, L., (1989), 43.
43. Case 229/83, Association des Centres distributeurs Édouard Leclerc and other v. SARL 'Au blé vert' and others, [1985] ECR 1, para. 18 to 20.
44. Case 231/83, Henri Cullet and Chambre syndicate des réparateurs automobiles et détaillants de produits pétroliers v. Centre Leclerc à Toulouse and Centre Leclerc à Saint-Orens-d-Gameville, [1985] ECR 305. The same line of argumentation was employed in a later case. See Case 188/86, Ministère public v. Régis Lefèvre, [1987] ECR 2963, para. 7.
45. Case 188/86, Ministère public v. Régis Lefèvre, [1987] ECR 2963, para. 17.
46. This was confirmed by the Court in Case 188/86, Ministère public v. Régis Lefèvre, [1987] ECR 2963, para. 7.
47. Case 231/83, Henri Cullet and Chambre syndicate des réparateurs automobiles et détaillants de produits pétroliers v. Centre Leclerc à Toulouse and Centre Leclerc à Saint-Orens-d-Gameville, [1985] ECR 305, para. 17.
48. Gyselen, L., (1989), 44.

contrary to Article 81 EC Treaty. The Court held that the *effet utile*[49] of Competition law could not be safeguarded if a Member State were to require or favour the adoption of agreements, decisions or concerted practices contrary to Article 81 or to reinforce their effects.[50] Consequently it ruled that it would run counter Member States' obligations under Article 10 *juncto* 3(g) and 81 EC Treaty to approve and thus to reinforce the effects of such tariffs if they resulted from infringements of Article 81 EC Treaty. The interesting element of this case is that State measures reinforcing existing illegal practices are caught by the jurisprudence. This is therefore a reaffirmation of the Courts decision in *BNIC* where it stated that infringements of 81(1) EC Treaty remain illegal even if supported by government.[51]

For a short period of time, the Court took a very broad perspective and was willing to condemn Member State legislation. The first case in which the ECJ did not leave it to the discretion of other institutions was *Vlaamse Reisbureaus*.[52] In this decision the Court appeared to be building upon its methodology established in *Asjes*.[53] It first required the examination of the presence of infringements under Article 81(1) EC Treaty before assessing whether national provisions were capable of reinforcing their effects.[54] Particularly noteworthy is not only that the Court assessed both elements by itself but also the outspoken clarity with which it condemns anticompetitive practices.[55] Furthermore the Court uses three criteria. Firstly, whether an originally contractual prohibition was transformed into a legislative provision, secondly the availability of legal remedies and thirdly the existence of effective sanctions.[56]

In another case, *BNIC II*,[57] the Court ruled upon the legality of a production quota system which was made binding through national legislation. In assessing the compatibility of the agreement under Article 81(1) EC Treaty, the Court

49. Case 209/84–213/84, Lucas Asjes and others, Andrew Gray and others, Andrey Gray and others, Jacques Maillot and others and Léo Ludwig and others, [1986] ECR 1425, para. 71, citing again Walt Wilhelm and Inno.
50. Case 209/84–213/84, Lucas Asjes and others, Andrew Gray and others, Andrey Gray and others, Jacques Maillot and others and Léo Ludwig and others, [1986] ECR 1425, para. 72.
51. Case 123/83, Bureau national interprofessionnel du cognac v. Guy Clair, [1985] ECR 391, para. 16 ff.
52. Case 311/85, ASBL Vereniging van Vlaamse Reisbureaus v. ASBL Sociale Dienst van de Plaatselijke en Gewestelijke Overheidsdiensten, [1987] ECR 3801, para. 24. See also Case 66/86, Ahmed Saeed Flugreisen and Silver Line Reisebüro GmbH v. Zentrale zur Bekämpfung unlauteren Wettbewerbs e.V., [1989] ECR 803.
53. Case 209–213/84, Lucas Asjes and others, Andrew Gray and others, Andrey Gray and others, Jacques Maillot and others and Léo Ludwig and others, [1986] ECR 1425, para. 72.
54. Case 311/85, ASBL Vereniging van Vlaamse Reisbureaus v. ASBL Sociale Dienst van de Plaatselijke en Gewestelijke Overheidsdiensten, [1987] ECR 3801, para. 11.
55. Case 311/85, ASBL Vereniging van Vlaamse Reisbureaus v. ASBL Sociale Dienst van de Plaatselijke en Gewestelijke Overheidsdiensten, [1987] ECR 3801, para. 17.
56. Case 311/85, ASBL Vereniging van Vlaamse Reisbureaus v. ASBL Sociale Dienst van de Plaatselijke en Gewestelijke Overheidsdiensten, [1987] ECR 3801, para. 23.
57. Case 136/86, Bureau national interprofessionnel du cognac v. Yves Aubert, [1987] ECR 4789.

identified restrictions of competition on the basis that the agreement penalized any increase in production, the agreement tended to freeze the existing situation and make it more difficult for a producer to improve its competitive position on the market.[58] The Court held that Member State legislation violated the joint application jurisprudence in particular in those cases where it reinforced the effects of an agreement contrary to Article 81(1) EC Treaty by making it generally binding.[59]

Even though the Court condemned State measures under the joint application jurisprudence in *Ahmed Saeed*,[60] about half a year after its ruling in *van Eycke*,[61] the latter is to be regarded as a conservative restatement of the Court's existing jurisprudence and has since served as an important point of reference. In *van Eycke* the Court had to decide on the interaction of Articles 3(g), 10(2) and 81 EC Treaty. Questionable was whether a decree continued in legislative form previously existing self-regulatory agreements or concerted practices among banks which had the effect of restricting interest rates and making them compulsory.

In its assessment the Court stressed that Articles 81 and 82 per se only applied to the conduct of undertakings[62] and summarized previous case law by holding that Member States were under the obligation not to introduce or maintain in force measures which may render ineffective competition rules applicable to undertakings[63] by creating (1) new infringements, (2) reinforcing infringements or by (3) delegating authority.[64]

With regard to the first criterion, measures that require or favour the adoption of agreements, decisions or concerted practices contrary to Article 81,[65] the Court merely holds that it is not apparent from the facts of the case that the legislation in question was intended to require or favour the adoption of new restrictive agreements or the implementation of new practices. The Court does thus not offer additional information. With regard to the second element, more guidance is offered.

---

58. Case 136/86, Bureau national interprofessionnel du cognac v. Yves Aubert, [1987] ECR 4789, para. 17–19.
59. Case 136/86, Bureau national interprofessionnel du cognac v. Yves Aubert, [1987] ECR 4789, para. 24.
60. Case 66/86, Ahmed Saeed Flugreisen and Silver Line Reisebüro GmbH v. Zentrale zur Bekämpfung unlauteren Wettbewerbs e.V, [1989] ECR 803.
61. Case 267/86, Pascal Van Eycke v. ASPA NV., [1988] ECR 4769.
62. This is to be regarded as placing additional emphasis on what the Court has been consistently holding since its decision in Leclerc, Case 229/83, Association des Centres distributeurs Édouard Leclerc and other v. SARL 'Au blé vert' and others, [1985] ECR 1, para. 14.
63. Case 267/86, Pascal Van Eycke v. ASPA NV., [1988] ECR 4769, para. 16.
64. Case 267/86, Pascal Van Eycke v. ASPA NV., [1988] ECR 4769, para. 16.
65. This criterion has been expressed in Case 209–213/84, Lucas Asjes and others, Andrew Gray and others, Andrey Gray and others, Jacques Maillot and others and Léo Ludwig and others, [1986] ECR 1425, para. 72. Neergard U., (1998), 72, maintains that it can also be found in Leclerc, Case 229/83, Association des Centres distributeurs Édouard Leclerc and other v. SARL 'Au blé vert' and others, [1985] ECR 1, para. 15.

With regard to the second criterion, the Court gives guidance to the interpretation of 'reinforcing effects'[66] of agreements. As in previous cases,[67] the Court was not concerned with public authorities encouraging undertakings to set up anticompetitive agreements but rather whether the State measure incorporates either wholly or in part the terms of agreements concluded between undertakings and requires or encourages compliance on the part of those undertakings. It is therefore to be viewed as a stricter criterion than expressed in *Vlaamse Reisbureaus.*[68] Consequently State measures which have the effect to stabilise cartels but are not directly part of an existing agreement are not caught by this interpretation.[69]

With regard to the third criterion concerning measures that deprive themselves of their official character by delegating responsibility to private traders for taking decisions affecting the economic sphere,[70] the Court ruled that the legislation retained an official character even though consultations with representatives of associations of credit establishments occurred.

Subsequent case law appeared merely to emphasise that the term 'State' was to be interpreted broadly as an emanation of the State[71] and that the jurisprudence could only be applied if the trade between Member States was impeded.[72] In *Marchandise* the Court, however, seemed to imply that the third criterion of the *van Eycke* test regarding the delegation of public authority could be applied independently.[73]

---

66. As introduced in Case 209–213/84, Lucas Asjes and others, Andrew Gray and others, Andrey Gray and others, Jacques Maillot and others and Léo Ludwig and others, [1986] ECR 1425, para. 72.

67. Case 209–213/84, Lucas Asjes and others, Andrew Gray and others, Andrey Gray and others, Jacques Maillot and others and Léo Ludwig and others, [1986] ECR 1425; Case 136/86, Bureau national interprofessionnel du cognac v. Yves Aubert, [1987] ECR 4789.

68. Case 311/85, ASBL Vereniging van Vlaamse Reisbureaus v. ASBL Sociale Dienst van de Plaatselijke en Gewestelijke Overheidsdiensten, [1987] ECR 3801, para. 23.

69. Neergaard, U., (1998), 72, argues that it is difficult to see how legislation can reinforce an agreement if the legislation does not contain, in whole or in part, terms of that agreement. The dilemma is however, that Member States can change the market structure by creating barriers to entry and thereby stabilize cartels. Furthermore, the criterion of including wholly or in part elements of a cartel agreement is far to a legalistic an approach to be practicable in real life situations. Illegal cartel agreements are neither published nor expressly drafted in contract form. And given that – unlike in other fields of law where one at least is able to detect infringements – Competition law is characterized by illicit infringements where parts of the violations remain unnoticed, this is expected to constitute a very restrictive benchmark.

70. This has been applied by the Court in Leclerc and Cullet. Case 229/83, Association des Centres distributeurs Édouard Leclerc and other v. SARL 'Au blé vert' and others, [1985] ECR 1, para. 15; Case 231/83, Henri Cullet and Chambre syndicate des réparateurs automobiles et détaillants de produits pétroliers v. Centre Leclerc à Toulouse and Centre Leclerc à Saint-Orens-d-Gameville, [1985] ECR 305, para. 17.

71. C 339/89, Alsthom Atlantique SA v. Companie de Construction Mechanique Sulzer SA., [1991] ECR I-107, para. 11.

72. Case C-60/91, José António Batista Morais, [1992] ECR I-02085, para. 12–13.

73. C 332/89, André Marchandise, Jean-Marie Chapuis and SA Trafitex, [1991] ECR I-01027, para. 23. See Neergard U., (1998), 75.

The new more restrictive approach of the Court can be seen in a series of rulings (*Reiff, Meng, Ohra*), which were taken all on the same day. In its assessment whether there was a price agreement between undertakings contrary to Article 81 EC Treaty, the Court decided in *Reiff*[74] that the representatives of a tariff board were acting upon their own behalf and were not bound by their respective institutions. Even though in a very similar situation as in *BNIC* the Court reached a different conclusion.[75] The decisive difference[76] was that in *Reiff*, the Court was satisfied with the national requirement that board members had to 'discharge their duties in an honorary capacity and were not to be bound by orders or instructions',[77] while in *BNIC* the mere fact that board members were proposed for appointment by the trade organisations concerned was deemed sufficient to regard them as representatives of undertakings and hence to reach the conclusion that there was an agreement between undertakings.[78]

Even though it may not be surprising that the Court does not evaluate the effectiveness of national law and readily accepts a national law obligation that board members have to be independent, it is nevertheless doubtful if a mere obligation suffices to ensure that honorary representatives do not act by taking in particular the concerns of their undertakings into account.[79]

With regard to the 'delegation of power' criterion, the Court stated in *Reiff* that the Minister's competence to participate in board meetings and the power to overrule the board's decision in agreement under particular circumstances (agreement with another minister) was sufficient to conclude that the State did not delegate its authority.[80] Even though the 'delegation of power' criterion established in *van Eycke* did not exist when the *BNIC* ruling was taken, it is interesting to notice that in the latter case the Court held that the mere fact that an agreement had to be approved by public authorities in order to be binding upon all market participants was immaterial to the question whether an agreement would be caught by Article 81 EC Treaty.[81]

In *Meng* the Court had to decide whether a national law extending a cartel-based prohibition to pass on agent's commissions to insurance customers from the

---

74. Case C-185/91, Bundesanstalt für Güterfernverkehr v. Gebrüder Reiff GmbH & Co. KG., [1993] ECR I-05801, para. 17–19.
75. Case 123/83, Bureau national interprofessionnel du cognac v. Guy Clair, [1985] ECR 391, para. 18–20.
76. Other differences include that the Court assessed the facts in BNIC within the framework of Art. 85 (now 81 EC Treaty) while Reiff was assessed within the framework of the joint application doctrine.
77. Case C-185/91, Bundesanstalt für Güterfernverkehr v. Gebrüder Reiff GmbH & Co. KG., [1993] ECR I-05801, para. 4.
78. Case 123/83, Bureau national interprofessionnel du cognac v. Guy Clair, [1985] ECR 391, para. 3, 19–20.
79. For a critique on this point see Bach, A., (1994), 1360 ff.
80. Case C-185/91, Bundesanstalt für Güterfernverkehr v. Gebrüder Reiff GmbH & Co. KG., [1993] ECR I-05801, para. 20–23.
81. Case 123/83, Bureau national interprofessionnel du cognac v. Guy Clair, [1985] ECR 391, para. 21–23.

life insurance sector to the health and legal expense insurance sectors constituted a reinforcement of the effects of a pre-existing cartel.[82] The Court further narrowed the scope of the jurisprudence by requiring that national legislation could not be regarded as reinforcing the effect of a pre-existing agreement, decision or concerted practice unless they simply reproduce the elements of an agreement, decision or concerted practice between economic agents in *that* sector.[83] In comparison to the *van Eycke* test, the Court required the agreement to be in the same sector and disregarded the possibility that legislative measures could encourage compliance on the undertakings concerned.[84]

The Court's finding and the narrowing of the scope of the joined application jurisprudence is particularly lamentable from an economic perspective. Theory predicts that multi-market contacts[85] of undertakings provide for more possibilities to retaliate against cartel violators and stabilize cartels. Extending the scope of a cartel agreement to other sectors gives undertakings a legal and therefore potentially more credible[86] enforcement mechanism. Consequently the extension of the scope of an agreement from life insurance to health insurance encourages compliance to the existing cartel in life insurance.

Advocate General Tesauro is stated to be critical about the time gap between the introduction of a private sector cartel and the entering into force of regulation.[87] The Advocate General must have overlooked the fact that even well established business practices with an anticompetitive effect need to be enforced if individual parties can benefit from deviating from them. Therefore regulation intervening eleven or even forty-four years after the transformation of the original cartel rules restricting retrocession into binding ordinances must be regarded as reinforcing the original cartel.

In *Ohra*[88] the Court was questioned whether Articles 3(g), 10(2) and 81(1) would preclude State rules which prohibit insurance companies from granting financial advantages to policy holders and where thus producing effects similar to those of a private cartel. The Court's ruling is interesting because it expressly states what has already been inferred from the Court's jurisprudence, namely that for the second criterion of the *van Eycke* test of 'measures reinforcing the effects' of an agreement etc., one has to establish its existence in the first place.[89] Since also the other two criteria were not fulfilled, the Court did not condemn the anticompetitive effect of Member State legislation.

---

82. Case 2/91, Wolf W. Meng, [1993] ECR I-05751, para. 3, 18–19.
83. Case 2/91, Wolf W. Meng, [1993] ECR I-05751, para. 19. Emphasis added. See also Reich, N., (1994), 472.
84. Case 267/86, Pascal Van Eycke v. ASPA NV., [1988] ECR 4769, para. 18.
85. Here it is expressly assumed that a subset of undertakings is operating in more than one insurance market and that multi-market contacts do indeed exist.
86. Threats to punish cartel violators are only effective if they are credible. To the extent that the costs of legal enforcement are lower than the costs of retaliation, a cartel will benefit from the introduction of legal enforcement mechanisms.
87. Advocate Genaral Tesauro's opinion (of 14 Jul. 1993) is cited in Bach, A., (1994), 1365.
88. Case 245/91, Ohra Schadeverzekeringen NV., [1993] ECR I-05851.
89. Case 245/91, Ohra Schadeverzekeringen NV., [1993] ECR I-05851, para. 9 and 12.

Despite its more restrictive approach, the Court in *CNSD*[90] was still prepared to condemn national legislation which constituted a flagrant violation of the EC Treaty. Not only did the measure require the formation of an agreement and relinquish public authority to private entities but it also provided tools to coerce cartel violators into compliance.

In summary, this section has shown that the Court did recognise the anticompetitive detriment that can result from national measures. In the course of evolution of the jurisprudence, the ECJ emancipated its application from both, the minimalist approach of adhering to the exact wording of the Treaty and the dependence of provisions governing the four freedoms. After alternating indications for a broadening and closing of the jurisprudence, the Court holds currently a strict interpretation which is, however, not free from economic critique.

3.2   JOINED APPLICATION JURISPRUDENCE TODAY UNDER ARTICLE 81

This section of the chapter summarizes the historic evolution presented in the preceding section by reciting the joined application jurisprudence as it is currently applied by the Court. After presenting the legal criteria established by the case law, the link to the European Emissions Trading System is made. This is done by analysing how national measures contained in NAPs such as grandfathering and PSR are to be assessed in light of the established criteria.

A State measure will only be caught by the joined application jurisprudence if it (i) introduces new infringements, (ii) reinforces infringements or (iii) delegates authority to private entities. Each will be taken in turn. Yet it should be noticed that a *conditio sine qua non* for the application of European Competition law is that trade between Member States must be affected.[91]

(i) The measures require or favour the adoption of agreements, decisions or concerted practices contrary to Article 81[92]

Due to the small number of cases which discuss this issue, the Court's application is still unclear. The Court did, however, condemn national legislation that required

---

90. C-35/96, Commission v. Italy (CNSD), [1998] ECR I-03851, para. 52–60.
91. For a statement of this criterion w.r.t. State measures see Case 136/86, Bureau national interprofessionnel du cognac v. Yves Aubert, [1987] ECR 4789, para. 16; Case 311/85, ASBL Vereniging van Vlaamse Reisbureaus v. ASBL Sociale Dienst van de Plaatselijke en Gewestelijke Overheidsdiensten, [1987] ECR 3801, para. 18; Case C-60/91, José António Batista Morais, [1992] ECR I-02085, para. 12; Case C-35/96, Commission v. Italy (CNSD), [1998] ECR I-03851, para. 48; and Case C-35/99, Manuele Arduino, [2002] ECR I-01529, para. 33.
92. This criterion has been expressed in Case 209–213/84, Lucas Asjes and others, Andrew Gray and others, Andrey Gray and others, Jacques Maillot and others and Léo Ludwig and others, [1986] ECR 1425, para. 72. Neergard, U., (1998), 72, maintains that it can also be found in Leclerc, Case 229/83, Association des Centres distributeurs Édouard Leclerc and other v. SARL 'Au blé vert' and others, [1985] ECR 1, para. 15.

the conclusion of an agreement contrary to Article 81, declined to influence its terms and assisted to ensure the compliance of the agreement.[93] Furthermore, the Court held that a national measure cannot require or favour the adoption of an agreement, decision or concerted practice if the obligations these stipulations create are self-sufficient or self-contained.[94]

(ii) The measures reinforce the effects of a violation of Article 81 EC Treaty[95]

Provided that the conduct of undertakings infringes Article 81 EC Treaty[96] and a pre-existing agreement, decision or concerted practice could be identified in that economic sector,[97] the Court examines whether Member State legislation reinforces the effect thereof. In doing so, the Court firstly scrutinizes the relationship between the experts advising government on the legislation and the regulated undertakings, and secondly the Court examines the link between the illegal practices of undertakings and the national legislation.

With respect to the first element, the Court held that the link between such experts and the regulated sector must be limited: representatives, in particular when appointed by public authorities upon proposal of the undertakings they are charged to regulate,[98] must be acting upon their own behalf, not be bound by their respective undertakings, take interests of other sectors as well as the public[99] at large into account.[100] The Court has been accepting requirements for impartiality contained in national legislation.[101]

Concerning the second element, the Court considered not only if an original contractual prohibition was transformed into a legislative provision[102] with

---

93. Case C-35/96, Commission v. Italy (CNSD), [1998] ECR I-03851, para. 55.
94. Case 2/91, Wolf W. Meng, [1993] ECR I-05751, para. 15; Case 245/91, Ohra Schadeverzekeringen NV., [1993] ECR I-05851, para. 11.
95. As introduced in Case 209–213/84, Lucas Asjes and others, Andrew Gray and others, Andrey Gray and others, Jacques Maillot and others and Léo Ludwig and others, [1986] ECR 1425, para. 72.
96. Case 245/91, Ohra Schadeverzekeringen NV., [1993] ECR I-05851, para. 12.
97. Case 2/91, Wolf W. Meng, [1993] ECR I-05751, para. 19.
98. Case 123/83, Bureau national interprofessionnel du cognac v. Guy Clair, [1985] ECR 391, para. 3, 19–20.
99. In Case C-38/97, Autotransporti Librandi, [1998] ECR I 05955, para. 38–42, the Court held that the term public interest applied in Case C-96/94, Centro Servizi Spediporto, [1995] ECR I-02883, para. 42, corresponds to the term general interest as applied in Case C-185/91, Bundesanstalt für Güterfernverkehr v. Gebrüder Reiff GmbH & Co. KG., [1993] ECR I-05801, para. 17–19 and Case C-153/93, Delta Schiffahrts- und Speditionsgesellschaft mbH, [1994] ECR I-02517, para. 21–22. Public interest criteria are to be determined by national legislation and their application is observed by national courts. See Case C-38/97, Autotransporti Librandi, [1998] ECR I 05955, para. 47.
100. Case C-185/91, Bundesanstalt für Güterfernverkehr v. Gebrüder Reiff GmbH & Co. KG., [1993] ECR I-05801, para. 17–19.
101. Case C-185/91, Bundesanstalt für Güterfernverkehr v. Gebrüder Reiff GmbH & Co. KG., [1993] ECR I-05801, para. 4.
102. Case 136/86, Bureau national interprofessionnel du cognac v. Yves Aubert, [1987] ECR 4789, para. 24.

legal remedies and effective sanctions[103] but in particular if the State measure incorporated either wholly or in part the terms of agreements concluded between undertakings and requires or encourages compliance on the part of those undertakings.[104] Consequently State measures which have the effect to stabilise cartels but are not directly part of an existing agreement are not caught by this interpretation.[105]

(iii) The measures that deprive the legislation of its official character by delegating to private traders responsibility for taking decisions affecting the economic sphere[106] contrary to Article 81 EC Treaty

Provided that the conduct of undertakings infringe Article 81 EC Treaty and if private undertakings do have responsibility for taking decisions affecting the economic sphere,[107] the Court examines whether the State delegates its power to a private body. The Court has repeatedly held that this would be the case if government bodies did not have the final authority over the respective measures which are being promulgated.[108] Elements considered include the right to participate in board meetings,[109] the composition of the board[110] and the power to overrule decisions in particular circumstances.[111]

---

103. Case 311/85, ASBL Vereniging van Vlaamse Reisbureaus v. ASBL Sociale Dienst van de Plaatselijke en Gewestelijke Overheidsdiensten, [1987] ECR 3801, para. 23.
104. Case 267/86, Pascal Van Eycke v. ASPA NV., [1988] ECR 4769, para. 18.
105. See the economic appraisal for an elaboration for a critique on this point. In addition it should be noticed, that the ECJ has been willing to condemn infringements of Art. 81 also in cases where no formal written contract was found but other elements of proof suggested collusion. In Case T-41/96, Bayer AG v. Commission [2000] ECR II 3383, para. 69, the Court states that the form of an agreement is unimportant but that it represents the full expression of the parties' intention.
106. This has been applied by the Court in Leclerc and Cullet. Case 229/83, Association des Centres distributeurs Édouard Leclerc and other v. SARL 'Au blé vert' and others, [1985] ECR 1, para. 15; Case 231/83, Henri Cullet and Chambre syndicate des réparateurs automobiles et détaillants de produits pétroliers v. Centre Leclerc à Toulouse and Centre Leclerc à Saint-Orens-d-Gameville, [1985] ECR 305, para. 17.
107. Case 245/91, Ohra Schadeverzekeringen NV., [1993] ECR I-05851, para. 13.
108. Case 267/86, Pascal Van Eycke v. ASPA NV., [1988] ECR 4769, para. 19; Case C-153/93, Delta Schiffahrts- und Speditionsgesellschaft mbH, [1994] ECR I-02517, para. 21–22.
109. Case C-185/91, Bundesanstalt für Güterfernverkehr v. Gebrüder Reiff GmbH & Co. KG., [1993] ECR I-05801, para. 20–23.
110. See Case C-96/94, Centro Servizi Spediporto, para. 23; Case C-38/97, Autotrasporti Librandi, [1998] ECR I-05955, para. 33–34, Case C-250/03 Giorgio Emanuele Mauri v. Ministero della Giustizia, Comissione per gli esami di avvocato presso la Corte d'appello di Milano, [2005] n.y.r., para. 32.
111. Case C-185/91, Bundesanstalt für Güterfernverkehr v. Gebrüder Reiff GmbH & Co. KG., [1993] ECR I-05801, para. 20–23; Joined Cases C-140/94, C-141/94, C-142/94, DIP, [1995] ECR I-03257, para. 21–23, Case C-250/03 Giorgio Emanuele Mauri v. Ministero della Giustizia, Comissione per gli esami di avvocato presso la Corte d'appello di Milano, [2005] n.y.r., para. 33 ff., Joined cases C-94/04 and C-202/04, Federico Cipolla v. Rosaria Fazari and Stefano Macrino, Claudia Capodarte v. Roberto Meloni, [2006], n.y.r., para. 50 ff.

Private bodies could be the recipient of legislative authority if it could be ensured with reasonable probability that the experts it consists of, are functioning like an arm of the State working in public interest.[112] In this context the Court examined whether the advising experts were independent of the economic operators concerned, took account of the public interest and the interest of undertakings in other sectors.[113] In the case of an undue delegation of authority, the national measure in question is held to be infringing Community law. It should be borne in mind, however that the Court has repeatedly stated that the mere anticompetitive effect of national legislation is not counter the EC Treaty. Consequently rules that themselves forbid competition are not held to constitute delegation of authority.[114]

3.2.1   Analysis: The EU ETS and Article 81 EC Treaty

This section examines how anticompetitive measures taken within framework of the EU ETS are to be assessed in light of the foregoing discussion. The same structure as applied above will be used.

(i) The measure requires or favours the adoption of agreements, decisions or concerted practices contrary to Article 81

Questionable is whether measures taken by Member States within the framework of the European Emissions Trading System which – as presented in the beginning of the chapter – have the effect of impacting undertakings' propensity to collude are not only in fact but in law held to require or favour the adoption of collusive practices between undertakings.

In light of the legal assessment of the case law it is clear that obligations created by national measures that are self-sufficient or self-contained cannot be regarded to reinforce or favour cartelisation. Since neither Directive 2003/87/EC nor any of the National Allocation Plans submitted by the Member States obliges undertakings to collude, it can be concluded that national measures taken within the EU ETS framework are neither self-sufficient nor self-contained and can therefore not be excluded from the field of application of the joint application jurisprudence.[115]

---

112. Case C-35/99, Manuele Arduino, [2002] ECR I-01529, para. 39, but also Joined cases C-94/04 and C-202/04, Federico Cipolla v. Rosaria Fazari and Stefano Macrino, Claudia Capodarte v. Roberto Meloni, [2006], n.y.r., para. 49. See on this point also Vossestein A., J., (2002), 861.
113. See Case C-35/99, Manuele Arduino, [2002] ECR I-01529, para. 37–39. The criteria are essentially a restatement of the criteria used by the Court to examine the second criterion of the van Eycke test: see Case C-185/91, Bundesanstalt für Güterfernverkehr v. Gebrüder Reiff GmbH & Co. KG., [1993] ECR I-05801, para. 17 to 19 and 24; Case C-153/93, Delta Schiffahrts- und Speditionsgesellschaft mbH, [1994] ECR I-02517, para. 16 to 18 and 23; Joined Cases C-140/94, C-141/94, C-142/94, DIP, para. 18 and 19; Case C-35/96, Commission v. Italy (CNSD), [1998] ECR I-03851, para. 44, Case C-198/01 CIF, [2003] n.y.r., para 77.
114. Case 2/91, Wolf W. Meng, [1993] ECR I-05751, para. 20.
115. Case 2/91, Wolf W. Meng, [1993] ECR I-05751, para. 15; Case 245/91, Ohra Schadeverzekeringen NV., [1993] ECR I-05851, para. 11.

While it is clear that the Court requires the violation of Article 81 before it will condemn any State measure, it is, however, not at all clear how strong the link the ECJ requires between a State measure and undue behaviour of undertakings has to be. Unfortunately the Court failed to clarify how it interprets the terms 'requiring' or 'favouring'. The flagrant violation the ECJ condemned in *CNSD* expressly requiring an association of undertakings to form agreements, granting it relative decision-making powers and providing legal compliance rules is clearly a strict point of reference.

If the *CNSD* judgement were taken as a benchmark, barriers to entry established under allocation mechanisms would not be caught but escape legal sanctioning. They do not contain the express obligation to collude but nevertheless increase benefits from cartelisation. Therefore it is suggested to make use of economic tools to investigate to what extent observed price changes that cannot be explained on the basis of production cost increases originating in the introduction of the Emission Trading System, are related to governmental measures.

(ii) The measures reinforce the effects of a violation of Article 81 EC Treaty

Another branch of the ECJ's approach to State measures under Article 81 EC Treaty requires the presence of a pre-existing infringement. In its assessment, the Court would at first closely examine the relationship between advising experts and regulated undertakings. It would be examined whether the choice and design of the particular allocation method and the amount of allowances granted would favour the undertakings and sectors of the advising experts. If the Court would find that advisors were neither impartial nor obliged to take the interests of other sectors and the public at large into account, it would proceed to test a second criterion.

In a second step the Court would examine if a contractual prohibition was transformed into legislative provisions with legal remedies and effective sanctions and whether it would either wholly or in part contain the terms of the pre-existing cartel agreement. Since the creation of the EU ETS introduces a new production factor to the market process, it is rather unlikely that there are any prior collusive agreements between undertakings regulating prices or sectoral output based on the overall amount of CO_2 emissions. Consequently it is quite unlikely that the Court – if it were to adhere to its present interpretation of the joint application jurisprudence – would be able to condemn State measures taken within the framework of the EU ETS.

Yet irrespective of the fact that CO_2 emission allowances constitute new products in most Member States, the second criterion employed by the Court is too legalistic to be of practical relevance in most cartel cases. Illegal cartel agreements are rarely drafted in contract form and unless revealed by whistle blowers lured by effective leniency policies, not available to the Court. It is therefore unlikely that this criterion – if applied strictly by the Court – would permit the application of the joint application jurisprudence towards the EU Emission Trading System. It should be noticed, however, that the ECJ has been willing to condemn infringements of Article 81 also in cases where no formal written contract was found but other elements of proof suggested collusion.[116]

(iii) The measures that deprive the legislation of its official character by delegating to private traders responsibility for taking decisions affecting the economic sphere contrary to Article 81 EC Treaty

In the context of the European Emissions Trading System, it is clear that governments do not delegate their ultimate authority to allocate allowances to private undertakings. Therefore this criterion is not relevant for the assessment of such State measures.

This section has reviewed jurisprudence governing the application of the joint application jurisprudence with regard to Article 81 EC Treaty. It found little guidance as to how State measures taken within the framework of the EU ETS are to be assessed in the absence of pre-existing cartels. The Court is called upon to take due consideration of economic insights to assess the link between State measures and collusion. The benchmark employed by the Court that violations of Article 81 can only be reinforced if they do include part of a cartel agreement is criticized as very legalistic and as an element that will constitute a prohibitively high burden of proof in practice and appears to differ from the Court's case law under Article 81 procedures. Since also the delegation criterion employed by the Court cannot be employed to contain the State measures under consideration, it is concluded that the measures under considerations are unlikely to fall within the ambit of the joint application jurisprudence as it stands today.

3.3   JOINED APPLICATION JURISPRUDENCE TODAY UNDER ARTICLE 82

Member States are restrained in their liberty not only with regard to cartels but also with regard to abuse of dominant positions. It is settled case law that Article 82 EC Treaty read in conjunction with Article 3(g) or 10(2) EC Treaty, requires Member States to refrain from introducing or maintaining in force measures which may render ineffective the competition rules applicable to undertakings.[117] In contrast to the joined application jurisprudence under Article 81 EC Treaty however, considerable uncertainty remains because most cases regarding Article 82 were dealt with in combination with Article 86 rather than 3(g) and 10(2).

The Court expressly examined the possibility of a joint application under Article 82 but always failed to establish a dominant position which constitutes a

---

116. See Case T-41/96, Bayer AG v. Commission [2000] ECR II 3383, para. 69.
117. Case 13/77, N.V. GB-INNO-B.M. v. Association des détaillantes en tabac, [1977] ECR 2115, para. 31; Case C-96/94, Centro Servizi Spediporto, [1995] ECR I-02883, para. 20; Joined Cases C-140/94, C-141/94, C-142/94, DIP, [1995] ECR I-03257, para. 14; Case C-70/95, Sodemare SA, [1997] ECR I-03395, para. 41.

precondition for establishing a violation of the Article.[118] Nevertheless it follows that in principle national legislation can be declared to be incompatible with Community law provided that it places an undertaking in a dominant position[119] or creates sufficiently strong links between undertakings to give rise to collective dominance.[120] The later is characterized by the absence of competition among undertakings[121] and the adoption of the same market conduct.[122]

Legislation leading to full or partial market foreclosure through licensing[123] or discrimination between undertakings applying for financial reimbursements for social services rendered[124] is not held to create dominant positions nor lead to collective dominance. Similarly, the ECJ held that legislation fixing road-haulage tariffs by public authorities, does not create collective dominance characterized by the absence of competition between undertakings.[125] The restrictive nature of the Court's interpretation of collective dominance, shows itself in the denial that collective agreements that allow derogations from legally established and mandatory tariff schemes by extending and legally enforcing majority decisions upon non-collaborating undertakings, can give rise to anticompetitive effects.[126]

Under Article 82 the establishment of the existence of a dominant position is but one criterion for finding that there is an infringement of this provision. It therefore is reasonable to conclude that the Court also requires that the other requirements of this provision are met. Article 82 commonly is held only to be applicable to undertakings abusing a dominant position in a relevant market if there is an effect on trade between Member States. Furthermore, taking the logic of the joint application jurisprudence into consideration, one must expect that the Court requires a link between the State measure and the infringement of Article 82.

---

118. Joined Cases C-140/94, C-141/94, C-142/94, DIP, [1995] ECR I-03257, para. 20–26; Case C-96/94, Centro Servizi Spediporto, [1995] ECR I-02883, para. 31–35.
119. See Case C–85/76, Hoffmann-La Roche, [1979], ECR I-00461, para. 38 for dominance and Case C-96/94, Centro Servizi Spediporto, [1995] ECR I-02883, para. 31; Joined Cases C-140/94, C-141/94, C-142/94, DIP, [1995] ECR I-03257, para. 24; Case C-70/95, Sodemare SA, [1997] ECR I-03395, para. 44. Case C-38/97, Autotransporti Librandi Snc di Librandi F. & C., [1998] ECR I-05955, para. 27. It should be noted that this interpretation of Art. 82 in combination with Arts 3(g) and 5 EC Treaty, contrasts sharply the Court's finding in its traditional analysis under Art. 82 EC Treaty and the actual wording of the Article. Dominance as such is not condemned but its abuse.
120. Joined dominance: Joined Cases C-140/94, C-141/94, C-142/94, DIP, [1995] ECR I-03257, para. 26–27; Case C-70/95, Sodemare SA, [1997] ECR I-03395, para. 47–48.
121. Joined Cases C-140/94, C-141/94, C-142/94, DIP, [1995] ECR I-03257, para. 27
122. Case C-393/92, Almelo and others, [1994] ECR I-01477, para. 42; Case C-96/94 Centro Servizi Spediporto [1995] ECR I-02883, para. 33; Joined Cases C-140/94, C-141/94, C-142/94, DIP [1995] ECR I-03257, para. 26; Case C-70/95, Sodemare SA, [1997] ECR I-03395, para. 46.
123. Joined Cases C-140/94, C-141/94, C-142/94, DIP, [1995] ECR I-03257, para. 27.
124. Case C-70/95, Sodemare SA, [1997] ECR I-03395, para. 47.
125. Case C-96/94, Centro Servizi Spediporto, [1995] ECR I-02883, para. 34; Case C-38/97, Autotransporti Librandi Snc di Librandi F. & C., [1998] ECR I-05955, para. 32.
126. Case C-96/94, Centro Servizi Spediporto, [1995] ECR I-02883, para. 29; Case C-38/97, Autotransporti Librandi Snc di Librandi F. & C., [1998] ECR I-05955, para. 49–52. The Court argues that such derogations have the effect that they generate more competition.

3.3.1   Application

Whether national measures introduced within the framework of the EU ETS system could be declared to be incompatible with the common market on the basis of a joined application of Articles 3(g), 10(2) and 82 constitutes a non-trivial question. Particularly so in light of the considerable gaps existing in the Courts interpretations to this date.

It follows from the above summary of jurisprudence, that a national measure can only be declared incompatible with the Community law if it (1) gives rise to dominance, (2) there is an abuse by one or more undertakings and (3) there is a link between a national measure and the infringement of Community law. With regard to the first two elements, it is expected that the ECJ applies the same standards as under Article 82. Therefore NAPs or allocation mechanisms can only be subject to judicial scrutiny under the joint application jurisprudence if undertakings are abusing market power. The condemnation of national measures will crucially depend on the link between legislation and abuse.

The link between a Member State legislation and infringement is, however, not addressed by existing case law. Drawing parallels between the granting of statutory monopolies and abuse under the Court's case law regarding Article 86 is not only insightful but also warranted given the object and purpose of the joint application jurisprudence with regard to Article 82 EC Treaty.

In proceedings jointly applying Article 86 and 82, four approaches to State measures creating legal monopolies – epitomes of dominant firms – can be recognized.[127] Both the 'absolute-' and the 'limited sovereignty approach' view monopolies generally as favourable. The first appears to suggest that monopolies are compatible with EC legislation unless the Commission can prove the contrary,[128] while the limited sovereignty approach condemns monopolies only when they cannot avoid abusing their dominance when executing their normal operations.[129] By contrast, both the 'limited-' and the 'absolute competition approach' view legal monopolies rather critically. While in the first case, the ECJ appears to view the granting of exclusive rights creating a dominant position as impermissible unless a Member State can prove its necessity,[130] the Court's interpretation under the absolute competition approach[131] goes much further. Here the Court holds that the granting of rights is not permissible if it creates a situation where an undertaking is led to or induced to infringe Article 86. This comes very close to state that State measures

---

127. See Edward, D., Hoskins, M., (1995) for this typoloy.
128. Case C-202/88, French Republic v. Commission, [1991] ECR I-01223.
129. Case C-323/93, Crespelle, [1994] ECR I-05077, para. 18; Case C-41/90, Höfner v. Macrotron, [1991] ECR I-01979, para. 29.
130. Case C-155/73, Giuseppe Sacchi, [1974] ECR 00409; Case C-320/91, Corbeau, [1993] ECR I-02533.
131. See Case C-179/90, Merci convenzionali porto di Genova Spa v. Siderurgica Gabrielli SpA, [1991] ECR I-05889, para. 19; Case C-18/93, Corsica Ferries Italia Srl. V. Corpo dei Piloti del Porto di Genova, [1994] ECR I-01783; Case C-260/89, Elliniki Radiophonia Tiléorassi AE, [1991] ECR I-02925, para. 38.

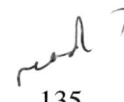

creating dominance are illegal per se.[132] Though it should be noticed however, that while cases are existing to support the extreme positions of the Court, in practices it appears that the limited competition and sovereignty approach are more widely applied.[133]

Nevertheless the developments under the absolute competition approach are noteworthy with respect to the assessment of the link between NAPs and allocation schemes and abuse offer interesting insights. Here the Court recognizes the inter-action between legislation placing undertakings artificially into a dominant position and the creation of incentives for undertakings to engage in abusive behaviour. This is also the case with regard to State measures taken within the framework of the European Emissions Trading System. It appears that such State measures could fall within the ambit of Competition law if the Court were to follow the footsteps of its absolute competition approach and if it were extending it from the realm of Article 86 to the joint application jurisprudence under Article 82 EC Treaty. In this way, undue distortions of competition arising from Member State interference with the market could be addressed.

From a legal point of view an important observation has to be made. The condemnation of dominance applied by the Court contrasts sharply with the tra-ditional approach it takes under Article 82. Here it holds that dominance as such is not against Community law. Despite the seeming paradox in legal terms, it reflects good economic judgment and deserves praiseworthy recognition. In economic terms, dominance can be a natural result of competition on the merits – as recog-nized by the ECJ under Article 82. Yet State interventions creating dominance in markets where it would not have been able to evolve nor survive is clearly dis-torting free competition and consequently to be rejected on economic grounds – this is recognized by the Court through application of the absolute competition approach under Article 86.[134] Thus from a conceptual point of view the proposed legal paradox is resolved by reference to State intervention.

This section has shown that even in the absence of case law guiding the application of the joint application jurisprudence under Article 82, some important findings can be derived through the drawing of legal parallels. If the Court would extend its absolute competition approach from Article 86 to Article 82 it would theoretically be able to close the existing legal gap and bring State measures distorting competition – such as are present under the current EU ETS – into the realm of the core provisions of Competition law. Thereby it would be enabled to safeguard competition on the merits and contain social welfare reducing distor-tions stemming from abusive behaviour.

Whether the ECJ is indeed willing to extend the scope of the joint application jurisprudence – and in particular strict approaches towards it – will depend on the impairment of the *effet utile* of Community law. This effectiveness is expected to depend on the perceived gravity of the anticompetitive effects and the possibility

132. See Craig P., De Búrca, G., (2003), 1129.
133. Craig P., De Búrca, G., (2008), 1078.
134. Edward, D., Hoskins, M., (1995), 162 ff.

that the infringements can satisfactorily be addressed through application of the existing provisions directed against undertakings. In the absence of empirical studies for abuse being directly attributable to the change in the propensity to abuse originating in State measures taken within the EU ETS framework, it is to be expected that the Court will not broaden the application of the joint application jurisprudence.

3.4   APPLICABILITY OF ARTICLES 82 AND 86 EC TREATY AGAINST
               STATE MEASURES

In contrast to Articles 81 and 82, which are directed against undertakings, Article 86 appears to be more directed towards the State. This Article has the objective of bringing State measures affecting public undertakings, and those granted special or exclusive rights within the field of application of EC law in general and Competition law in particular. Since many public undertakings and those undertakings which enjoy special or exclusive rights hold a dominant position, Article 82 is of particular relevance. Before assessing whether this provision would be applicable to State measures under the European Emissions Trading System, the term undertaking and subsequently the Court's application of the concepts of exclusive and special rights have to be examined.[135]

Even though the Court in principle applies the same broad[136] concept of undertakings under Article 86 as under Articles 81 and 82 EC Treaty, two observations have to be made. Firstly, while it is apparent from existing case law that the notion of undertakings also applies to public entities,[137] Article 86(1) expressly employs the term 'public undertaking', bringing those undertakings into the ambit of the provision that are subject to Member State control. Public undertakings are defined as undertakings over which public authorities exercise directly or indirectly dominant influence through ownership, financial participation or the rules that govern it.[138] In addition the ECJ considered government's ability to appoint more than half of the members of the undertakings administrative, managerial or supervisory body.[139] Secondly, entities exercising powers which form part of the prerogatives of the State[140] or base their operations on solidarity[141]

---

135. There are also other terms which need to be assessed to reach the conclusion that a State measure falls within the ambit of Art. 86 EC Treaty. Yet for the issue at hand it suffices to examine only undertakings, exclusive and special rights.
136. C-41/90, Höfner v. Macrotron, [1991] ECR I-01979, para. 21.
137. C-41/90, Höfner v. Macrotron, [1991] ECR I-01979, para. 22–23.
138. See Art. 2, Commission Directive 2006/111/EC of 16 Nov. 2006.
139. Joint Cases 188/80–190/80, France, Italy and the United Kingdom v. Commission, [1982] ECR I-02545, para. 25.
140. Case C-364/92, SAT Fluggesellschaft mbH v. Eurocontrol, [1994] ECR I-00043, para. 30; Case C-343/95, Diego Cali, [1997] ECR I-01547, para. 22–23.
141. Joined Cases C-159/91 and C-160/91, Poucet and Pistre, [1993] ECR I-00637, para. 18–19; C-70/95, Sodemare SA, [1997] ECR I-03395, para. 29; Joined cases C-264/01, 306/01, 354/01,

are not regarded as being economic in nature and hence are not classified as undertakings.[142]

The Court uses the concept of 'exclusive right' to describe a statutory monopoly[143] that enjoys a dominant position over a substantial part of the common market.[144] The Court has, however, been prepared to extend this concept to oligopolies.[145] Yet from the wording applied by the Court it is apparent that the number of undertakings which can enjoy such an exclusive right is intended to be limited[146] and closed.[147]

With regard to the concept of special rights, jurisprudence[148] does not offer sufficient insight and one has to turn to secondary law for interpretation.[149] Article 1 of Directive 2002/77/EC[150] defines 'special rights' as rights granted by Member States to a limited number of undertakings according to subjective and discretionary criteria. The above review is limited to a clarification of undertakings falling within the ambit of the Article. As explained below, the undertakings participating in the European Emissions Trading System do not meet these criteria and hence there is no need for further elaborating upon the legal functioning of this Article.

355/01, AOK Bundesverband [2004] ECR I-02493, para. 64; C-244/94, Fédération Française des Soci étés d'Assurance, [1995] ECR I-04013, para. 16–22; C-67/96, Albany International BV v. Stichting Bedrijfspensioenfonds Textielindustrie, [1999] ECR I-05751, para. 84–87; Joined Cases C-180/98–C-184/98, Pavel Pavlov and Others v. Stichting Pensioenfonds Medische Specialisten, [2000] ECR I-06451, para. 119; Case C-218/00, Cisal di Battistello Venanzio & C. Sas v. Istituto nazionale per l'assicurazione contro gli infortuni sul lavoro (INAIL), [2002] ECR I-00691, para. 45.

142. See Jones, A., Sufrin, B., (2004), 533–537.
143. C-323/93, Crespelle, [1994] ECR I-05077, para. 17; C-41/90, Höfner v. Macrotron, [1991] ECR I-01979, para. 29; C-260/89, Elliniki Radiophonia Tiléorassi AE, [1991] ECR I-02925, para. 38; Case C-320/91, Corbeau, [1993] ECR I-02533, para. 8; Joined Cases C-147/97 and C-148/97, Deutsche Post, [2000] ECR I-00835, para. 37.
144. C-179/90, Merci convenzionali porto di Genova Spa v. Siderurgica Gabrielli SpA, [1991] ECR I-05889, para. 14.
145. C-209/98, Entreprenørforeningens Affalds/Miljøsektion (FFAD) v. Københavns Kommune, [2000] ECR I-03743, para. 54.; see also C-155/73, Giuseppe Sacchi, [1974] ECR 00409, para. 14; Case 66/86, Ahmed Saeed Flugreisen and Silver Line Reisebüro GmbH v. Zentrale zur Bekämpfung unlauteren Wettbewerbs e.V., [1989] ECR 803, para. 50; C-260/89, Elliniki Radiophonia Tiléorassi AE, [1991] ECR I-02925, para. 36.
146. Case 66/86, Ahmed Saeed Flugreisen and Silver Line Reisebüro GmbH v. Zentrale zur Bekämpfung unlauteren Wettbewerbs e.V, [1989] ECR 803, para. 50; C-155/73, Giuseppe Sacchi, [1974] ECR 00409, para. 14.
147. See Jones, A., Sufrin, B., (2004), 545.
148. Case C-202/88, France v. Commission, [1991] ECR I-01223, para. 45; Joint Cases C-271, 281 and 289/90, Spain, Belgium & Italy v. Commission, [1992] ECR I-5833, para. 29–30; C-387/93, Banchero, [1995] ECR I-04663, para. 54.
149. This passage is based on See Jones, A., Sufrin, B., (2004), 545.
150. Article 1 of Directive 2002/77/EC of 16 Sep. 2002.

3.4.1   Application

In order to assess the applicability of Article 86 with regard to State measures giving rise to anticompetitive concerns within the framework of the European Emissions Trading System, one has to examine first whether the undertakings concerned meet the required criteria. Undertakings must be classified as public undertakings or as undertakings that have been granted exclusive or special rights.

As presented above undertakings will only be regarded as public if public authorities are able to exert influence over the commercial decisions of undertakings. With regard to Emissions Trading systems their number is likely to be limited and may include installations such as academic hospitals. If such undertakings engage in collusive or abusive activities they are subject to Community law.

Article 86 could be employed to condemn national measures if undertakings are unduly granted exclusive or special rights. With regard to the exclusive rights it has to be noted that the European Emissions Trading System applies to several sectors and is binding upon more than 5.000 undertakings. Even if an allocation contained in a NAP would be granted to a monopolist or to oligopolists on a relevant market, it is to be expected that the Court will take regard of the other allocations contained within the NAP. The Court would then question the existence of a closed group of undertakings and hence not view CO_2 emission allowances as exclusive rights.

In light of the preamble to Directive 2002/77/EC[151] special rights would only exist if a NAP would allocate emission allowances to a limited number of undertakings (two or more) within a geographical area otherwise than according to objective, proportional and non-discriminatory criteria. Given such criteria, it is extremely unlikely that the ECJ would conclude that any emission allowance allocation – irrespective of whether it involves grandfathering or PSR – constitutes a special right. This is because firstly, the Court would have to distinguish between the various sectors when counting the number of recipients under a NAP instead of emphasising the total number of undertakings which are granted emission allowances. Secondly, and perhaps more convincingly, if allocations were indeed made according to subjective, disproportional and outright discriminatory criteria, the NAPs would violate the criteria contained in Directive 2003/87/EC. This entails that the European Commission should not have been declaring this NAP compatible with Community law. In this case, the Commission's decision should be annulled under an Article 230 EC Treaty (review of legality) proceeding. Condemning discriminatory criteria in its totality is more effective than annulling a subset of specific State measures under Article 86.

It follows from the above that the ECJ is unlikely to condemn under Article 86 many State measures that give rise to anticompetitive concerns stemming from National Allocation Plans in general or the employed allocation formats in particular. This finding is based on the belief that there are no public undertakings

---

151.   Article 1 of Directive 2002/77/EC of 16 Sep. 2002.

concerned, and that State measures taken within the framework of the Emission Trading System do not qualify as exclusive or special rights. It can therefore be concluded that the interpretation of Article 86 EC Treaty is too narrow to address the myriad of anticompetitive measures Member States can employ.

4   ECONOMIC APPRAISAL AND SUMMARY

The previous sections of the chapter have presented distortions of competition that affect allocative efficiency and stem from anticompetitive State measures taken within the framework of the European Emissions Trading System. The chapter reviewed the Court's jurisprudence directed at containing distortive State measures in general and those resulting from NAPs and allocation formats under the EU ETS in particular. Special emphasis has been placed upon the evolution of the relevant case law.

Building upon the finding of the previous sections that the existing case law suggests that the ECJ is unlikely to find violations under the joint application jurisprudence, in particular with respect to Article 81 and 86, this section presents an economics based treatment of State measures under Articles 81 and 82 and concludes the analysis.

The joint application jurisprudence was established by the Court upon recognition of the gap in the EC Treaty that left Member States room to take anticompetitive measures that could not be appropriately addressed under competition law articles. Taking the EU ETS as an example, the ECJ's approach under the joint application jurisprudence under Article 81 appears too narrow to contain anticompetitive effects originating from such State measures. As for Article 82, there is the theoretical possibility for the Court to address such effects. Under Article 86 the Court has recognized the interaction between legislation placing undertakings artificially into a dominant position and the creation of incentives for undertakings to engage in abusive behaviour. This is also the case with regard to State measures taken within the framework of the EU ETS. If the Court were to follow the footsteps of its 'absolute competition approach' and extend it from the realm of Article 86 to the joint application jurisprudence under Article 82 EC Treaty, it might be able to address anticompetitive distortions originating in the EU ETS.

Despite this theoretical possibility that anticompetitive and allocative efficiency distorting State measures taken within the EU ETS framework could be brought into the realm of Competition law, and the economic insights that State measures impacting undertakings' propensity to collude and abuse will reduce social welfare, it is unlikely that such measures are contained from a legal perspective. The ECJ's ability to address 'legal gaps' is very much determined by the EC Treaty. Since the Court bases its jurisprudence upon the seemingly vague concept of *effet utile*, it is not surprising that there are natural limits to it. Unless the Court was convinced that the State measures altering undertakings' propensity to collude and abuse would endanger the effectiveness of the Treaty, the author does not expect extensions of the jurisprudence. It appears more likely that the Court would be welcoming a rise in cartelisation and abuse proceedings under competition Articles directed against undertakings as an indirect means to mitigate anticompetitive effects originating in State measures.[152] Even though the former is more desirable from a socio-economic and policy point of view, the latter is more in line with legal jurisprudence. It can therefore be concluded that EC Competition law does not appear to be well designed to contain anti-competitive and allocation distorting State measures stemming from the EU ETS.[153]

Before addressing another kind of distortion – State aid – that can be created through Member States' allocation of emission allowances in the subsequent chapter, the remainder of this section develops a series of critiques where the ECJ has disregarded industrial economic insights in relation to the joint application jurisprudence. Points of critique raised regard the discrimination between undertakings, private and public actions, collective agreements and the requirement to include part of an agreement to establish an infringement. Each is treated in turn.

In *Sodemare SA* the Court artificially distinguishes between undertakings (profit and non-profit organisations). It held the opinion that national legislation reserving public finance to non-profit organisations and setting profit companies in the same sector at a comparative disadvantage, did not constitute an anticompetitive concern that fell under the scope of EC cartel law.[154] Yet from an economic perspective it is clear that the artificial division of undertakings into profit seeking enterprises and non-profit organisations distorts competition on the merits. It sets certain operators at a financial comparative disadvantage, prevents them from entering the market and shields non-profit firms from competitive pressure. The overall social welfare is reduced by giving rise to suboptimal use of scarce resources and through the potential[155] creation of incentives to unduly inflate costs in order to be reimbursed for them.[156]

The Court does not only artificially distinguish between undertakings but also between private and State measures. By holding that a national measure cannot require or favour the adoption of an agreement, decision or concerted practice if the

---

152. When setting the level of penalty the conduct of undertakings may be assessed in light of national legislation; this can be viewed as a mitigating factor. See Joined cases 40 to 48, 50, 54 to 56, 111, 113 and 114–73, Coöperatieve Vereniging 'Suiker Unie' UA and others v. Commission of the European Communities [1975] ECR 01663, para. 620, Case C-198/01 CIF, [2003] n.y.r., para. 56.
153. It should be noticed that here EC Competition law is examined only. The complementary function of the four freedoms, in particular of Arts 28 (goods) and 49 (services) is not considered here. On this point see also Craig P., De Búrca, G., (2008), 1084.
154. Case C-70/95, Sodemare SA, [1997] ECR I-03395, para. 12–15 and 43 following and the fouth paragraph of the Court's decision which reads as follows: 'Articles 85 and 86, read in conjunction with Articles 3(g), 5 and 90 of the EC Treaty, do not apply to national rules which allow only non-profit making private operators to participate in the running of a social welfare system by concluding contracts wich entitle them to be reimbursed by the public authorities for the costs of providing social welfare services of a health-care nature.'
155. Whether this is indeed the situation in the case at hand cannot be determined without a close examination of the criteria applied by national administration.
156. Frank, R., (1997), 414, has termed this the gold plated water cooler effect.

obligations these stipulations create are self-sufficient or self-contained,[157] the Court draws an artificial differentiation between private and public action. This is analogous to the critique van Gyselen (1989)[158] formulated with regard to *Cullet*. In this case the State did not delegate the power to fix prices to private undertakings but to public authorities.[159] The Court concluded that it was unable to apply Article 81(1) to Member State measures. Only if a State delegates price fixing power to undertakings and thereby makes it superfluous for them to collude, Member State regulation is viewed to undermine the effectiveness of Article 81(1) and consequently to be held unlawful. If a Member State, however, establishes a retail price maintenance scheme by itself, it is not deemed to be incompatible with the common market. Thus the State can freely orchestra infringements which would be condemned if they were executed by undertakings. As noted correctly by van Gyselen (1989), such delineation is artificial because the regulatory choice will depend to a large degree on the particularities of the relevant product market.[160]

In *Sodemare* and *Servizi* the Court denies that anticompetitive effects can stem from collective agreements. Following its restrictive interpretation of collective dominance, the ECJ denies that collective agreements, which allow derogations from legally established and mandatory tariff schemes by extending and legally enforcing majority decisions upon non-collaborating undertakings, can give rise to anticompetitive effects.[161] Even though from an economic point of view such agreements influence the undertakings propensity to collude and abuse.

Even more crucial than failing to recognize economic theory in the above cited cases, the Court fails to realise the most fundamental differences between Competition law and other fields of law. Namely that uncertainty does not only relate to the offender but can also relate to the offence itself. The Court held in *van Eycke* that State measures could only reinforce the effects of existing agreements if they incorporate either wholly or in part the terms of agreements concluded between undertakings and requires or encourages compliance[162] and was – as in previous cases[163] – not concerned about public authorities encouraging undertakings to set up anticompetitive agreements. Even though this criterion was slightly

---

157. Case 2/91, Wolf W. Meng, [1993] ECR I-05751, para. 15; Case 245/91, Ohra Schadeverze-keringen NV., [1993] ECR I-05851, para. 11.
158. Gyselen, L., (1989), 44.
159. Case 231/83, Henri Cullet and Chambre syndicate des réparateurs automobiles et détaillants de produits pétroliers v. Centre Leclerc à Toulouse and Centre Leclerc à Saint-Orens-d-Gameville, [1985] ECR 305, para. 17.
160. Gyselen, L., (1989), 44.
161. Case C-96/94, Centro Servizi Spediporto, [1995] ECR I-02883, para. 29; Case C-38/97, Autotransporti Librandi Snc di Librandi F. & C., [1998] ECR I-05955, para. 49–52.
162. Case 267/86, Pascal Van Eycke v. ASPA NV., [1988] ECR 4769, para. 23.
163. Case 209–213/84, Lucas Asjes and others, Andrew Gray and others, Andrey Gray and others, Jacques Maillot and others and Léo Ludwig and others, [1986] ECR 1425; Case 136/86, Bureau national interprofessionnel du cognac v. Yves Aubert, [1987] ECR 4789.

loosened in *Vlaamse Reisbureaus*,[164] it must still be viewed as being the current interpretation of the Court.

From an economic point of view, there are two important flaws contained in this line of argumentation. Firstly, Member States can, by creating entry barriers, change the market structure and thereby facilitate collusion and stabilize cartels by reducing the threat of new entry. Through State measures agreements can be reinforced without employing their express wording. Secondly, the criterion of including wholly or in part elements of a cartel agreement is far to legalistic an approach to be practicable and has to be regarded as being more strict than the Court's case law under Article 81 would suggest. Illegal cartel agreements are neither published nor expressly drafted in contract form. So even if there is an illegal agreement, one does not know the wording. Furthermore Competition law enforcement is characterized by the difficulty of detection. Unlike in other fields of law where one at least is able to detect illegal trespass, under Competition law one first has to discover the existence of cartels. If one accepts that not all illegal cartels are discovered – even if one was able to acquire the exact terms of those agreements that are discovered – this approach is still too narrow to maintain the same level of Competition law enforcement persisting before the State measure creating undue incentives for undertakings was introduced.

5   JOINT APPLICATION POST 2012?

Taking the amendment of the EU ETS as a basis, it is noticeable that Member States are playing a much lesser role than in the current EU ETS. The National Allocation Plans that arguably granted Member States much discretion will be abolished. The allocation formats prescribed by the amendment, benchmarking and auctioning, will be further fleshed out by the Commission. To design the exact allocation rules, the European Commission has time until 30 June 2010[165] with regard to auctioning and until 31 December 2010[166] in the case of benchmarking.

With regard to the Community-wide *ex ante* benchmarking allocation scheme it can be observed that a high degree of harmonisation can be inferred from the wording of Article 10a(1) which speaks of fully harmonized implementation measures.[167] Allocation will be based upon a Commission regulation that takes as its defining allocation principle the average performance of the 10% most efficient installations in a sector or sub-sector during the period 2007–2008. The task of Member States therefore appears to be limited to an executive function. It may therefore be questioned if one can indeed speak of a State measure within the framework of the joint application jurisprudence. Given that the object and

---

164. Case 311/85, ASBL Vereniging van Vlaamse Reisbureaus v. ASBL Sociale Dienst van de Plaatselijke en Gewestelijke Overheidsdiensten, [1987] ECR 3801, para. 23.
165. Article 10(4), Directive 2009/29/EC.
166. Article 10a(1), Directive 2009/29/EC.
167. Article 10a(1), Directive 2009/29/EC.

purpose of the joint application jurisprudence is the maintenance of the *effet utile* of Competition law provisions of the EC Treaty, it may be argued that the Court might be inclined to interpret State measures broadly. It does, however, appear more expedient – if indeed the *effet utile* of a Competition law provision was undermined – for the Court to address the allocation decision of the Commission rather than an administrative act on Member State level. In this context it should be noticed that also the Commission is obliged to respect EC primary law.[168]

Regarding auctioning, a similar and yet more qualified finding can be made. Here the degree of Member State discretion is not yet known. The modalities of the Commission regulation on auctioning is still to be fleshed out. It appears that the Commission will select one or more auctioning systems and allow Member States to employ them subject to its monitoring.

A Member State may thus select an auctioning mechanism from the list presented by the Commission and set the relevant parameters in such a way that only a limited number of undertakings may find it attractive to participate in the bidding. If as a result they could collude successfully one could argue that the implemented State measure did facilitate a collusive process. Though before examining the application of the joint application jurisprudence further, one should note that it may be more easy for the Commission and the Court to establish a conflict with the auctioning criteria enlisted in Article 10(4) of the EU ETS amendment. These criteria specify that auctions shall be granted full, fair and equitable access, afford all participants equal access to information, cost efficiency and enable small emitters due access to auctions. It is therefore believed that it would be more expedient for the Court to review an auction that gave rise to concerted practices under the criteria of the EU ETS amendment than under the joint application jurisprudence.

It therefore appears that the role of Competition law within the framework of the joint application jurisprudence under the amendment is very remote. This appears the case in particular with regard to the harmonized *ex ante* benchmarking and perhaps to a somewhat lesser degree also under auctioning.

---

168. See Art. 211 EC Treaty.

Chapter 6

EU Emissions Trading System and State Aid

1   INTRODUCTION[1]

After analysing if Competition law is capable to contain government induced distortions stemming from the Emission Trading System, this chapter addresses allocation issues from a State aid perspective. The State aid provision is designed to prevent Member States from granting aid that sets undertakings at a comparative advantage and affects the internal market.

As presented in chapter three, not all allocation mechanisms grant undertakings a comparative advantage. Auctions in which undertakings compete for buying emission rights for example do not give rise to undue advantages as such. State aid involvement is unlikely to be present if auctioning is the only allocation format used. Later in this chapter it is examined to what extent auctions may give rise to undue advantages in practice during the third trading phase when auctioning is envisaged to co-exists with benchmarking. Here some references regarding auctions in the first and second trading period is made where necessary. At first it is examined to what extent free allocation mechanisms such as grandfathering or Performance Standard Rate (PSR) may lead to favourable treatment of undertakings. Both will be examined in this chapter from a legal and industrial economic perspective. The chapter analyses the methods' proneness to give rise to anticompetitive distortions of the level playing field in the EU and examines their compatibility with the European common market. In doing so, the chapter is subdivided into four sections. After this introduction section two presents State aid legislation and spells out relevant criteria. Section three is subdivided into three parts. Parts one and two apply the developed State aid criteria to grandfathering and the PSR

---

1. A version of this chapter is also be published in Weishaar, S. (2007c).

system. Part three represents a conclusion of the section. Section four examines to what extent Competition law is able to contain distortions stemming from allocation mechanisms under the post 2012 regime.

2   EUROPEAN STATE AID REGULATION APPLICABLE
           TO EMISSION TRADING SYSTEMS

This section reviews the substantive State aid requirements applicable to Emission Trading Systems. While the first section reviews the necessary requirements which have to be fulfilled in order to establish the existence of State aid, section two presents derogations under which Emission Trading Systems can be declared to be nevertheless compatible with the common market.

2.1   THE EXISTENCE OF STATE AID: ARTICLE 87(1)

Article 87(1) EC Treaty declares incompatible with the common market:

> any aid granted by a Member State or through State resources in any form whatsoever which distorts or threatens to distort competition through favouring certain undertakings or the production of certain goods [. . .] in so far as it affects trade between Member States.

From a legal point of view, a number of criteria need to be fulfilled in order to fall within the ambit of Article 87(1) EC Treaty:

- Transfer of a benefit or an advantage (notion of aid).
- Aid favouring a certain undertaking over others (selectivity principle).
- Granted by the State or through State resources.
- It should be an undertaking or . . . production.
- Distorts or threatens to distort competition.
- Community dimension: aid which is capable of affecting trade between Member States.

Only if all these criteria are fulfilled does State aid exist within the meaning of Article 87(1) EC Treaty. Before taking each criterion in turn, it should be noticed that criteria v and vi are in fact inseparably linked in their legal assessment. It is only for didactical reasons, namely to explain anticompetitive distortions from an economic side, that they are treated separately.

After reviewing each criterion the relevant Emissions Trading decisions will be reviewed. Although decided before the establishment of a Community legal framework, the Commission's decisions concerning early emission trading schemes can offer interesting insights.[2] Subsequently, the following section will

---

2. See Merola, M. & Crichlow, G., (2004), 31.

discuss two relevant derogations. Under these derogations State aid may still be declared compatible with the common market.

2.1.1 Transfer of a Benefit or an Advantage (Notion of Aid)

The first criterion to be examined concerns the notion of aid. The Treaty contains no express definition of the concept of aid as referred to under Article 87(1). Evans (1997) states that any attempt by the legislator to give a definition would neither have been feasible nor useful, since it would have been liable to restrict the scope of the concept.[3] The Commission claims that the concept of aid falls exclusively within the ambit of Community law and that it therefore has the power to interpret the concept of aid under Article 87 EC Treaty[4] and indeed the Commission does so very broadly in order to ensure that competition, in accordance with Article 3(g) EC Treaty, is not distorted.[5]

In light of this objective it is not surprising that the European Court of Justice (ECJ) does not distinguish between measures of State intervention by reference to their causes or aims but determines aid solely based on their effects.[6] Transferral of an advantage is the objective test employed to determine its existence.[7]

In *Banco Exterior* the Court confirmed that the concept of aid is:

> wider than that of a subsidy because it embraces not only positive benefits, such as subsidies themselves, but also interventions which, in various forms, mitigate the charges which are normally included in the budget of an undertaking and which, without therefore being subsidies in the strict meaning of the word, are similar in character and have the same effect.[8]

Examples of practices constituting State aid are tax exemptions,[9] exemption from para-fiscal charges, preferential interest rates, loan guarantees,[10] granting

---

3. Evans, A., (1997), 27.
4. Case T-459/93 *Siemens SA v. Commission* [1995] ECR II-1675, para. 52.
5. Evans, A., (1997), 27.
6. Case 173/73 *Italy v. Commission* [1974] ECR 709, para. 13; Case C-241/94 *France v. Commission* [1996] ECR I-4551, para. 19–20; joined Cases T-228/99 and T-233/99, *Westdeutsche Landesbank Girozentrale and Land Nordrhein-Westfalen v. Commission*, [2003] ECR II-435, para. 180.
7. Case T-67/94 *Ladbroke Racing Ltd. v. Commission* [1998] ECR II-00001, para. 52; Case T-46/97 *SIC v. Commission* [2000] ECR II-2125, para. 83; joined Cases T-228/99 and T-233/99, *Westdeutsche Landesbank Girozentrale and Land Nordrhein-Westfalen v. Commission*, [2003] ECR II-435, para. 180.
8. Case C-387/92 *Banco de Credito Industrial SA (Banco Exterior de España SA) v. Ayuntamiento de Valencia* [1994] ECR I-877, para. 13; see also Case 30/59 *De Gezamenlijke Steenkolenmijnen Limburg v. High Authority* [1961] ECR 1, para. 19; C-156/98 *Germany v. Commission* [2000] ECR I-6857, para. 25.
9. Case C-387/92 *Banco de Credito Industrial SA (Banco Exterior de España SA) v. Ayuntamiento de Valencia* [1994] ECR I-877, para. 14; Case C-6/97, *Italy v. Commission* [1999] ECR I-2981, para. 16.
10. Even when they are not relied upon, see Case C-404/97 *Commission v. Portugal* [2000] ECR I-4897, para. 44–45.

preferential terms for land or real estate, provision of goods, services or personnel on preferential terms,[11] indemnities against operating losses and the purchase of shares of a company in financial difficulties,[12] preferential terms for public orderings, deferment of collection of fiscal or social contributions[13] and dividend grants.[14] It is therefore to be concluded that not only direct benefits accruing to an undertaking but also cost reductions constitute aid.[15]

In the context of the U.K. Emission Trading Systems, the Commission has held that incentive money paid to market participants through an auction in exchange for emission reductions constitutes an advantage.[16] Not only the granting of funds but also the allocation of transferable emission allowances free of charge to participants of the U.K. and Danish trading system was held to constitute a transfer of intangible assets that could be traded and hence as amounting to the granting of an undue advantage.[17] Arguments that recipients had to undertake investments in order to be able to realize benefits were not entertained.[18] In practice this is tantamount to imply that a reduction in production as a precondition to be able to sell emission allowances on the market constituted State aid. This is a legal extension of the Commission's argument which calls into question the economic efficiency arguments which have led to the introduction of an Emission Trading System. Yet this point has not been addressed in either decision. While it was still unclear from the U.K. and Danish decision, the Commission's decision in Belgium Green Electricity Certificates clarifies that the mere fact that there is a market for tradable emission allowances, is not sufficient to assert that the transfer of an intangible asset constitutes aid.[19] In this case the Commission recognized that the Belgium certificates were only a recognition of actual emission reductions and hence undertakings would not benefit from an undue advantage.

---

11. Case C-241/94 *France v. Commission* [1996] ECR I-4551, para. 37; Case C-39/94 SFEI [1996] ECR I-3547, para. 59–60; Case C-143/99 *Adria-Wien Pipeline GmbH and Wietersdorfer & Peggauer Zementwerke GmbH v. Finanzlandesdirektion fuer Kaernten* [2001] ECR I-8365, para. 40.
12. Case 323/82 *SA Intermills v. Commission* [1984] ECR 3809, para. 31; Joint Cases C-278/92, C-279/92 and C-280/92 *Spain v. Commission* [1994] ECR I-4103, para. 20 and 41; Case C-305/89, *Italy v. Commission* [1991] ECR I-1603, para. 18.
13. Case C-75/97, *Belgium v. Commission* [1999] ECR I-3671, para. 44 and 53.
14. This passage is based on Craig, P. & De Burca G., (2003), 1141 and Fairhurst, J. & Vincenzi, C., (2003), 431.
15. Case 30/59 *De Gezamenlijke Steenkolenmijnen Limburg v. High Authority* [1961] ECR 1, para. 19.
16. European Commission, (2001), State aid No N 416/2001 United Kingdom Emission Trading Scheme, C(2001) 3739 fin, 28 Nov. 2001, 9.
17. European Commission, (2001), State aid No N 416/2001 United Kingdom Emission Trading Scheme, C(2001) 3739fin, 28 Nov. 2001, 9; European Commission, (2000), Statsstøttesag Nr. N 653/1999 – CO_2-kvoter, SG (2000) D/, 12.04.2000, 5–6.
18. European Commission, (2001), State aid No N 416/2001 United Kingdom Emission Trading Scheme, C(2001) 3739fin, 28 Nov. 2001, 9.
19. European Commission, (2001), Steunmaatregel nr. N 550/2000 België Groenestroomcertificaten, SG(2001) D/290545, 25 Jul. 2001, 10.

2.1.2 Aid Favouring a Certain Undertaking over Others (Selectivity Principle)

This criterion is applied in order to differentiate between State aid favouring certain undertakings or the production of goods[20] and general State interventions regarding fiscal rules, social security measures, etc. Selectivity is thus a precondition for creating a distortion of competition in the first place.[21] The ECJ looks at the effects of the measure to distinguish between a selective State aid measure directed to aid certain undertakings and general economic measures.[22]

The Commission considers that measures that have neither the objective nor the effect of favouring certain undertakings or the production of certain goods, or measures applicable to persons in accordance with objective criteria, that is, irrespective of the location, the sector, or undertaking in which the beneficiary is employed, do not involve aid.[23]

This, however, does not imply that all undertakings are to be affected equally by a general measure. Differential impacts on some sectors will be held to be selective[24] and thus constitute State aid within the meaning of Article 87(1) as long as the differential effects are not inherent in the system or justified by the nature or the reasons relating to the general measure.[25] Measures allowing authorities too much discretion in the granting of aid will be judged to be selective.[26] Recipients do not need to be identifiable *ex ante*, it is sufficient if the scheme could favour certain undertakings. In the context of assessing the selective nature of advantages the Court has also employed the term 'specificity' which appears to address the differential impact of aid on a subset of the beneficiaries.[27]

To summarize the above, measures of a general economic nature that are objective, non-discriminatory and non-discretionary will not be adjudged as selective and hence not fall within the ambit of Article 87(1) EC Treaty. Central to the selectivity criterion is not if due to the measure at hand the beneficiary is better or worse but rather if the measure is such as to favour certain undertakings or production of certain goods in comparison with other undertakings which are in a legal

---

20. Case T-55/99 *CETM v. Commission* [2000] ECR II-3207, para. 39; Joined Cases T-92/00 and T-103/00 *Territorio Histórico de Álava v. Commission* [2002] ECR II 1385, para. 48, Case C-143/99 *Adria-Wien Pipeline GmbH and Wietersdorfer & Peggauer Zementwerke GmbH v. Finanzlandesdirektion fuer Kaernten* [2001] ECR I-8365, para. 41.
21. Frenz, W., (2007), 191.
22. Case 173/73 *Italy v. Commission* [1974] ECR 709, para. 13; Case C-241/94 *France v. Commission* [1996] ECR I-4551, para. 19–20; Case 310/85 *Deufil GmbH & Co. KG v. Commission* [1987] ECR 901, para. 8.
23. EC Commission, (1995) Notice on cooperation between national courts and the Commission in the State aid field, OJ C 312/8, para. 7.
24. Case C-75/97, *Belgium v. Commission* [1999] ECR I-3671, para. 28–31.
25. See Bacon, K., (2003), 59.
26. Case C-200/97 *Ecotrade Srl v. AFS* [1998] ECR I-7907, para. 40; Case C-241/94, *France v. Commission* [1996] ECR 4393, para. 23–24.
27. See Frenz, W., (2007), 224 ff. and Case C-501/00, *Spain v. Commission* [2004] n.y.r., para. 120.

and factual situation that is comparable in light of the object and purpose of the measure's objective.[28]

Regarding Emission Trading Systems, it is noticeable that the Commission's decision on the UK scheme is silent about this criterion. Merola and Crichlow (2004), however, argue that it can be inferred from the Commission's reference to direct participants of the scheme[29] and therefore has been assessed by the Commission. In any event, selectivity has been important in the Danish CO_2 emission quota system.[30] This system was held to constitute a selective measure because it was only binding upon the electricity sector.

2.1.3   Granted by the State or through State Resources

The third criterion which needs to be examined is whether the alleged aid is granted by the State or through State resources. The ECJ held that for the purpose of Article 87(1) the meaning of State is broad. It does not only include national or central authorities but also regional and local authorities, as well as public or private bodies established or appointed by the State.[31] Hence any aid given by such institutions falls within the meaning of 'State' used in Article 87(1) provided that the aid given is imputable to the State. 'Imputability' takes a central role in the assessment of State involvement.[32] It not only seeks to determine the States active involvement in the measure through for example, appointments in private bodies by the State, but also extends to the discretion enjoyed by Member States to design or chose measures. Case law clarifies that the Article refers to decisions of Member States by which, in pursuit of their own economic and social objectives, they give, by unilateral and autonomous decisions, resources to undertakings or other persons or procure for them advantages intended to encourage the attainment of the economic or social objectives sought.[33]

As regards the second part of this criterion 'through State resources', the ECJ held that there has to be a financial burden for the State which directly or indirectly affects a public account in order to constitute aid within the meaning

---

28. Case C-143/99 *Adria-Wien Pipeline GmbH and Wietersdorfer & Peggauer Zementwerke GmbH v. Finanzlandesdirektion fuer Kaernten* [2001] ECR I-8365, para. 41. Case C-75/97, and *Belgium v. Commission* [1999] ECR I-3671, para. 28 ff.
29. Merola, M. & Crichlow, G., (2004), 39 and European Commission, (2001), State aid No N 416/2001 United Kingdom Emission Trading Scheme, C(2001) 3739 fin, 28 Nov. 2001, 9.
30. European Commission, (2000), Statsstøttesag Nr. N 653/1999 – CO_2-kvoter, SG (2000) D/, 12 Apr. 2000, 5–6.
31. Case 78/76 *Steinike v. Bundesamt für Ernährung und Forstwirtschaft* [1977] ECR 595, para. 21; Joint Cases 67, 68 and 70/85, *Van der Kooy BV v. Commission* [1988] ECR 219, para. 35; Case C-303/88 *Italian Republic v. Commission* [1991] ECR I-1433, para. 11; Case C-305/89, *Italy v. Commission* [1991] ECR I-1603, para. 13; Case C-482/99 *France v. Commission* [2002] ECR I-4397, para. 48; Case C-200/97 *Ecotrade Srl v. AFS* [1998] ECR I-7907, para. 35.
32. Case C-482/99 *France v. Commission* [2002] ECR I-4397, para. 24.
33. Case 61/79 *Amministrazione delle finanze dello Stato v. Denkavit italiana* [1980] ECR 1205, para. 31, Case T-351/02 *Deutsche Bahn AG v. Commission* [2006] n.y.r., para. 100.

of Article 87(1).[34] Thus an actual or immediate transfer of resources from the State to the beneficiary is not necessary.[35] The granting of loan guarantees or preferential charges in which the State foregoes profits falls within the interpretation of 'State resources'.[36]

In *Preussen Elektra* the ECJ clarified that State aid had to be granted directly or indirectly through State resources[37] and that Article 10 EC Treaty could not be used to extent the scope of Article 87 EC Treaty to support measures which are decided upon by a Member State but financed by private undertakings.[38] Thus, even if there is an advantage granted through State action, as long as no extra financial burden is placed upon public authorities, there is no aid within the meaning of the Article.[39]

Concerning Emission Trading Systems the Commission has interpreted this criterion restrictively. The payment of incentive money from State resources in the U.K. Emission Trading System was held to be State aid.[40] In the context of both the Danish and the U.K. Emission Trading Systems, the Commission held that by not auctioning tradable emission allowances the State was forgoing revenue it could have raised.[41] The Commission particularly emphasized the voluntary nature of the U.K. system which foreclosed the State's possibility to gain revenue from an auction.[42] In the Belgium Green Certificate case, however, the Commission followed the line of argumentation of the Belgium government which cited the *Preussen Elektra* case[43] in order to justify the obligation on energy distributors to buy a certain amount of green energy certificates.[44] The Commission also accepted that even though the government did grant intangible assets in the form of green energy

34. Case 82/77, *Opebaar Ministerie v. Van Tiggele* [1978] ECR 25, para. 23–25; Joined Cases C-72/91 and C-73/91 *Sloman Neptun v. Bodo Ziesemer* [1993] ECR I-887, para. 19 and 21; Case C-189/91 Kirsammer-Hack [1993] ECR I-6185, para. 16; Joint Cases 213–215/81 *Norddeutsches Vieh- und Fleischkontor v. BALM* [1982] ECR 3583, para. 22 and 23; Joint Cases C-52/97, C-53/97 and C-57/94 *Viscido, Scandella, Terragnolo and Others v. Ente Poste Italiane* [1998] ECR I-2629, para. 13, C-295/97, *Industrie Aeronautiche e Meccaniche Rinaldo Piaggio SpA v. International Factors Italia SpA* [1999] ECR I-3735, para. 35.
35. See Quigley, C. & Collins, A., (2003), 26.
36. Joint Cases T-204/97 and T-270/97 *EPAC v. Commission* [2000] ECR II-2267, para. 80; Case C-200/97 *Ecotrade Srl v. AFS* [1998] ECR I-7907, para. 43; Joint Cases 67, 68 and 70/85, *Van der Kooy BV v. Commission* [1988] ECR 219, para. 28.
37. Case C-379/98 *Preussen Elektra v. Schleswag AG* [2001] ECR I-2099, para. 58.
38. C-379/98 *Preussen Elektra v. Schleswag AG* [2001] ECR I-2099, para. 63–65.
39. Case T-613/97 *Ufex v. Commission* [2000] ECR II-4055, para. 108–110.
40. European Commission, (2001), State aid No N 416/2001 United Kingdom Emission Trading Scheme, C(2001) 3739fin, 28 Nov. 2001, 9.
41. European Commission, (2001), State aid No N 416/2001 United Kingdom Emission Trading Scheme, C(2001) 3739fin, 28 Nov. 2001, 9; European Commission, (2000), Statsstøttesag Nr. N 653/1999 – CO$_2$-kvoter, SG (2000) D/, 12 Apr. 2000, 5–6.
42. European Commission, (2001), State aid No N 416/2001 United Kingdom Emission Trading Scheme, C(2001) 3739fin, 28 Nov. 2001, 9. On this point also see Merola, M., Crichlow, G., (2004), 37.
43. C-379/98 *Preussen Elektra v. Schleswag AG* [2001] ECR I-2099.
44. European Commission, (2001), Steunmaatregel nr. N 550/2000 België Groenestroomcertificaten, SG(2001) D/290545, 25 Jul. 2001, 6 and 10.

certificates to green energy producers, it did not incur a budgetary loss since the certificate merely constitutes prove of the actual production of green energy.[45]

2.1.4   It Should be an Undertaking or . . . Production

It is believed that the notion of undertaking under Articles 87 and 88 EC Treaty is identical to the one applied under Articles 81 and 82 EC Treaty. In *Höfner*[46] the ECJ held that any entity engaged in an economic activity regardless of its legal status and the way it was financed amounted to an undertaking.

2.1.5   Distorts or Threatens to Distort Competition

Aid within the meaning of Article 87 EC Treaty must distort or threaten to distort competition. In order to determine anticompetitive distortions, the ECJ assesses the direct and immediate effects of aid on the competitive position of the recipient. A comparison of the market position prior to and after the granting of aid is conducted and distortion is established if the position of the undertaking is more favourable *ex post*.[47] Any aid granted to an undertaking is treated as giving it an advantage in relation to actual or potential competitors and as affecting competition.[48] Even though actual proof of an anticompetitive distortion is not required[49] and its potential presence is sufficient, the Commission must at least set out those circumstances in the statement of reasons for its decision.[50] Thus unlike under Articles 81 and 82 EC Treaty, under Article 87(1) EC Treaty the Commission is under no obligation to prove anticompetitive distortions on relevant product markets and the Court appears to be reluctant to do so.[51]

So far, the ECJ has not been willing to entertain arguments that the particular aid before it lowered the relatively higher costs of a sector or that aid should be justified on the grounds that other States made similar payments within their territory.[52] This implies that the ECJ looks at the anticompetitive effects without taking any grounds for motivation of the aid into account.

---

45. European Commission, (2001), Steunmaatregel nr. N 550/2000 België Groenestroomcertificaten, SG(2001) D/290545, 25 Jul. 2001, 6 and 10.
46. Case C-41/90 *Höfner and Elser v. Macroton GmbH* [1991] ECR I-1979, para. 21.
47. Case 173/73 *Italy v. Commission* [1974] ECR 709, para. 17.
48. Evans, A., (1997), 77.
49. Case T-288/97 *Regione Friuli Venezia Giulia v. Commission* [2001] ECR II-1169, para. 49–50; Case T-35/99 *Keller SpA v. Commission* [2002] ECR II-261, para. 85; Case T-214/95, *Vlaamse Gewest v. Commission* [1998] ECR II-717, para. 67.
50. Joined Cases 296/82 and 318/82, *Netherlands and Leeuwarder Papierwarenfabriek v. Commission* [1985] ECR 809, para. 24, Joint Cases C-329/93, C-62/95 and C-63/95 *Germany, Hanseatische Industrie-Beteiligungen GmbH and Bremer Vulkan Verbund AG v. Commission* [1996] ECR I-5151, para. 52.
51. Case 730/79 *Philip Morris v. Commission* [1980] ECR 2671, para. 9–13; Case 53/00 *Ferring SA v. Agence centrale des organismes de sécurité sociale* [2001] ECR I-9067, para. 21.
52. Case 78/76 *Steinike v. Bundesamt für Ernährung und Forstwirtschaft* [1977] ECR 595, para. 24; Case T-214/95, *Vlaamse Gewest v. Commission* [1998] ECR II-717, para. 54.

As the guardian of the Treaty with respect to State aid measures, the Commission has promulgated Commission Regulation (EC) No. 69/2001 on threshold levels below which all cumulated aid not exceeding EUR 200.000, granted within a period of three years, is adjudged not to be capable of distorting competition – the so-called *de minimis* rule.[53] Therefore only aid exceeding this threshold can fall within the ambit of Article 87(1) EC Treaty.

With regard to Emission Trading Systems, the Commission has held in the UK trading system that incentive money paid to market participants through an auction in exchange for emission reductions, constitutes an advantage which distorts competition between the recipient companies and their competitors.[54] Similarly the granting of an undue advantage in the form of an intangible asset was held to constitute a distortion of competition in the Belgium,[55] UK,[56] and Danish[57] schemes.

Besides such legal considerations, the determination of the monetary value of emission allowances granted under a National Allocation Plan (NAP) to a particular entity is a non-trivial issue. From an economic point of view 'aid' should include all economically quantifiable advantages accruing to an entity. Not only does the market price of an emission allowance vary considerably over time but also do particular industries face very different demand elasticities. Consumers of some sectors will tolerate price increases while others may not. The marginal benefit from an emission allowance accruing to an undertaking thus also depends on its ability to pass increased production costs on to consumers. By contrast, those sectors that are unable to do this will experience a larger cut in profits. Thus in summary the actual effects of aid granted to undertakings may vary and are not easily determined.

2.1.6 Community Dimension: Aid Which is Capable of Affecting Trade between Member States

In *Philip Morris v. Commission* the ECJ presumed that trade between Member States was affected if the given aid was capable of strengthening the position of an undertaking *vis-à-vis* undertakings competing in intra-community trade.[58]

---

53. The value was increased from EUR 100.000 to EUR 200.000 by Commission Regulation 1998/2006 of 15 December which replaced Commission Regulation 69/2001 of 12 Jan. 2001. A transitionary period ended 30 Jun. 2007.
54. European Commission, (2001), State aid No N 416/2001 United Kingdom Emission Trading Scheme, C (2001) 3739 fin, 28 Nov. 2001, 9.
55. European Commission, (2001), Steunmaatregel nr. N 550/2000 België Groenestroomcertificaten, SG (2001) D/290545, 25 Jul. 2001, 5–6.
56. European Commission, (2001), State aid No N 416/2001 United Kingdom Emission Trading Scheme, C (2001) 3739fin, 28 Nov. 2001, 9.
57. European Commission, (2000), Statsstøttesag Nr. N 653/1999 – CO_2-kvoter, SG (2000) D/, 12 Apr. 2000, 5–6.
58. Case 730/79 *Philip Morris v. Commission* [1980] ECR 2671, para. 11; Case T-214/95, *Vlaamse Gewest v. Commission* [1998] ECR II-717, para. 50; Case T-288/97 *Regione Friuli Venezia Giulia v. Commission* [2001] ECR II-1169, para. 41; Case T-152/99 *Hijos de Andrés Molina,*

Even though also limited amounts of aid or aid granted to small recipients can affect intra-community trade,[59] the aforementioned *de minimis* regulation[60] is applicable. As is the case with regard to competitive distortions, the relevant criterion here is not the actual effect on intra-community trade but its potential detriment.[61] As was the case under competitive distortions, the Commission must at least set out those circumstances in the statement of reasons for its decision.[62]

With regard to Emission Trading Systems, the Commission has held in the UK scheme that incentive money paid to market participants through an auction in exchange for emission reductions constitutes an advantage which distorted trade between Member States.[63] Similarly the granting of an undue advantage in the form of an intangible asset was generally held to distort trade between Member States.[64]

2.2   Derogations of Article 87(1)

If all of the above criteria have been fulfilled, State aid may still be permissible and be adjudged as being compatible with the common market. Article 87(2) lists a number of types of aid which are automatically compatible with the common market while Article 87(3) lists types of aid that may be deemed to be compatible. It is therefore evident that with regard to the latter category the European Commission has more discretion. The Commission's wide discretion to allow aid under

---

SA v. Commission [2002] ECR II-3049, para. 220; Joint Cases T-298/97, T-312/97, T-313/97, T-315/97, T-600 to 607/97, T-1/98, T-3/98 to T-6/98, T-23/98, *Alzetta Mauro and Others v. Commission* [2000] ECR II-2319, para. 81. In Joint Cases C-278/92, C-279/92 and C-280/92 *Spain v. Commission* [1994] ECR I-4103, para. 40, the Court clarified that the beneficiary undertaking itself does not need to engage in exports. See also Case 102/87 *France v. Commission* [1988] ECR 4067, para. 19; Case C-75/97, *Belgium v. Commission* [1999] ECR I-3671, para. 47.

59. Case C-142/87 *Belgium v. Commission* [1990] ECR I-959, para. 43.
60. Commission Regulation 1998/2006 of 15 December which replaced Commission Regulation 69/2001 of 12 Jan. 2001. A transitory period ended 30 Jun. 2007.
61. Joint Cases T-298/97, T-312/97, T-313/97, T-315/97, T-600 to 607/97, T-1/98, T-3/98 to T-6/98, T-23/98, *Alzetta Mauro and Others v. Commission* [2000] ECR II-2319, para. 78.
62. Joined Cases 296/82 and 318/82, *Netherlands and Leeuwarder Papierwarenfabriek v. Commission* [1985] ECR 809, para. 24, Joint Cases C-329/93, C-62/95 and C-63/95 *Germany, Hanseatische Industrie-Beteiligungen GmbH and Bremer Vulkan Verbund AG v. Commission* [1996] ECR I-5151, para. 52.
63. European Commission, (2001), State aid No N 416/2001 United Kingdom Emission Trading Scheme, C (2001) 3739fin, 28 Nov. 2001, 9.
64. European Commission, (2001), State aid No N 416/2001 United Kingdom Emission Trading Scheme, C (2001) 3739fin, 28 Nov. 2001, 9; European Commission, (2000), Statsstøttesag Nr. N 653/1999 – CO$_2$-kvoter, SG (2000) D/, 12 Apr. 2000, 5–6; European Commission, (2001), Steunmaatregel nr. N 550/2000 België Groenestroomcertificaten, SG (2001) D/290545, 25 Jul. 2001, 5–6.

Article $87(3)^{65}$ is only subject to the marginal control of the ECJ which ruled in *Italy v. Commission* that the Commission had to take all relevant factors into account.[66]

Because of the complexity of the derogations, this author will restrict the analysis to the review of those provisions that may be applicable in Emission Trading Systems. Looking at the object and purpose of the Emission Trading Program it is clear that environmental considerations are an important element. Consequently the Environmental aid guidelines,[67] seeking to integrate environmental policy and protection (Article 6 EC Treaty) with the objectives of the Treaty (Article 3 EC Treaty), are of relevance. In particular, the Environmental guidelines pursue the dual objective of ensuring competitive markets[68] and of bringing environmental protection and cost internalization into the realm of competition policy.[69] They seek to determine whether and under what conditions State aid may be regarded as *necessary* to ensure environmental protection and sustainable development without having *disproportionate* effects on competition and economic growth.[70]

The Environmental State aid guidelines that came into force in 2001 were applicable to the first EU ETS trading period. As of 02 April 2008 they were replaced by new guidelines. These new provide more transparency regarding the applicable State aid derogations and the structure of the legal test conducted by the European Commission. Before examining how they address State aid derogations under the second and third trading phase of the EU ETS, the legal framework under the former State aid guidelines that were applicable to the first trading period is treated.

Even though the former environmental state aid guidelines predated the EU ETS and thus did not duly address emission trading systems they nevertheless allowed for the application of Articles 87(3)(b) (aid for an important project of common European interest) and (c) (aid to facilitate the development of certain economic activities or economic areas) of the EC Treaty in their context. Both

---

65. Case 730/79 *Philip Morris v. Commission* [1980] ECR 2671, para. 17; Joint Cases 62/87 and 72/87 *Exécutif Régional Wallon v. Commission* [1988] ECR 1573, para. 21; Case T-152/99 *Hijos de Andrés Molina, SA v. Commission* [2002] ECR II-3049, para. 48; Case C-142/87 *Belgium v. Commission* [1990] ECR I-959, para. 56; Case C-39/94 SFEI [1996] ECR I-3547, para. 36; Case 78/76 *Steinike v. Bundesamt für Ernährung und Forstwirtschaft* [1977] ECR 595, para. 8; Case C-156/98 *Germany v. Commission* [2000] ECR I-6857, para. 67; Case C-303/88 *Italian Republic v. Commission* [1991] ECR I-1433, para. 34.
66. Case C-261/89 *Italy v. Commission* [1991] ECR I-1437, para. 20.
67. OJ C 37, 03 Feb. 2001, 3–15. The interested reader is referred to European Commission (2007). These guidelines were applicable to the first trading period. The larger part of the second trading period is falling under the new environmental State aid guidelines of OJ C 82, 01 Apr. 2008, 1–33. Specific provisions of the text are applicable to trading phases post 2012.
68. OJ C 37, 03 Feb. 2001, para. 14(a).
69. OJ C 37, 03 Feb. 2001, para. 14(b). The objective of cost internalization is softened in para. 18 in which it is deemed that if cost internalization is not fully possible a suboptimal implementation should be used as a temporary second best solution.
70. OJ C 37, 03 Feb. 2001, para. 5, emphasis added.

derogations will therefore be considered in detail below. The current guidelines contain specific guidance for the application of Article 87(3)(c) in the context of the EU ETS but also enables the application of Articles 87(3)(b) where applicable.

Derogations to the principle described in Article 87(1) that aid schemes should not contain elements which render it incompatible with the common market may only be permitted if they contribute to the Community interests described in Article 87(3). This so-called 'compensatory justification' principle has been upheld by the ECJ in *Philip Morris v. Commission.*[71] These exemptions are interpreted narrowly when individual, regional or sectoral aid is examined, and are only applied if the Commission is convinced that the market would not work to attain the envisaged objectives of the aid to be granted.[72] The Commission scrutinizes individual aid awards under aid schemes in order to determine whether aid is necessary and whether it can be justified by the recipient's contribution to the objectives of Article 87(3) (proportionality).

Consequently, the first element which needed to be examined was the necessity of the aid at hand for the realization of the project. Evans (1997) states that necessity is determined by assessing the seriousness of the problem and if it could be addressed without this aid.[73] The author postulates that aid for profitable[74] investments that would have been made in the absence of any governmental support, particularly so if excess demand exists, as well as aid *ex post*, should not be granted.[75] Aid is not deemed to be necessary if the undertaking or its parent company[76] has the means to invest or if its financial position is not different from that of its competitors.[77]

The second element which needed to be addressed is proportionality. The proportionality test consists of two considerations. Firstly, the European Commission conducted a balancing act between the adverse competitive effects of granting aid and its positive Community impacts. Secondly, it was examined whether the advantage could be reached in a less distortionary manner. Since section F of the Environmental guidelines, governing policies, measures and instruments for reducing greenhouse gases did not contain specific conditions for authorizing State aid,[78] the European Commission had more discretion in drawing up its criteria to assess Nation Allocation Plans. It was only 'limited' by specific requirements already contained in the Environmental guidelines.[79]

---

71. Case 730/79 *Philip Morris v. Commission* [1980] ECR 2671, para. 16 and 17.
72. This passage is based on 84/499/EEC 1984 OJ L 276/43.
73. Evans, A., (1997), 110.
74. Commission Decision 84/499/EEC 1984, OJ L 276/43.
75. This passage is based on Evans, A., (1997), 110 ff.
76. Commission Decision 88/468/EEC 1988 OJ L229/37, Commission Decision 89/254/EEC OJ L 106/34, Case 323/82 *SA Intermills v. Commission* [1984] ECR 3809, para. 39.
77. Commission Decision 82/776/EEC 1982 OJ L323/37.
78. OJ C 37, 03 Feb. 2001, para. 71.
79. With regard to the application of Art. 87(3)(b) EC Treaty, para. 73 of the Environmental guidelines give more specific criteria. With regard to Art. 87(3)(c) EC Treaty, more uncertainty remains.

The criteria for the allocation of emission allowances are presented in Annex III of the Directive 2003/87/EC. Particularly criteria one (Assessment of emissions development) and two (Non-discrimination between companies or sectors) could have been serving as a point of reference. Within the framework of the Directive, NAPs that allocated more than 95% of emission allowances for free, must have yielded sufficient environmental benefits in order to be judged to be compatible with the common market.[80]

2.2.1   Article 87(3)(b)

The first limb of Article 87(3)(b) permits aid to promote the execution of an important project of common European interest. An aid scheme will only fall within the scope of this provision if it forms part of a transnational European program, which is supported by a number of Member States or if it stems from a concerted action.[81]

The former Environmental guidelines stated in paragraph 73 that aid falling within the ambit of Article 87(3)(b) had to be necessary for the project to proceed, and that the project under consideration had to be specific, well-defined, qualitatively important and has to make an exemplary and clearly identifiable contribution to the common European interest. Furthermore, they stated that additional aid could be granted under Article 87(3)(c) EC Treaty. This implies that both a necessity and a proportionality test were to be applied.

Regarding Emission Trading Systems it is clear that climate change is a priority of environmental policy of the European Union[82] and that it forms part of the Sixth Environmental Action Programme.[83] Before the EU ETS Emission Trading Systems were national and have consequently not been exempted under Article 87(3)(b). The Commission could, however, have used this derogation in its assessment of NAPs created pursuant to Directive 2003/87/EC. Merola and Crichlow (2004) point out that if 87(3)(b) would be used for the whole NAP, there would be:

> the possibility of avoiding the precise verification of the proportionality of the aid to the required *quid pro quo*, because the compensatory justification would be implicit in the fulfilment of the specified requirements of Article 87(3)(b), as indicated in paragraph 73 of the Guidelines.[84]

---

80. The Commission holds the opinion that free allocations of emission allowances beyond the amount that Member States are obliged to allocate for free constitute State aid. See European Commission (2004b), 2.
81. Case 730/79 *Philip Morris v. Commission* [1980] ECR 2671, para. 25; Joint Cases 62/87 and 72/87 *Exécutif Régional Wallon v. Commission* [1988] ECR 1573, para. 22–23.
82. European Commission, (2001), State aid No N 416/2001 United Kingdom Emission Trading Scheme, C(2001) 3739fin, 28 Nov. 2001, 10.
83. European Commission, (2001) On the sixth environment action programme of the European Community, 'Environment 2010: Our future, Our choice', Brussels, 24 Jan. 2001, COM (2001) 31 final, 2001/0029 (COD), 24.
84. Merola, M. & Crichlow, G., (2004), 47.

While Farnsworth (2005) dismisses this argument as being counter to State aid rationale,[85] this author believes that the Commission pursuant to its competition assessment under Article 87(1) has developed an understanding of the problematic allocations and will concentrate its limited resources on examining whether they could be justified by the environmental benefits of the NAP. The environmental benefit would be subject to severe scrutiny if more than the maximum average annual emission is proposed to be allocated to industry[86] or if the NAP unduly favours industrial activities.[87]

2.2.2   Article 87(3)(c)

Under Article 87(3)(c) the European Commission can authorize aid to facilitate the development of certain economic activities or of certain economic areas in those cases where such aid does not negatively affect trading conditions to an extent contrary to common interest.[88]

Independent of Community averages,[89] Article 87(3)(c) allows for aid to be legitimated by addressing the needs of a particular economic area.[90] The intensity of aid is measured by a complex formula determining the 'net grant equivalent' and taking into consideration the regional handicap.[91] Sectoral aid embraces aid for the protection of the environment, aid for rescuing and restructuring of firms in difficulty provided that there is a restructuring plan aiming at the reduction and reorientation of activities.[92] Operating aid cannot be granted.[93]

In its assessment whether trading conditions are affected to an extent contrary to common interest, the Commission compares the beneficial effects of aid to their adverse effect on trading conditions and competition.[94] It thereby considers the effects on competitors and product markets,[95] market

---

85. Farnsworth, N., (2005), 299.
86. See Commission Decision on the Polish National Allocation Plan COM (2005) 549 final, para. 4. In para. 9 it is stated that non-compliance with criterion two 'fundamentally jeopardises the overall environmental objective' of an emission trading scheme.
87. See Commission Decision on the French National Allocation Plan COM (2004) 3982/7 final, para. 4.
88. OJ C 37, 03 Feb. 2001, para. 72.
89. While the application of Art. 87(3)(a) is contingent upon the Community average, under 87(3)(c) aid for economic activities or areas is dependent upon national criteria.
90. See Commission Guidelines on National Regional Aid, (1998) OJ C74/9, para. 3.6.
91. For a review of the Commission Guidelines on National Regional Aid see Wishlade, F., (1998).
92. Joint Cases 278/92, C-279/92 and C-280/92 *Spain v. Commission* [1994] ECR I-4103, para. 67; Joint Cases T-126/96 and T-127/96 *BFM and EFIM v. Commission* [1998] ECR II-3437, para. 98.
93. Case T-459/93 *Siemens SA v. Commission* [1995] ECR II-1675, para. 48; Case C-288/96 *Germany v. Commission* [2000] ECR I-8237, para. 90; Joint Cases 62/87 and 72/87 *Exécutif Régional Wallon v. Commission* [1988] ECR 1573, para. 29; Case T-55/99 *CETM v. Commission*, [2000] ECR II-3207, para. 83–84.
94. This passage is based on Quigley, C. & Collins A. M., (2003), 89 ff.
95. Joined Cases 62/87 and 72/87 *Exécutif Régional Wallon v. Commission* [1988] ECR 1573, para. 30–32.

demand,[96] profit margins,[97] intensity of trading on the common market,[98] and the particular circumstances of the recipient.[99]

This derogation of Article 87(1) is not only frequently used[100] but has also been applied in the context of Emission Trading Systems. All schemes discussed earlier fell under Article 87(3)(c), provided of course that the Commission came to the conclusion that State aid existed.[101] In the U.K. scheme the Commission concluded that the Emission Trading System was in conformity with section F of the former Environmental guidelines, generated a net environmental benefit, and did not adversely affect trading conditions to an extent contrary to the common interest.[102]

Since the former Environmental guidelines predated the Directive 2003/87/EC and were not offering concrete guidance on the assessment criteria to be applied under Article 87(3)(c), the Commission has enjoyed considerable discretion in setting them for assessing NAPs.[103] It is most likely that the Commission has turned to criteria one (Assessment of emissions development) and two (Non-discrimination between companies or sectors) of Directive 2003/87/EC for guidance and even looked to paragraph 18 (Temporary second best solution and Incentive measure) of the former guidelines for inspiration to assess the net environmental benefits.

In its assessment under 87(3)(c) the Commission may have taken a sectoral approach in considering economic activities and could have been inclined to examine the voluntary introduction of stringent measures that go beyond the scope of the Community action in the context of both economic areas or sectors.[104]

The new Environmental guidelines clarify that Article 87(3)(c) is used for assessing aid in the context of emission trading systems. The applied methodology is similar to the former State aid guideline. Here too a balancing test is executed. The first two steps of the test address the positive effects of the aid while the third element compares the positive against the negative effects. At the first level of the test it is examined if the aid addresses a well-defined objective of common interest such as the environment. In the second step it is examined if the aid addresses a

---

96. Case 310/85 *Deufil GmbH & Co. KG v. Commission* [1987] ECR 901, para. 16; Joint Cases 62/87 and 72/87 *Exécutif Régional Wallon v. Commission* [1988] ECR 1573, para. 17.
97. Case 259/85 *France v. Commission* [1988] ECR 1573, para. 17.
98. Case 730/79 *Philip Morris v. Commission* [1980] ECR 2671, para. 26.
99. Case 323/82 *SA Intermills v. Commission* [1984] ECR 3809, para. 39; Joint Cases 296/82 and 318/82, *Netherlands and Leeuwarder Papierwarenfabriek v. Commission* [1985] ECR 809, para. 26.
100. Merola, M. & Crichlow, G., (2004), 48.
101. Even though the Commission concluded that the Belgium Green Certificate scheme did not involve State resources and could therefore not constitute State aid, it stated that it would also have been compatible with 87(3)(c). European Commission, (2001), Steunmaatregel nr. N 550/2000 België Groenestroomcertificaten, SG(2001) D/290545, 25 Jul. 2001, 10.
102. European Commission, (2001), State aid No N 416/2001 United Kingdom Emission Trading Scheme, C(2001) 3739fin, 28 Nov. 2001, 10 ff.
103. The interested reader is referred to European Commission (2007).
104. See also Merola, M. & Crichlow, G., (2004), 48.

market failure, if it is an appropriate policy instrument, if it gives an incentive effect (if aid is necessary[105]) and proportionality. The third step entails the balancing act.

The environmental guidelines further specify in section 3.1.12 that aid in the context of tradable permit schemes may be declared compatible with the common market under Article 87(3)(c) if a number of requirements are satisfied. Aid can only be declared permissible if it attains additional environmental protection beyond what is ensured on the basis of Community legislation,[106] the allocation is executed in a transparent manner, based on objective criteria and reliable data and does not give rise to over allocation.[107] Furthermore, the system should be of a general nature and not favour certain undertakings or sectors unless justified by the logic of the scheme[108] nor put new entrants at a comparative advantage or disadvantage *vis-à-vis* incumbents.[109]

When the third trading phase commences the new guidelines contain additional criteria further explaining necessity and proportionality of State aid under the emission trading permits.[110] Pursuant to these criteria the choice of aid beneficiaries must be based on objective and transparent criteria and aid must be granted in an uniform way ensuring that similar installations are treated similarly. Furthermore it is made clear that auctioning should lead to substantial increases in production costs without the consumers having to bear the costs.[111] Criterion 'd' seems to entail that aid for undertakings should be proportional to its environmental performance taking due consideration of the best performing techniques employed in the European Economic Area.[112]

As under its predecessor also Article 87(3)(b)[113] might be declared applicable for aid schemes that promote the execution of an important project of common European interest. This derogation could for example be considered in cases of a pan-European project promoting carbon capture and storage. For national allocation schemes, however, Article 87(3)(c) continues to appear to be the most promising alley.

Pursuant to the new Environmental State aid guidelines four conditions need to be fulfilled:[114] Firstly the project needs to be specific and clearly defined in terms of participants, objectives and its effects. Secondly, it must make a significant contribution to a common European interest in a concrete and practical way and the benefits must accrue to the Community as a whole. Furthermore it must be necessary and contain incentive effects, involve a high level of risk and yield

---

105. See OJ C 82, 01 Apr 2008, 1–33, s. 3.2.
106. OJ C 82, 01 Apr. 2008, 1–33, recital 140(a).
107. OJ C 82, 01 Apr. 2008, 1–33, recital 140(b).
108. OJ C 82, 01 Apr. 2008, 1–33, recital 140(c).
109. OJ C 82, 01 Apr. 2008, 1–33, recital 140(d).
110. OJ C 82, 01 Apr. 2008, 1–33, recital 141.
111. OJ C 82, 01 Apr. 2008, 1–33, recital 141(b and c).
112. OJ C 82, 01 Apr. 2008, 1–33, recital 141(d).
113. OJ C 82, 01 Apr. 2008, 1–33, s. 3.3.
114. OJ C 82, 01 Apr. 2008, 1–33, recital 147–148.

substantial environmental effects. Albeit not a requirement under this section of the guidelines, if follows from the general description of the balancing test and the individual assessment of the cases that the proportionality requirement has to be complied with as well. Aid granted under Article 87(3)(b) EC Treaty entitles the Commission to authorize aid at higher rates than otherwise laid down under the Environmental guidelines.[115]

3 STATE AID ASSESSMENT OF FREE
   ALLOCATION MECHANISMS

After having reviewed the relevant State aid norms of the EC Treaty, the following part analyses the compatibility with the common market of two forms of allocation mechanisms that Member States have taken or could have taken pursuant to Directive 2003/87/EC. At first the grandfathering allocation mechanism is defined and analysed in light of the proceeding part. Subsequently, in part four, the PSR system is examined along the same lines.

3.1 STATE AID ASSESSMENT OF GRANDFATHERING

This part examines how grandfathering will be assessed under European Competition law as presented in section two. The same structure will be applied. First the existence of State aid within the meaning of Article 87(1) EC Treaty is examined by checking how grandfathering relates to the six State aid criteria. Subsequently, it is examined how grandfathering could benefit from one of the two derogations applicable to Emission Trading Systems.

3.1.1 Transfer of a Benefit or an Advantage (Notion of Aid)

In order for grandfathering allocation mechanisms to fall within the ambit of Article 87(1) EC Treaty, there has to be a transfer of a benefit or an advantage accruing to an undertaking or a sector. Cost internalization of negative externalities such as CO_2 emissions under the European Emissions Trading System implies that an undertaking will face inflated costs of production. If CO_2 emission allowances need to be purchased on the market and be submitted on the basis of actual emission, they inflate the operating costs of the firm and consequently also the marginal costs of production.

If, by contrast, government does not decide to sell allowances but to distribute them on the basis of historic emissions for free, the same behavioural incentives are created. Entrepreneurs receiving emission allowances will consider the market value of the received allowances as well as their abatement costs, and take their production decision on the basis of their particular opportunity costs. Nevertheless

---

115. OJ C 82, 01 Apr. 2008, 1–33, recital 150.

it must be noted that from a strategic firm behavioural point of view the decision to invest in abatement technology is not straightforward. The updating of the historical reference period and uncertainty about future reference periods can lead to the time shifting of investments.[116]

Because in theory the costs considered are the same in both allocation methods[117] (with the only distinction that they are direct expenditures for the undertaking in the first case and opportunity costs in the second[118]), one can conclude that the allocation format does not have a bearing on the actual production decision.[119] This implies that grandfathering does not constitute an operative aid to the undertaking but rather a windfall profit in the form of a lump sum transfer of emission allowances which do have a market value. Clearly windfall profits fall within the definition of a subsidy as a special case of 'aid' and hence, are to be adjudged as satisfying the first State aid criterion of Article 87(1) EC Treaty.[120] The Commission has been reaching the same conclusion in its assessment of the Danish emission quota system.[121]

In the literature it has been argued[122] that since the production decision are not influenced by grandfathering emission allowances for free, there is no distortion of competition. The authors conclude that from an efficiency point of view there could therefore not be State aid. They do, however, acknowledge that there would be State aid from an equity point of view for a firm that has to pay for allowances, will face direct capital expenses, and thus will be set at a disadvantage *vis-à-vis* a grandfathered firm.

While in principle one can distinguish between equity and efficiency arguments that guide policy making,[123] this author beliefs that such a differentiation is not helpful with regard to the interpretation of the State aid concept which embraces both equity and efficiency considerations.[124] If distortions of competition would be found that could be associated with either approach, this State aid element would be fulfilled.

---

116. In the case of updating, incentives to time shift cannot be neutralized by taking reference periods that are distant enough in history. For an early account recognizing the importance of inter-temporal adaptations see Lyon, R. M., (1982), 30.

117. Because such subsidies are inherently pecuniary in nature and do not affect the actual nor the marginal emission reduction costs and hence do not distort the product markets, they can be described as lump sum subsidies. See also Jensen, J. & Rasmussen, T. N., (1998).

118. See also Woerdman, E., (2003), 111.

119. This will be the case if one assumes that firms are not credit rationed and do not incur financing costs.

120. In contrast to Woerdman, E., (2003), 113, this author does not view the definition of State aid to be insufficient to decide whether grandfathering should be regarded as State aid or not. Also in contrast to this Zwingmann, K., (2007), 166, asserts that a benefit within the meaning of Art. 87(1) could not be derived from the free allocation of emission allowances as such since their benefits could only be realized through firm actions oriented towards emission reductions.

121. European Commission, (2000), Statsstøttesag Nr. N 653/1999 – CO$_2$-kvoter, SG(2000) D/, 12 Apr. 2000, 5–6.

122. See Woerdman, E., (2003), 113, but also Woerdman, E. & Arcuri, A., (2006), 12.

123. See Van der Laan, R. & Nentjes, A., (2001) but also Woerdman, E., (2002), 255 ff.

124. See Van der Laan, R., & Nentjes, A., (2001), 139 but also Woerdman, E., (2002), 320 ff.

Furthermore, it is also questionable if there are indeed no economic efficiency arguments against grandfathering. At first it has to be noticed that the term economic efficiency used in the above cited literature remains unclear. If the efficiency concept is tantamount to allocative efficiency[125] or efficient use of resources, it has been shown elsewhere in this book that auctions and grandfathering schemes do differ with regard to their ability to attain allocations that bring recipients directly to their contract curve and that a reduction of competitive pressure – stemming from entrenched market shares for example – can lead to x-inefficiency.[126]

In addition, the reference to opportunity costs and the link to economic efficiency is not clear for it should be noticed that a firm's ability to pass on costs to consumers does not depend upon costs that are levied from either trading partner but will always fall most heavily on that side that can least avoid it.[127] The legal incidence of a levy has no effect on its economic incidence. Who eventually bears the cost burden is an issue of demand elasticities and not of opportunity costs and consequently independent of the nature (opportunity or actual costs) of the costs. This author therefore concludes that under both efficiency and equity approaches to competitive distortions, grandfathering does constitute State aid and that reference to opportunity costs as a guide to anticompetitive distortions to State aid legislation is not utterly convincing.

3.1.2 Aid Favouring a Certain Undertaking over others (Selectivity Principle)

The selectivity principle is applied in order to differentiate between selective aid granted to particular undertakings giving rise to anticompetitive distortions and general economic measures. The selectivity principle requires that emission allowances are to be allocated on the basis of objective, non-discriminatory and non-discretionary criteria. Yet, under Directive 2003/87/EC Member States do have a margin of discretion to decide how many emission allowances they allocate to a particular entity. Consequently one can conclude that grandfathering, as applied under this Directive, is potentially selective and liable to constitute State aid. Besides the discretion afforded to Member States there are other factors reconfirming this finding. Grandfathering is likely to be selective because abatement costs differ not only between individual undertakings but in particular also between industries leading to discriminatory variations in the cost burden placed upon particular undertakings which can give rise to a discriminatory effect. The discrimination under grandfathering stems in particular from the setting of the historical standards used for allocation. Unrepresentative temporary downturns or

---

125. Woerdman, E. & Arcuri, A., (2006), 12 do refere to the work of Van der Laan, R. & Nentjes, A., (2001) who associate the 'economic' interpretation of European State aid with allocative efficiency. See Van der Laan, R. & Nentjes, A., (2001), 136.
126. Frank, R., (1997), 412, defines x-inefficiency as a condition in which a firm fails to obtain maximum output from a given combination of inputs.
127. Frank, R., (1997), 52 ff.

increases in production lead to a discriminatory bias in the data used to determine emission allowance requirements which are used as a basis for allocation. Last but not least past Commission decisions offer guidance on selectivity and grand-fathering by indicating that the selectivity principle is not only concerned with the very group of companies that receive rights for free. The Commission adjudged the Danish system to be selective because it was only binding upon the electricity sector.[128] The underlying differentiation between trading and non-trading sectors, could be extended and brought into the realm of the discussion of covered and uncovered sectors and incumbents and new entrants.

3.1.3   Granted by the State or through State Resources

The third criterion to be assessed is whether the measure at hand is granted by the State or through Member State resources. This is of course only relevant with regard to those allocations where no obligation to that effect obliges Member States to grant resources. In light of the ECJ's ruling in *Preussen Elektra*, as discussed before, these two elements contained in Article 87(1) EC Treaty are cumulative rather than alternative. At first it will be assessed if grandfathering is granted by the State and subsequently, a financial burden is placed upon the Member State. Only if both elements are confirmed, can the measure constitute aid within the meaning of Article 87(1) EC Treaty.

3.1.3.1   *By the State*

The relationship between institutions that grandfather emission allowances to undertakings and the State can take many forms. According to ECJ case law not only direct allocation through a State institution or public bodies but also through private bodies which are either established or appointed to administer aid[129] constitute an 'emanation' of the State and hence fall within the meaning of 'State' as used in Article 87(1) EC Treaty. Thus any institution granting or administering CO_2 emission allowances does meet this criterion.

The examination of imputability of the granted benefit to the State is, however, more complex. With regard to the European Emissions Trading System, one has to observe the wording of the legal text. Directive 2003/87/EC states in Article 10 that Member States shall allocate at least 95% of the allowances for the three-year period beginning 1 January 2005 free of charge.[130] Member States are obliged to transpose directives into national law[131] and since the wording of Article 10 of

---

128. European Commission, (2000), Statsstøttesag Nr. N 653/1999 – CO₂-kvoter, SG(2000) D/, 12 Mar. 2000, 5–6.
129. Case 78/76 *Steinike v. Bundesamt für Ernährung und Forstwirtschaft* [1977] ECR 595, para. 21.
130. This amount is reduced to 90% for the five-year period beginning 1 Jan. 2008, see Art. 10, Directive 2003/87/EC.
131. Article 10 EC Treaty.

Directive 2003/87/EC does not grant Member States any room for discretion, the mere compliance with their obligations under the Treaty cannot be used against them.[132] Therefore, the forgone revenue which constitutes a financial burden on the Member State with regard to these 95% of emission allowances, cannot constitute aid within the meaning of Article 87(1).

Since Member States are not under a duty to allocate more than the amount of emission allowances prescribed in Directive 2003/87/EC free of charge, their decision to do so constitutes an act which is at their own discretion. Member States are thus bound by their obligation to respect the EC Treaty, including Article 87(1).[133] Since in practice it is not discernable which emission allowances are derived from the permissible 95% and which from the questionable excess amount of allowances, all proposals containing such allocations have to be adjudged to contain aid granted through Member State resources.

3.1.3.2   *Through Member State Resources*

With regard to the second element, the granting of aid through Member State resources, it is questionable whether grandfathering directly or indirectly affects a public account. Given the fact that the Member States allocate emission allowances on the basis of historic figures for free to undertakings and that those allowances do constitute an asset for the undertaking and even have a market price, it has to be concluded that the State foregoes revenue by not selling emission allowances.[134]

In addition to the above it should be noted in passing that competitive distortions and burdens for Member States can also stem from other sources that are not directly related to allocation mechanisms. Distortions stem form the discretion of Member States to allocate relatively more or less allowances to the EU ETS trading sectors, from differences in the quality of law enforcement and from allowing the use of project based emissions under the EU ETS. Since these possibilities fall beyond the sphere of interest of this text, they are only discussed briefly.

With regard to the first element, it is known that in the second trading period most[135] Member States are subject to the obligations of the Kyoto Protocol to meet particular national targets. Within the EU some Member States are also bound to emission targets within the framework of the EU Burden Sharing Agreement. Member States do, however, enjoy discretion to allocate relatively more or less allowances to the trading sector as a whole. Favouring the trading sectors on the expense of transportation for example, will directly influence the amount of emissions that is being allocated within the framework of the Directive and hence have an immediate impact on State resources.

---

132.   On this point see also Frenz, W., (2007), 210.
133.   See also European Commission (2004b), 2.
134.   A similar observation is made by Welch, W. P., (1983).
135.   It is noticeable that Cyprus and Malta do participate in the EU ETS but have allocated their emission allowances on the basis of business as usual. See Malta (2004) and Malta (2006) as well as Cyprus (2004) and Cyprus (2007).

Differences in the quality of enforcement of the EU ETS Directive could lead to distortions of competition. Such distortions could stem from difference in the coverage of undertakings,[136] or in the quality of monitoring and enforcement requirements. Even though many of the provisions concerning enforcement are harmonized by Directive 2003/87/EC[137] prevailing differences could lead to biased data and hence to distorted accounting.[138]

With regard to the third element, project mechanisms, Member States enjoy discretion[139] to allow operators to use Certified Emission Reductions (CERs)[140] and Emission Reduction Units (ERUs)[141] from project activities in the Community scheme up to a percentage of the allocation of allowances to each installation.[142] It is noticeable that distortions can take effect already in the first EU ETS trading period (2005–2007) with regard to CERs and from 2008 also with regard to ERUs and that the setting of percentages is only mandatory from 2008 onwards.[143] According to Article 30(3) of Directive 2003/87/EC as amended, the amount of CERs and ERUs Member States intent to allocate has to be specified in their NAPs. If NAPs differ with regard to the percentage of project based credits that can be exchanged for EU ETS allowances, undertakings' competitive position is impacted to the extent that price differentials between the EU ETS allowances, CERs and ERUs exist.[144]

The above can be summarized as follows. Only if both elements of this criterion, namely the granting of aid by the State as well as the granting of aid through Member State resources are fulfilled, can the measure be judged to fall within the scope of Article 87(1) EC Treaty with regard to those emission allowances that a Member State freely chooses to allocate for free. This was up to 5% of the emission allowances during the first trading period and up to 10% during the second trading period. To the extent that a NAP fulfils these two elements, the grandfathering mechanisms contained therein are liable to constitute State aid within the meaning of Article 87(1). This finding is consistent with the Commission's finding in the

---

136. Schmitt-Rady, B., (2006), 84 finds substantial variations in the definitions of installations that are subject to the EU ETS. It should be noticed that a preliminary ruling question has been posed to the ECJ that addresses the validity of Directive 2003/87/EC under the principle of equal treatment with regard to the exclusion of the aluminum and plastic industries from the scope of the Directive. See OJ C 117/8 of 26 May 2007.
137. Peeters, M., (2006), 185. Monitoring for example is subject to Commission Guidelines drawn up in accordance with Art. 14 and subject to Annex IV of Directive 2003/87/EC.
138. On the importance of monitoring, accounting and enforcement see Peeters, M., (2006) and Peterson, S., (2003).
139. Member States' discretion is limited by the criteria in Annex III of Directive 2003/87/EC (as amended) and by requirements that the total use of ERUs and CERs must be consistent with relevant supplementarity obligations under the Kyoto Protocol.
140. From Clean Development Mechanism (CDM) projects.
141. From Joint implementation (JI) projects.
142. Article 11(a)(1), Directive 2003/87/EC as amended by Directive 2004/101/EC.
143. See recital 5, Directive 2004/101/EC.
144. On this point see also De Cendra de Larragán, J., (2006), 106, but also STEM (2005), 12–13.

Danish emission quota system which predates the emission trading Directive by several years.[145]

3.1.4   It Should be an Undertaking or ... Production

Article 87(1) EC Treaty limits the scope of incompatible aid by referring expressly to undertakings or the production of certain goods. Hence part of the legal exercise is to determine whether the recipients of CO_2 emission allowances are to be considered to be undertakings. Only if this is assessed in the affirmative, there can be State aid.

From the foregoing legal assessment, it is apparent that the ECJ applies a broad interpretation of the concept of undertaking. Whether a particular grandfathering allocation fulfils this criterion obviously depends on the recipient. Directive 2003/87/EC restricts the allocation of emission allowances to operators[146] which are defined as any person who operates or controls an installation.[147] Therefore, any recipient of grandfathered emission allowances of the European Emissions Trading System will be found to be engaged in an economic activity and be identified as an undertaking. Thus any person – legal or natural alike – being allowed to receive emission allowances will automatically fulfil this State aid criterion.

3.1.5   Distorts or Threatens to Distort Competition

With regard to grandfathering and the actual or potential distortion of competition, it is questionable whether anticompetitive distortions exist in the first place and secondly, whether the aid upon which they are based exceeds the *de minimis* threshold as required by Commission Regulation (EC) No. 69/2001.[148] Each will be discussed in turn.

3.1.5.1   *Firstly, Competitive Distortions*

Allocating emission allowances to undertakings for free can give rise to competitive distortions if there is dissimilar treatment of undertakings leading to comparative advantages. Both from a firm as well as a social welfare point of view, particular regard of the undertakings' marginal abatement cost structures, including expected technological developments and growth forecasts, have to be taken into account in order to ensure a non-distorted competitive market equilibrium.

---

145. European Commission, (2000), Statsstøttesag Nr. N 653/1999 – CO₂-kvoter, SG(2000) D/, 12 Apr. 2000, 5–6.
146. See Art. 11(1), Directive 2003/87/EC.
147. See Art. 3(f), Directive 2003/87/EC.
148. The *de minimis* threshold was increased from EUR 100.000 to EUR 200.000 by Commission Regulation 1998/2006 of 15 December which replaced Commission Regulation 69/2001 of 12 Jan. 2001. A transitionary period ended 30 Jun. 2007.

Based on the particular constellations between the involved undertakings, one can distinguish four kinds of competitive relationships that can be distorted. These are firstly, relations between incumbents and new entering firms, secondly, between trading and non-trading sectors, thirdly, relationships between competing firms of the same Member State in particular regarding covered and non-covered parts of a sector as well as undertakings within a covered sector and fourthly between trading sectors. Each will be discussed in turn. If any of these relationships demonstrates actual or potential competitive distortions, the measure at hand is liable to fall within the ambit of Article 87(1) EC Treaty, as long as the granted aid exceeds the *de minimis* threshold level which is assessed in part two of this passage.

3.1.5.1.1   Between Incumbent and New Entering Firms
Competitive distortions could arise from the Member State's NAPs if they would not afford potential competitors equal treatment. Granting new entrants relatively less CO_2 emission allowances, sets them at a comparative disadvantage *vis-à-vis* incumbent undertakings. This results in an increase of the barriers to entry. In the absence of perfectly competitive markets, this will lead to higher consumer prices and to potential x-inefficiency.[149] Thus, competition will be distorted if new entrants were not grandfathered adequate amounts of emission allowances. The examination of a sector would be required to determine if the actual allocation is in fact liable to give rise to such distortions. A full industrial economics analysis would have to be undertaken in order to determine the substance of such a claim. Yet in practice, however, the Commission is not obliged to prove the actual competitive distortion, its mere potential detriment is sufficient.[150]

It should be noted that the wording of the Directive 2003/87/EC is relatively weak and affords Member States some discretion by merely obliging them to take the need to provide access to allowances for new entrants into account, in particular with regard to Articles 87 and 88 EC Treaty.[151] It seems that equally authentic language versions of the Directive differ in clarity. While the English version speaks of the obligation to take into account the need of new entrants, the Dutch version of the text speaks of the necessity to keep emission allowances available for new entrants.

This ambiguity can translate into severe disadvantages when national reserves for new entrants are depleted. In such a case new entrants would not be entitled to allowances in most Member States.[152] New entrants would clearly be disadvantaged because they would have to acquire emission allowances on the market and face direct expenses which immediately lead to an increase of the barrier to entry.

---

149. Frank, R., (1997), 412, defines x-inefficiency as a condition in which a firm fails to obtain maximum output from a given combination of inputs.
150. Since the analysis of anticompetitive distortions and the Community dimension are inseparably linked, the findings made in section two are applicable. See for example Case 730/79 *Philip Morris v. Commission* [1980] ECR 2671, para. 11.
151. See Art. 11(3) and Annex III criteria 6 of Directive 2003/87/EC.
152. Notable exceptions include the German and the Polish NAP for 2005–2007, see Germany (2004), 37 and Poland (2004), 39.

Interestingly enough, while the Commission acknowledges that it is crucial that new entrants have access to allowances, it at the same time states that the interests of new entrants are sufficiently safeguarded by affording new entrants the possibility to buy allowances on the market.[153] From an economic perspective equality of treatments cannot be safeguarded because the differences in production costs will immediately set new entrants at a comparative disadvantage.

Negative discrimination of new entrants does not only arise from resource depletion and the possibility to have new entrants pay for their allowances but they do also exist in those cases where they are awarded allowances for free. Under the Dutch NAP new entrants can never receive more allowances than needed while incumbent top performers can benefit from up to 10% more allowances.[154]

In general it has to be observed that barriers to entry will be created if the introduction of a particular environmental standard implies that the investments of a newly entering firm are higher than those of incumbents. To the extent that the potential competitor is credit rationed,[155] investments in the market are sunk,[156] or the financial burden of lending money is positive, new entrants need – *ceteris paribus* – a higher level of profits to find it profitable to invest. To the extent that the particular market under consideration is not perfectly competitive, such barriers to entry give incumbent firms the possibility to raise prices and reap positive economic profits.[157] A reduced threat of new entries does not only have positive effects on cartel stabilization but is also liable to give rise to x-inefficiency due to reduced competitive pressure. In the absence of empirical evidence, it is impossible to determine the importance of these effects.

In lieu of the above it can be concluded that not only under pure grandfathering schemes but also under the existing NAPs, new entrants can suffer from comparative disadvantages which distorts competition on the merits.

3.1.5.1.2   Trading Sectors and Non-trading Sectors

Anticompetitive distortions can also arise if competing sectors do not fall under the Emission Trading System. In this case the increase in production costs which is associated with grandfathering, despite the fact that emission allowances are allocated for free, leads to distortions of relative prices. This implies that non-covered sectors attain a comparative advantage from the introduction of an Emissions Trading System. Such advantages – even if not directly considered within a State aid framework – still puts sectors in a stronger position and distorts competition on the merits. One example is the private housing construction market which

---

153. COM (2003) 830 final, 12.
154. See the Dutch NAP, The Netherlands (2004a), 27.
155. This implies that a potential entrant is unable to attain sufficient funds to make the necessary investments to enter the market.
156. Sunk costs refer to irrecoverable costs once invested. They do not have a bearing on a firm's decision to exit a market but constitute a decisive factor for market entry.
157. Positive economic profit is defined as profit above a normal rate of return which merely reflects the opportunity costs of capital. Positive economic profit is logically separable from accounting profit.

uses both cement and wood as production inputs. If cement prices increase due to enhanced environmental standards, wood becomes relatively cheaper and thus more attractive as a construction material. Wood producers enjoy a comparative advantage and benefit from the introduction of the European Emissions Trading System. This will increase the social welfare only to the extent that such price increases will reflect the internalization of negative externalities.

3.1.5.1.3   Anticompetitive Distortions between Competing Firms
              of the Same Member State

When assessing anticompetitive distortions between competing firms, one has to distinguish between competitors who themselves fall under the Emission Trading Directive and competitors that do not fall under it. A third element to be considered in this context regards the reduction of competitive pressure which could lead to the entrenchment of market shares.

Firstly, anticompetitive distortions are created if an undertaking receives relatively more or less emission allowances under grandfathering than its competitors. Even in the absence of an administrative intent to favour particular undertakings, the fixing of the historical basis for allocation of emission allowances can coincide with a temporal downturn of a company's sales, or simply ignore the entity's strong economic growth. Similarly, fixing the wrong historical basis can ignore meaningful environmental investments which – if taken at a later stage – would have yielded a comparative advantage.[158] The punishment of early movers will not have general efficiency effects with regard to past investments but creates distributive effects which impact the financial position of undertakings and therefore their competitive position. From this it appears that grandfathering schemes are directly liable to distort competition on the merits. In addition it should be noticed that equal allocation rules may also develop a discriminatory effect if the production mix, that is, the economic reality of undertakings, substantially differs.[159]

Secondly, anticompetitive distortions can arise if only a number of undertakings of a given sector are obliged to participate in the Emission Trading System while others are not. To the extent that the non-participating entities are not subject to additional and presumably costly legislative requirements, they will enjoy a comparative advantage with regard to those firms which are participating in the

---

158. One example that underlines the self defeating rationale behind the setting of historic standards are Carbon Capture and Storage projects. Electricity producers that could capture and store part of their CO_2 emission have an incentive to first pollute in order to be eligible for the grandfathering of emission allowances before they can actually benefit from their investments in CO_2 abatement. For a description of such a project see the Vattenfall's newsletter on the CO_2 free power plant project, No. 3, November 2005.

159. In Case T-387/04 *EnBW Energie Baden-Würtemberg AG v. European Commission* [2007], para. 34 ff. it is alleged that the German electricity producer RWE is set at a comparative advantage *vis-à-vis* EnBW because RWE has a large number of outdated coal power plants and will thus benefit disproportionally from the German transfer rules applicable when old plants are closed.

Emission Trading System.[160] Grandfathering emission allowances for free will mitigate the direct financial cost burden of the participating entities and thus alleviate the competitive pressure.[161] Boom and Nentjes (2003) assert that the Commission appears to be thinking that such effects can be easily contained.[162] It should be noted, that the grandfathering does not take into account the changing of market shares of firms, but only their historic emissions and thus has an inherently static focus which is not designed to address the needs of undertakings operating in a rapidly changing environment. Perhaps it is for this reason that some NAPs do contain variables to approximate sectoral developments and to allow for an emphasis on dynamic changes.

Thirdly, competitive pressure on less efficient producers operating on a market characterized by a given market demand is reduced under a capped grandfathering system. The rationale behind this is that part of the comparative production cost advantage of more competitive producers is absorbed by the necessity to buy additional emission allowances. Undertakings loosing market share are able to withstand competitive pressure better due to an influx of capital when they sell emission allowances they are grandfathered on a historical basis while actually producing less than before. Consequently, gaining market share on the expense of less efficient producers is rendered more difficult since less efficient producers are able to cross-subsidize their products. Such entrenchment of market shares does not only distort competition on the merits, give rise to real welfare losses but is also at odds with the polluter pays principle.

3.1.5.1.4   Between Trading Sectors

Anticompetitive distortions can also occur if specific trading sectors are grandfathered relatively more emission allowances than others. The existing market equilibrium will also be distorted if the burden of the Emission Trading System falls more heavily on one sector. This would be the case if a sector would receive an equal amount to another sector but has very strong growth prospects or very high abatement costs. In such cases the operating costs would increase more strongly in this sector. As a consequence, prices would be increased to the extent possible for the given demand elasticities and profits would decrease. Relative prices would change unequally and the general market equilibrium would be distorted.

3.1.5.2   *Secondly, Does It Exceed the* de minimis *Threshold?*

If, in any of the above explained relationships of undertakings anticompetitive effects have been identified, the aid which generates them in the first place will only fall within the ambit of Article 87(1) EC Treaty if it surpasses the *de minimis* threshold. The threshold will be met if the total amount of all aid received by the

---

160. This is based on the presumption that fewer emission allowances are being allocated to each installation than are actually needed.
161. See also Woerdman, E., (2003), 111.
162. See Boom, J. T. & Nentjes, A., (2003), 52 and COM (2000) 87 final, 13.

undertaking exceeds EUR 200.000 in the last three consecutive years.[163] Any aid below this threshold is presumed not to be sufficient to distort competition and will therefore not be subject to proceedings and directly adjudged to be compatible with the common market.

3.1.6   Community Dimension: Aid Which is Capable of Affecting Trade between Member States

In *Philip Morris v. Commission* the ECJ ruled that the mere presumption of the existence of a competitive distortion was sufficient to constitute State aid.[164] With regard to the allocation of emission allowances through the grandfathering allocation mechanism, such potential anticompetitive distortions of the level playing field stem from the dissimilar treatment of comparable undertakings. If an undertaking in one Member State is grandfathered relatively more emission allowances than its competitors in another Member State, the level playing field is distorted and competition on the merits is undermined. Therefore, unless it could be proven that there is no undue advantage accruing to a particular enterprise, grandfathering systems are liable to affect trade between Member States.

Under the European Emissions Trading System anticompetitive distortions between incumbents in different Member States are expected to arise from the selection of different historical standards and diverging closure and transfer rules. Each will be discussed in turn.

The selection of different reference periods for the setting of historical standards will give rise to competitive distortions if the selection of the base period would lead to differential treatment of comparable undertakings. That the historical standards differ can easily be seen by comparing the various NAPs.

The free movement of production capacity and competition on the merits is distorted by the regulatory differences in Member States' site closure and production transfer rules. In the Netherlands undertakings that terminate their operation at a particular production site in order to scrap capacity or in order to shift production to a more efficient installation, are allowed to retain all allocated emission allowances during the whole trading period (2005–2007).[165] In the absence of other transfer rules, the undertaking will benefit because it receives allowances on the basis of production statistics which do not reflect the technology applied. An undertaking closing a plant in the Netherlands and building additional capacity in another Member State would not only be able to retain emission allowances for idled production in the Netherlands but may even benefit from new entrance emission allowance in other Member States, setting it at a comparative advantage *vis-à-vis* other potential entrants and due to temporarily reduced production costs

---

163. The value was increased from EUR 100.000 to EUR 200.000 by Commission Regulation 1998/2006 of 15 December which replaced Commission Regulation 69/2001 of 12 Jan. 2001. A transitory period ended 30 Jun. 2007.
164. Case 730/79 *Philip Morris v. Commission* [1980] ECR 2671, para. 11.
165. See the Dutch NAP, The Netherlands (2004a), 41.

even *vis-à-vis* incumbent competitors. In Germany by contrast scrapping of capacity is only 'compensated' by the right to retain the allowances allocated during this particular year and transfer rules do only apply to replacement installations in Germany.[166] Yet here too, anticompetitive effects between Member States could result if particular undertakings would benefit from transfer rules while other competing companies would not.[167]

To the extent that firms' decisions are unduly influenced by incentives established by NAPs, barriers to entry are created. These curtail the effective functioning of competition as the 'natural' sanctioning mechanism of markets. A reduction of competitive pressure in a shielded environment creates real welfare losses in the form of x-inefficiency and an environment in which cartels could be more stable and dominant firms may have a propensity to abuse their position. The effective application of Competition law rules such as Articles 81 and 82 EC Treaty is necessary to contain such tendencies. To what extent differences in NAPs do actually lead to the partitioning of the internal market through the erection of barriers of entry and give rise to cartelization and abuse is subject to further research.

The examination of the six State aid criteria has shown that grandfathering systems can be liable to constitute State aid within the meaning of Article 87(1) EC Treaty. Whether this is indeed the case depends on the particular allocation made to each installation under a NAP and the characteristics of the market the entity is operating on, and cannot be answered on any level of generality.

3.1.6.1 Compatible with the Common Market

As outlined earlier, a NAP in the first trading phase could have been falling under the derogation of Article 87(3)(b) EC Treaty to the extent that it strictly implements the requirements spelt out by Directive 2003/87/EC. The Article's applicability is subject to the necessity criterion and the proportionality test. As for the second and third trading period the criteria enlisted in the new environmental guidelines[168] are applicable. In light with the logic of the balancing test also the proportionality test would have to be complied with. While their interpretation is contingent upon the particular allocations contained in the NAP, some observations regarding the grandfathering allocation mechanism can still be made.

The compensatory justification principle under Article 87(3)(b) requires that anticompetitive distortions can be outweighed by positive environmental benefits. While it is not necessary to compensate distortions and benefits on the basis of an individual allocation, the average balance has to be positive. It therefore suffices to

---

166. See Dutch NAP, The Netherlands (2004a), 41. and the German Nap, Germany (2004), section C2 and C3.3.
167. In Case T-387/04 *EnBW Energie Baden-Würtemberg AG v. European Commission* [2007], para. 34 ff., it is alleged that the German electricity producer RWE is set at a comparative advantage *vis a vis* EnBW. This may also lead to distortions of community trade.
168. See OJ C 82, 01 Apr. 2008, 1–33, s. 3.3

highlight the anticompetitive distortions that are likely to be created by the application of a grandfathering system.

As presented above, competitive distortions can stem from several sources. New entries will suffer from barriers to entry while trading sectors incur disadvantages from undue changes in relative prices *vis-à-vis* other trading sectors or non-trading sectors. Competing firms can suffer from dissimilar treatment stemming from the inadequate selection of the historic basis for emission allowance (early investments or temporary downturns in production etc.) or the fact that only some but not all members of a sector are obliged to participate in the trading system. The Commission would have had to review such distortions in the light of the compensatory justification principle.

With regard to the derogation provided for by Article 87(3)(c), the effect of the trading conditions is the relevant benchmark. As discussed earlier, while the Commission was enjoying large discretion to set the assessment criteria during the first trading phase, it would have had to take the effects on competitors and product markets, market demand, profit margins, the intensity of trading on the common market and the particular circumstances of the recipients into account. The measure had to be justified by the net environmental benefit it created and could be used to justify actions going beyond the scope of Directive 2003/87/EC. It is important to note that this derogation is subject to the same anticompetitive distortions as Article 87(3)(b).

Under the second trading phase similar rules are employed. It is, however stressed that aid can only be declared permissible if it attains additional environmental protection beyond what is ensured on the basis of Community legislation,[169] the allocation is executed in a transparent manner, based on objective criteria and reliable data and does not give rise to over allocation.[170] It is noteworthy that the aid scheme should be of a general nature and not favour certain undertakings or sectors unless justified by the logic of the scheme[171] nor put new entrants at a comparative advantage or disadvantage *vis-à-vis* incumbents.[172] It is not clear how the generality requirement of the scheme is interpreted in practice since aid must be selective in order to qualify as aid within the meaning of Article 87(1) EC Treaty in the first place.

3.2   STATE AID ASSESSMENT OF A PSR SYSTEM

This part examines how the PSR system will be assessed under European Competition law as presented in chapter two. Following the same structure, at first the existence of State aid within the meaning of Article 87(1) EC Treaty is examined by analysing how PSR relates to the six State aid criteria. Where the findings differ between a PSR on national and on European level, they will be expressly addressed. Particular attention will be paid to case T-233/04 on the Dutch NOx

---

169. OJ C 82, 01 Apr. 2008, 1–33, recital 140(a).
170. OJ C 82, 01 Apr. 2008, 1–33, recital 140(b).
171. OJ C 82, 01 Apr. 2008, 1–33, recital 140(c).
172. OJ C 82, 01 Apr. 2008, 1–33, recital 140(d).

emission trading system that incorporates a PSR system and is presently subject to appeal before the ECJ.[173] Subsequently it is examined how the PSR system could benefit from one of the two derogations applicable to Emission Trading Systems.

3.2.1 Transfer of a Benefit or an Advantage (Notion of Aid)

In order that the PSR system falls within the ambit of Article 87(1) EC Treaty, there has to be a transfer of a benefit or an advantage accruing to an undertaking or a sector. Undertakings will benefit under the PSR system if their production is more CO_2 efficient than the best-practice benchmark fixed by the government. In this case an undertaking is able to have its relative 'CO_2 savings' accredited by a third party and registered with the government. These CO_2 savings are termed 'credits' and constitute an asset with a market value. Therefore the European Commission is correct in its assessment that tradable emission documents constitute intangible assets[174] and that the PSR credits do constitute profit opportunities[175] for these undertakings. In its ruling 233/04 on the Dutch NOx system the Court of First instance confirms this line of argumentation and extends this finding by pointing out that undertakings may in addition benefit from the possibility to rely upon a trading market in order to avoid the payment of a fine.[176] Therefore it can be concluded that this criterion is met with regard to those cases.

3.2.2 Aid Favouring a Certain Undertaking over Others (Selectivity Principle)

How the PSR System is assessed with regard to the selectivity principle is a non-trivial question. The answer does necessarily depend upon the objectiveness of the employed criteria and the scope of the governmental benchmarks. The broader their scope and the more objective their criteria, the higher the likelihood that they are judged as being of a general economic nature. Yet, to the extent that covered sectors have heterogeneous abatement cost structures and distinct demand curves, the more likely it is that a measure even though it is not directly discriminatory, will still have such an equivalent effect. If for example large undertakings were favoured by a PSR scheme, it would be selective[177] unless it could be justified by the nature or generality of the scheme.[178] Therefore the result is ambiguous.

---

173. Case 279/08 is still pending before the ECJ at the time of writing. See OJ 223/30 of 30 Aug. 2008.
174. European Commission, (2003), Steunmaatregelen van de Staten N35/2003 – Nederland Systeem van verhandelbare emissierechten voor NOx, C(2003) 1761 fin, 24 Jun. 2003, s. 3.2.
175. Profit opportunities would exist if the investment cost for additional NOx emission abatement would be lower than the willingness to pay of those undertakings who are unwilling or unable to meet the governmental benchmark.
176. Case 233/04, Case T-233/04 *The Netherlands v. European Commission*, [2008] n.y.r., para. 73.
177. Case C-143/99 *Adria-Wien Pipeline GmbH and Wietersdorfer & Peggauer Zementwerke GmbH v. Finanzlandesdirektion fuer Kaernten* [2001] ECR I-8365, para. 41.
178. Case C-143/99 *Adria-Wien Pipeline GmbH and Wietersdorfer & Peggauer Zementwerke GmbH v. Finanzlandesdirektion fuer Kaernten* [2001] ECR I-8365, para. 42.

With regard to the specific case of the Dutch NOx scheme, the Court of First Instance held that the coverage of large industrial facilities under this scheme was based upon the objective criteria without any sectoral or geographic bias and that – to the extent that the measure was aimed at the biggest emitters – it could be motivated by the goals of the measure.[179]

With regard to the objective, obligations and fines imposed on large facilities the Court held that the legal and factual situation was different from that of smaller installations that were subject to different legal schemes.[180] In the eyes of the Court the Commission failed to establish that the smaller installations were subject to similar obligations, that stipulations on fines differed, or that the position of uncovered facilities was comparable to that of large covered facilities.[181] Likewise the Court did not reach the conclusion that the limited number of 250 covered installations allowed the conclusion that there was a selective measure.[182] Based on the ecological considerations the Court found that it was justified to distinguish between small and large emitters and that the selectivity criterion was not satisfied. Consequently no State aid was involved.[183]

3.2.3   Granted by the State or through State Resources

The two elements of aid granted by the State or through State resources require a cumulative assessment.[184] At first it will be examined whether the PSR system fulfils the criteria of aid granted by the State and subsequently, the financial burden on the Member State will be assessed. Only if both elements can be established, the system is liable to constitute State aid within the meaning of Article 87(1) EC Treaty.

3.2.3.1   *By the State*

Questionable here is whether a PSR system involves the State as a participant and whether the State is engaging in the transfer of a benefit. As presented above, the ECJ consistently held that for the purpose of Article 87(1) the meaning of State aid does include national, central, regional and local authorities but also public or private bodies established or appointed by the State. In light of this it is clear that any entity charged with the allocation of State aid in form of emission allowances would fall within the ambit of this Article.

It is thus important to assess whether the State or an emanation thereof is actually involved in the process of granting aid. The recently established Dutch NOx trading system is chosen as an epitome of a PSR Emission Trading System in order to address this question.

---

179. Case 233/04, Case T-233/04 *The Netherlands v. European Commission*, [2008] n.y.r., para. 88.
180. Case 233/04, Case T-233/04 *The Netherlands v. European Commission*, [2008] n.y.r., para. 89–90.
181. Case 233/04, Case T-233/04 *The Netherlands v. European Commission*, [2008] n.y.r., para. 91 ff.
182. Case 233/04, Case T-233/04 *The Netherlands v. European Commission*, [2008] n.y.r., para. 95.
183. Case 233/04, Case T-233/04 *The Netherlands v. European Commission*, [2008] n.y.r., para. 99 ff.
184. C-379/98 *Preussen Elektra v. Schleswag AG* [2001] ECR I-2099, para. 58.

The question of State involvement can be decomposed into two elements. Firstly, the question whether the entity charged with monitoring of the PSR trading system constitutes a State entity or emanation thereof and secondly, whether the State takes an active role in the distribution of aid.

The first question can be answered in the affirmative. The Dutch Emission Authority (Nederlandse Emissieautoriteit) can be classified as a part of the State because it forms part of the Dutch Ministry of Housing, Spatial Planning and the Environment.

With regard to the second element, the State's involvement in the distribution of aid, a number of observations have to be made. Before an undertaking, falling within the scope of application of the NOx trading scheme, is allowed to emit NOx, a license certifying the undertaking's ability to measure and monitor actual emissions has to be granted.[185] Monitoring reports have to be submitted to inform the Dutch Emission Authority about the emissions that have taken place. The actual emission is compared to the benchmark created by Administrative law. If less than the allowed amount has been emitted per unit of output, the undertaking is able to save or sell the credits. This is possible because these credits do have a property right characters.[186] One could therefore argue, as is done by the Dutch government, that in the NOx trading scheme emission allowances are not distributed to firms but that undertakings are obliged by administrative order to meet pre-determined production efficiency benchmarks.[187] The underlying intuition is of course that certification is solely based on a self-selection process. Furthermore, since only those undertakings which choose to over comply with the benchmark are able to receive and market credits, one could argue that the NOx credits are a mere administrative certification of actual emissions of an undertaking *vis-à-vis* the benchmark. Following this line of argumentation in this author's view they should not be considered State aid but be regarded as proof of earned rights by the undertakings as has been the case for the Belgium Green Certificates.[188] Hence it could be argued that the entity charged with controlling the compliance of private parties and maintaining the emission registry is not charged with the actual distribution of allowances to firms, calling into question the imputability of aid to the Member State.

Looking at prominent case law, however, it appears that the ECJ does not emphasize the involvement of the State but the question whether there is a direct or

185. VROM, (2005), 10.
186. VROM, (2005), 44. For a discussion of property rights within the context of Emission Trading Systems, see Koster, M. H., (2005) and Kortmann, J. S., (2005). Woerdman, E., (2005) notes that in the EU ETS allowances that are defined as an authorization to emit gases during a specified period of time are more akin to an authorization that can be terminated by the legislator rather then property rights.
187. European Commission, (2003), Steunmaatregelen van de Staten N35/2003 – Nederland Systeem van verhandelbare emissierechten voor NOx, C(2003) 1761 fin, 24 Jun 2003, under 3.2.
188. European Commission, (2001), Steunmaatregel nr. N 550/2000 België Groenestroomcertificaten, SG(2001) D/290545, 25 Jul. 2001, 5–6.

indirect financial burden on the State.[189] In *Preussen Elektra* support was granted through legislation and in the absence of direct State involvement, the Court was not concerned with the position of the State but with the financial burden placed upon it.[190] Therefore it would appear that the Court would not entertain the presented line of argument but concentrate upon the question whether alleged aid constitutes a burden upon State resources. While the Court of First Instance (CFI) in its ruling distinguished the Dutch NOx system from both the Belgium Green Certificates case and the *Preussen Elektra* case,[191] it is noticeable that the judgment does not expressly address State involvement in the granting of aid as such but is more concerned with its budgetary effects.[192] The element of imputability of aid to the State is discussed below in the context of Member State discretion in order to avoid repetition.

3.2.3.2   *Through State Resources*

If one continues to consider the second element of 'financing through State resources', one has to determine if a PSR system such as the Dutch NOx system does indeed have a negative budget implication for the State. In its decision on the Dutch NOx trading system[193] the European Commission argued that this condition was satisfied. It contended that the emission allowances were to be adjudged as intangible assets,[194] that they had a market value and that the Dutch government was deliberately foregoing revenue because it was at its full discretion to sell them to the operators.[195] The Court of First Instance endorsed this position in full.[196] Both the Commission and the Court were therefore distinguishing the Dutch NOx system from the Belgian Green Certificate case where the Commission ruled that even though the certificates granted by the government constituted intangible assets the government has not agreed to forego income because the certificates were mere proof of actual production and it did not provide the certificates free of charge.[197]

---

189. See Case 53/00 *Ferring SA v. Agence centrale des organismes de sécurité sociale* [2001] ECR I-9067, para. 27, for direct sales taxes falling within the ambit of 'State aid'.
190. C-379/98 *Preussen Elektra v. Schleswag AG* [2001] ECR I-2099, para. 58–61.
191. Case 233/04, Case T-233/04 *The Netherlands v. European Commission*, [2008] n.y.r., para. 76 and 77.
192. Case 233/04, Case T-233/04 *The Netherlands v. European Commission*, [2008] n.y.r., para. 69 ff., 74 and 75.
193. European Commission, (2003), Steunmaatregelen van de Staten N35/2003 – Nederland Systeem van verhandelbare emissierechten voor NOx, C(2003) 1761 fin, 24 Jun. 2003.
194. European Commission, (2003), Steunmaatregelen van de Staten N35/2003 – Nederland Systeem van verhandelbare emissierechten voor NOx, C(2003) 1761 fin, 24 Jun. 2003, under 3.1.
195. European Commission, (2003), Steunmaatregelen van de Staten N35/2003 – Nederland Systeem van verhandelbare emissierechten voor NOx, C(2003) 1761 fin, 24 Jun. 2003, under 3.2.
196. Case 233/04, Case T-233/04 *The Netherlands v. European Commission*, [2008] n.y.r., para. 75.
197. European Commission, (2001), Steunmaatregel nr. N 550/2000 België Groenestroomcertificaten, SG(2001) D/290545, 25 Jul. 2001, 5–6. Case 233/04, Case T-233/04 *The Netherlands v. European Commission*, [2008] n.y.r., para. 76.

Below the author examines the Court's and Commission's argument that the Dutch government is forgoing revenues by not employing auctions from two perspectives. Firstly, it is analysed within the PSR system itself and secondly by taking EC law as a framework of reference. The former presupposes that Member States do have a margin of discretion in choosing the regulatory measure they would like to take and examines the possibility of introducing auctioning into the PSR system, the later examines if such discretion actually exists.

At first the possibility to introduce auction mechanisms or to sell emission allowances has to be examined. The introduction of an auction mechanism into the PSR system would differentiate between those undertakings which are eligible to place a bid and those which are actually rewarded a NOx credit. Given the increased uncertainty whether investments in NOx abatement technology will actually be profitable, undertakings will *ceteris paribus* invest less in abatement technology. This does not only reduce the number of available NOx credits and negatively impact the Emission Trading System's liquidity but also leads to an increase in the overall abatement costs. Besides such strong economic reservations, it is also legally questionable under the protection of confidence concept because the PSR system is based on the accreditation of additional emission reductions.

In contrast to auctions, however, payments for NOx credits could more easily be levied. The government could reap part of the profits undertakings would generate from over complying with the government's benchmark and selling NOx credits on the market. The necessary result of the introduction of such levies, is of course the reduction of the profitability of investments in NOx abatement technology with similar yet potentially less grave economic impacts than under auction systems.[198] With regard to the issue of protection of confidence, the same result prevails.

Since the PSR system could be changed to incorporate selling or even auctioning of credits, the government could potentially forego revenue. It should however be stressed that the introduction of selling and auctioning goes counter the system's design, its *raison d'être* and is at odds with the former Commission decision in Belgium Green Certificates.[199] In this case – as is the case under the PSR system – a private undertaking eligible to receive government accreditation for its production would have been willing to pay for a certificate provided that the price would be lower than the expected profit that could be generated by selling the certificate on the market. Since the willingness to pay has been present in the Belgium Green Certificates case as well as in the PSR case but only been condemned in the latter, it has to be concluded that the Commission decision is either incoherent with its former decision or to be explained by the Commission's misconception of the PSR system.

---

198. This will be the case when the selling price of the credits lies below the auction price.
199. European Commission, (2001), Steunmaatregel nr. N 550/2000 België Groenestroomcertificaten, SG(2001) D/290545, 25 Jul. 2001, 5–6.

After examining if the PSR system could be redesigned in such a way as to verify the Commission's claim that the government was foregoing revenue by introducing the NOx trading system, we now turn to the second relevant dimension to examine the Commission's and Court's position. Since the EC Treaty does not compel Member States to levy taxes or to generate profits from its operations, it is clear that the European Commission's argument of foregone revenue seeks to extent the scope of Article 87(1). Yet in *Preussen Elektra* the ECJ did not condemn the employment of statutory provisions by a Member State to fix minimum prices above real market prices and to oblige private parties to bear the costs[200] despite the fact that tax revenues would be lower.[201] The CFI seems to circumvent this issue by emphasizing that the Dutch NOx case and the *Preussen Elektra* case were logically separable since it imposed an obligation to purchase among operators themselves and that it was not based on emission or pollution allowances.[202]

In addition, it has to be noticed that the ECJ did not entertain previous attempts of the European Commission to extent the field of application of Article 87(1). In *Commission v. France* the ECJ ruled that the scope of Articles 87 and 88 EC Treaty did not leave sufficient room for a competing concept of 'measures having equivalent effect' to State aid.[203]

That the argument of the European Commission constitutes an undue extension of the scope of Article 87(1) EC Treaty is also evident from the wording of Directive 2003/87/EC. If Member States were under Article 87(1) EC Treaty or any other Treaty provision under the duty to select the measure which generates most revenue, the Directive's inclusion of Article 10 requiring Member States to forgo revenue with respect to the allocation of emission allowances, would render the Directive null and void and would constitute a breach of primary EC law.

In the view of the author it can thus be argued that the State's budged was diminished by a policy decision taken at a time when the Netherlands had full discretion to choose between various allocation options and in the absence of any obligation to maximize State revenue, and that it cannot be taken as proof of a direct or indirect budget implication within the meaning of Article 87(1). Consequently it is this authors opinion that the second element 'financing through State resources' is not satisfied and accordingly there cannot be State aid.

If, however, the Court's and the European Commission's view point were motivated by the perception that the measures taken by the Dutch government amounted to *fraude a la loi*,[204] reflecting the malicious intent to circumvent

---

200. C-379/98 *Preussen Elektra v. Schleswag AG* [2001] ECR I-2099, para. 66.
201. C-379/98 *Preussen Elektra v. Schleswag AG* [2001] ECR I-2099, para. 62.
202. Case 233/04, Case T-233/04 *The Netherlands v. European Commission*, [2008] n.y.r., para. 77.
203. C-290/83 *Commission v. France* [1985] ECR 439, para. 18, and Quigley, C., Collins, A. M., (2003), 17.
204. Both the Commission and the Court of First Instance do not look at the possibility to extend the PSR system by auctioning but seem to take as a starting point that the Dutch government should have been selling or auctioning allowances in order to prevent budgetary effects.

Article 87(1) EC Treaty,[205] more appropriate measures[206] than contending that the PSR system constituted State aid should have been taken. If this were indeed the case, the more appropriate reaction should have been the initiation of an Article 226 EC Treaty proceeding for an alleged breach of the Netherlands's obligation under Article 10 EC Treaty to refrain from any measure that could jeopardize the attainment of the objectives of the Treaty.

While the above discussion is relevant for a national PSR system, the system's application in the particular multilateral context of the European Emissions Trading context is complicated by the existence of the Burden Sharing Agreement which is binding upon the former EU fifteen Member States. PSR systems could be set at European levels according to objective criteria and with the aim to minimize competitive distortions for particular product groups. To the extent that the national allocation under the PSR system would not be identical to the number of allowances granted under the Burden Sharing Agreement, transfers between national governments would be required. Such transfers would then imply budgetary consequences entailing the need for further investigations under State aid regulations. Following the same line of argumentation as presented under the Dutch NOx system above, Member States would not necessarily face budgetary consequences if a PSR system would be established on an EU wide level since they would not forgo revenue and consequently not satisfy this criterion.

3.2.4   It Should be an Undertaking or . . . Production

The scope of incompatible aid under Article 87(1) EC Treaty is restricted to undertakings or the production of certain goods. Thus to fall within the ambit of the Article, the recipients of CO_2 emission allowances are to be considered undertakings. Only if this is assessed in the affirmative, Article 87(1) is satisfied and there can be State aid.

The findings presented above with regard to grandfathering are the same for the PSR system and the same logic applies with regard to the line of argumentation and will therefore not be repeated. It suffices to notice that only operators are allowed to benefit under Directive 2003/87 EC and that consequently this criterion is satisfied.

3.2.5   Distorts or Threatens to Distort Competition

In assessing the existence of State aid within the ambit of Article 87(1) EC Treaty it has to be examined if the PSR system is liable to create actual or potential anticompetitive distortions and whether these are generated by aid which exceeds the

---

205. Especially since it is clear from C-379/98 *Preussen Elektra v. Schleswag AG* [2001] ECR I-2099, para. 65 that Art. 10 EC Treaty may not be used to extent the scope of Art. 87(1) EC Treaty.
206. C-290/83 *Commission v. France* [1985] ECR 439, para. 17.

de minimis threshold as required by Commission Regulation (EC) No. 69/2001.[207] Each will be discussed in turn.

3.2.5.1   *Firstly, Is There Distortion?*

The PSR Emission Trading System can give rise to distortions of competition if different sectors are subject to dissimilar treatment and are able to benefit from undue comparative advantages. As was the case under the grandfathering emission allowance allocation mechanism, particular regard of the undertakings' marginal abatement cost structures, including expected technologic developments and growth forecasts are important to ensure a non-distorted competitive market equilibrium.

As already stated above, based on the particular constellations between the involved undertakings, one can distinguish four kinds of competitive relationships that can be distorted. These are firstly, relations between incumbents and new entering firms, secondly, between trading and non-trading sectors, thirdly, relationships between competing firms of the same Member State in particular regarding covered and non-covered parts of a sector as well as undertakings within a covered sector and fourthly between trading sectors. Each will be discussed in turn. If any of these relationships demonstrates actual or potential competitive distortions, the measure at hand is liable to fall within the ambit of Article 87(1) EC Treaty, if the granted aid exceeds the *de minimis* threshold level assessed in part two of this passage.

3.2.5.1.1   Between Incumbent and New Entering Firms

Competitive distortions could arise from Emission Trading Systems if they would not afford potential competitors equal treatment. Granting new entrants relatively less CO_2 emission allowances sets them at a comparative disadvantage, *vis-à-vis* incumbent undertakings, and results in an increase of the barriers to entry. In the absence of perfectly competitive markets, this will lead to higher consumer prices and to potential x-inefficiency.[208] The PSR Emission Trading System does not generate such undesirable effects. By way of construction of the system, new entrants are subject to the same government benchmarks and hence are allowed to emit as much as incumbent firms.

While new entrants cannot be disadvantaged with regard to potential scarcity once they are on the market, new entering firms are subject to similar comparative disadvantages as mentioned under the grandfathering system. To the extent that the potential competitor is credit rationed,[209] investments in the market

---

207. Commission Regulation 1998/2006 of 15 December replaced Commission Regulation 69/2001 of 12 Jan. 2001. A transitory period ended 30 Jun. 2007.
208. Frank, R., (1997), 412 defines x-inefficiency as a condition in which a firm fails to obtain maximum output from a given combination of inputs.
209. This implies that a potential entrant is unable to attain sufficient funds to make the necessary investments to enter the market.

are sunk,[210] or the financial burden of lending money is positive, new entrants need – *ceteris paribus* – a higher level of profits to find it profitable to invest. To the extent that the particular market under consideration is not perfectly competitive, such barriers to entry give incumbent firms the possibility to raise prices and reap positive economic profits.[211] A reduced threat of new entries does not only have positive effects on cartel stabilization but is also liable to give rise to x-inefficiency due to reduced competitive pressure. In the absence of empirical evidence, it is impossible to determine the importance of these effects. What however is certain is that Competition law is essential to contain any cartelization and abuse that stems from Emission Trading Systems.

Besides the above mentioned comparative disadvantages of new entering firms *vis-à-vis* incumbents one could also expect that new undertakings benefit from investment in new and more efficient technology which may give them an advantage over incumbents and enhance their competitive position. To what extent such an effect exists and how it compares to the above mentioned negative effects is subject to further research and cannot be answered at any level of generality.

3.2.5.1.2   Trading Sectors and Non-trading Sectors
Anticompetitive distortions can also arise if competing sectors do not fall under the Emission Trading System. The same finding as explained with regard to grand-fathering holds true for PSR Emission Trading Systems. Here too, the increase in production costs, which is associated with the introduction of production standards, leads to distortions of relative prices. Only to the extent that such price increases reflect the internalization of negative externalities will this change in relative prices lead to an increase in social welfare.

3.2.5.1.3   Anticompetitive Distortions between Competing Firms
                 of the Same Member State
When assessing anticompetitive distortions between competing firms one has to distinguish between competitors who themselves fall under the PSR Emission Trading System and competitors that do not fall under it.

Firstly, as has already been argued above, anticompetitive distortions are created if an undertaking receives relatively more or less emission allowances than its competitors. Since under a PSR system entities producing similar goods are subject to the same legislative requirements, they are subject to the same cost burden. While this does not create anticompetitive distortions as such, comparative disadvantages can stem from the size and investment capabilities of firms. If abatement costs are not linearly increasing, small firms will have to spread their investment costs over fewer output and hence have relatively higher costs of

---

210. Sunk costs refer to irrecoverable costs once invested. They do not have a bearing on a firm's decision to exit a market but constitute a decisive factor for market entry.
211. Positive economic profit is defined as profit above a normal rate of return which merely reflects the opportunity costs of capital. Positive economic profit is logically separable from accounting profit.

production. While this sets them at a comparative disadvantage *vis-à-vis* large firms, this result would not be different if other allocation mechanisms were used.[212] With regard to investment capabilities, a similar observation has to be made. Only firms that are credit rationed will suffer from comparative disadvantages but this will not be different under PSR than under other allocation mechanisms.[213] From this it appears that PSR Emission Trading Systems are not liable to distort competition on the merits within a sector as long as PSR benchmarks reflect the equilibrium market price.

Secondly, anticompetitive distortions can arise if only a number of undertakings of a given sector are obliged to participate in the Emission Trading System while others are not. To the extent that the non-participating entities are not subject to additional and presumingly costly legislative requirements, they will enjoy a comparative advantage with regard to those firms which are participating in the Emission Trading System. This is based on the presumption that fewer emission allowances are being allocated to installations than are actually needed.

Legislation obliging some operators to attain production standards but not others, clearly creates a differential cost burden. To the extent that nearly all market players in a given industry are covered, competitive distortions are expected to be low. The same legal framework is applicable to them and competition on the merits would not be distorted. Whether a PSR system gives rise to State aid considerations is thus dependent upon its particular scope of application and not upon the PSR system itself.

One element which distinguishes a PSR system from grandfathering regards the entrenchment of market shares. While under grandfathering losers of market share are able to gain from selling allowances and winning market share is rendered more difficult by the obligation to buy additional allowances,[214] such problems are absent under PSR systems. Here there is no compensation for losers of market shares and winners are not burdened with the obligation to buy additional allowances.

3.2.5.1.4   Between Trading Sectors
Anticompetitive distortions can also occur if specific trading sectors have relatively more stringent government benchmarks under a PSR system. In contrast to

---

212. This finding is based upon the express assumption that the government benchmark under a PSR system would be set at that level that yields identical emission abatement incentives as would be generated by the equilibrium market price. Since abatement cost structures of firms are exogenously determined, the comparative disadvantages are necessarily identical. Besides from an economic efficiency point of view, losses of inefficient operators only constitute pecuniary effects and the result of a competitive selection process which only allows the fittest market participants to stay on the market.
213. This finding is based upon the express assumption that the government benchmark under a PSR system would be set at that level that yields identical emission abatement incentives as would be generated by the equilibrium market price. Since abatement cost structures of firms are exogenously determined, the comparative disadvantages are necessarily identical.
214. Here a market with a stable market size is implicitly assumed.

allocation mechanisms that use historic emission or production figures as a point of reference, the PSR system does not suffer from excessive burdens being placed upon sectors which have different growth or technical innovation potentials. Undue burdens will, however, be created every trading period anew if government fails to set production benchmarks in such a way as to equalize marginal sector abatement costs across different sectors. Failure to achieve this would distort the whole market equilibrium and consequently lead to inefficient resource allocation. The existence of State aid is thus dependent on the particular benchmarks set by the government and cannot be answered at any level of generality.

3.2.5.2 Secondly, Does It Exceed the de minimis Threshold?

As has already been mentioned above, if in any of the above explained relationships of undertakings anticompetitive effects have been identified, the aid which generates them in the first place, will only fall within the ambit of Article 87(1) EC Treaty if it surpasses the *de minimis* threshold.

3.2.6 Community Dimension: Aid Which is Capable of Affecting Trade between Member States

As already stated under the grandfathering system, the mere presumption of a distortion of competition suffices to constitute State aid. Since a PSR system applied only in one Member State cannot ensure equal treatment of similar enterprises in a multilateral environment, the PSR system is liable to constitute State aid with regard to this criterion unless it could be proven that firstly, no actual or potential undue advantages are being created and secondly, if they would not fall under the *de minimis* regulation.

If a PSR system would be introduced on a European level, an important difference to the grandfathering system discussed earlier needs to be highlighted. Such a PSR system would not be liable to generate intra-community competitive distortions stemming from the selection of historical standards and it is not subject to diverging closure and transfer rules.[215] Reservations regarding cartelization and abuse falling under Articles 81 and 82 EC Treaty appear at first sight to be mainly connected to new entrants.[216]

The above analysis has shown that PSR systems are not liable to constitute State aid within the meaning of Article 87(1) EC Treaty. If, nevertheless, the ECJ reached the conclusion that a particular PSR system would amount to State aid, the

---

215. It should be noticed that a grandfathering system with equal rules applying in all Member States would by construction also not be subject to this problem. It would, however, still lead to potential distortions regarding the choice of historical standards and create distortions between market share winners and losers. A PSR system could give rise to distortions of the general market equilibrium if the benchmarks were not set on appropriate levels.
216. A full analysis of the PSR and Competition law aspects falling under Arts 81 and 82 EC Treaty in connection to an industrial economics case study is required in order to establish robust findings.

aid could still be compatible with the common market if one of the derogations discussed below were applicable.

3.2.6.1   *Compatible with the Common Market*

As outlined previously, a NAP containing a PSR system in the first trading phase could have been falling under the derogation of Article 87(3)(b) EC Treaty only to the extent that it strictly implements the requirements spelled out by Directive 2003/87/EC. The Article's applicability is subject to the necessity criterion and the proportionality test. As for the second and third trading period the criteria enlisted in the new environmental guidelines[217] are applicable. In light with the logic of the balancing test also the proportionality test would have to be complied with. While their interpretation is dependent upon the particular allocations contained in the NAP and the particular approach to emission allowances remains largely uncertain, some general observations regarding the PSR system can nevertheless still be made.

As already explained above, in accordance with the compensatory justification principle under Article 87(3)(b), anticompetitive distortions can be outweighed by positive environmental benefits. The balance does not have to be positive for the individual allocations but on average. It therefore suffices to highlight the anticompetitive distortions that are likely to be created by the application of the PSR system.

As presented before competitive distortions can stem from several sources. New entries may suffer from barriers to entry and trading sectors may incur disadvantages from undue changes in relative prices *vis-à-vis* other trading sectors or non-trading sectors. On the other hand, new entrants may more easily apply low polluting technology, which compensates for the mentioned entry barriers. Competing firms will only suffer from dissimilar treatment if size or investment capabilities of undertakings differ or due to the fact that not all members of a sector are obliged to participate in the trading system. The Commission would have had to review such distortions in the light of the compensatory justification principle.

With regard to the derogation provided for by Article 87(3)(c), the effect of the trading conditions is the relevant benchmark. As discussed earlier, while the Commission was enjoying large discretion to set the assessment criteria during the first trading phase, it would have had to take the effects on competitors and product markets, market demand, profit margins, the intensity of trading on the common market and the particular circumstances of the recipients into account. The measure had to be justified by the net environmental benefit it created and could be used to justify actions going beyond the scope of Directive 2003/87/EC. It is important to note that this derogation is subject to the same anticompetitive distortions as Article 87(3)(b).

---

217.   See OJ C 82, 01 Apr. 2008, 1–33, s. 3.3.

Under the second trading phase similar rules are employed. It is, however stressed that aid can only be declared permissible if it attains additional environmental protection beyond what is ensured on the basis of Community legislation,[218] the allocation is executed in a transparent manner, based on objective criteria and reliable data and does not give rise to over allocation.[219] It is noteworthy that the aid scheme should be of a general nature and not favour certain undertakings or sectors unless justified by the logic of the scheme[220] nor put new entrants at a comparative advantage or disadvantage *vis-à-vis* incumbents.[221] It is not clear how the generality requirement of the scheme is interpreted in practice since aid must be selective in order to qualify as aid within the meaning of Article 87(1) EC Treaty in the first place.

3.3   CONCLUSION

This section has reviewed the assessment of both the grandfathering allocation mechanisms and the PSR Emission Trading System under European State aid legislation. The overall finding is that grandfathering systems can be liable to constitute State aid and that PSR systems are less liable to fall within the ambit of Article 87(1) EC Treaty. Clearly, the Commission has the power to interpret the concept of aid and has held that the Dutch NOx system, as an epitome of a PSR system, constitutes State aid. Yet the Court of First Instance did not follow the Commission's line of argumentation regarding the selectivity criterion and ruled that the Dutch NOx system does not constitute State aid. The Commission appealed this case in particular regarding the admissibility of the case before the CFI.[222] This case, case 279/08 is still pending before the ECJ. As has been presented in this chapter, it is this authors opinion that there are no State aid resources involved in a PSR system and consequently not all of the necessary criteria of Article 87(1) EC Treaty are fulfilled. In addition, a PSR system established on an EU wide level would not impede trade between Member States, undermining a further criterion for State aid. Therefore, it must be concluded that PSR systems do not constitute State aid. With great interest is the decision of the Court on this matter awaited.

If the Court concludes that the PSR systems – like grandfathering systems – fall within the ambit of Article 87(1) EC Treaty, they may still be declared to be compatible with the common market as falling under one of its derogations.

The brief discussion about the derogations of Article 87(1) EC Treaty has shown that it is difficult to establish concrete findings as to when a scheme should be declared compatible with the common market. This is related to the fact that the

---

218. OJ C 82, 01 Apr. 2008, 1–33, recital 140(a).
219. OJ C 82, 01 Apr. 2008, 1–33, recital 140(b).
220. OJ C 82, 01 Apr. 2008, 1–33, recital 140(c).
221. OJ C 82, 01 Apr. 2008, 1–33, recital 140(d).
222. Case C-279/08 P, OJ C 223/30 of 30 Aug. 2008.

overall level of anticompetitive distortions is inseparably linked to the individual emission allowance allocations. Nevertheless some differences regarding the choice of the allocation format can be established. It has been shown, that in both derogations an environmental compensatory justification is needed in order to justify State aid measures. A comparison of the grandfathering allocation mechanisms and the PSR system shows, however, that the anticompetitive distortions that are likely to be created by the PSR system are lower and that consequently the environmental net benefit required to justify the application of a measure will be lower too. This is tantamount to state that with regard to the proportionality test, a PSR system is preferable to grandfathering and this will be particularly so if a PSR system is introduced on an EU wide level because intra-community trade will not be distorted.

The disadvantage placed on new entries under the grandfathering system – provided of course that they are not afforded equal treatment – is mitigated by the fact that new entrants are per definition subject to the same benchmarks. New entrants would – as is the case under the grandfathering system – still be disadvantaged if barriers to entry would rise as a result of the introduction of the trading system. Such barriers will exist if competitors are credit rationed, investments are sunk or if financing costs are positive.

Unlike grandfathering mechanisms, PSR schemes are not liable to distort competition on the merits with regard to intra-industry competition within and even between Member States if a PSR is introduced on an EU wide level. Undertakings covered by the scheme are subject to the same legislative framework. The cost burden may differ to the extent that investment capabilities or returns to scale differ between enterprises. While investment capabilities would give rise to socially undesirable market distortions, disadvantages based on the mere size of the firm may reflect the natural play of market forces. Regarding covered and non-covered parts of a sector, it appears that grandfathering mechanisms are capable to partially compensate covered undertakings by the mere fact that they receive intangible assets for free. This clear mitigation of the financial cost burden is not possible under the PSR system. The lesson to be learned from a competition point of view is of course not the increase of transfers to undertakings but to create a level playing field in the sense that the same rules are applicable to all market participants. Furthermore, it has to be noticed that unlike grandfathering mechanisms, PSR systems do not support the entrenchment of market shares.

Last but not least it has to be observed that while grandfathering has an inherently static focus and needs particular correcting factors in order to accommodate for different sector growth rates or technical innovation potential, the PSR system is subject to continuous adjustments and therefore not prone to generate similar inter-sectoral distortions of competition.

Consequently, due to its lower level of anticompetitive distortions, the overall conclusion is that a PSR system will require a lower level of compensatory justification than a grandfathering allocation system and thus be preferred under the proportionality test – provided of course that it is justly adjudged to constitute State aid in the first place.

4   AUCTIONING UNDER THE EU ETS AMENDMENT

The recent amendment envisages a strong emphasis on auctioning as an allocation mechanism. This means that the current decentralized Member State oriented approach to emissions trading is substituted by an EU-wide harmonized one based on auctioning. The EU ETS amendment delegates legislative power, subject to the comitology procedure (regulatory procedure with scrutiny[223]), to the Commission. The Commission shall adopt a Regulation on the timing, administration and other aspects of auctioning by 30 June 2010 to ensure that auctions are conducted in an open, transparent and non-discriminatory way safeguarding cost efficiency, market access and information access.[224] In particular electricity producers are in principle subject to auctioning.[225] The transitory period originally envisaged by the European Commission to introduce auctioning is prolonged. Under the EU ETS amendment the 80% *ex ante* benchmark based free allocation is calculated on the average consumption of the top 10% installations during the period 2007–2008 and reduced to 30% by 2020 with a view on a total phase out by 2027.[226]

In addition to these temporary derogations to auctioning, installations in sectors or sub-sectors exposed to a significant risk of carbon leakage can receive 100% free allocation in accordance with Article 10a(12).[227] By 31 December 2009 – and then every five years – the Commission shall, after discussing with the Council, identify exposed sectors.[228] Furthermore, for the *energy-intensive* sectors or sub-sectors exposed to a significant risk of carbon leakage, the Commission analyses their position in the light of the outcome of international negotiations by 30 June 2010 and submits proposals, including the *adjustment of the proportion* of allowances received free from charge.[229] In addition it has to be noticed that the overall community-wide quantity of allowances is reduced annually by 1.74%[230] to ensure environmental effectiveness.

This section of the chapter examines the role of State aid legislation in the context of auctions under the EU ETS amendment. The choice not to closely examine State aid issues under the harmonized free allocation benchmarks is based upon the understanding that they will be set by the Commission and merely be employed by the Member States. It is noticeable that the EU ETS amendment employs the word 'calculation' in the context of free allocation.[231] This shows that

---

223. See Art. 10(4), Directive 2009/29/EC; Craig P. & De Búrca, G., (2003), 150, and Art. 5a Council Decision 2006/513/EC, (2006) amending Decision 1999/468/EC laying down the procedures for the exercise of implementing powers conferred on the Commission, OJ L 200/11.
224. Article 10(4), Directive 2009/29/EC.
225. Article 10a(6), COM (2008) 16 final.
226. Article 10a(11), Directive 2009/29/EC.
227. Article 10a(12), Directive 2009/29/EC.
228. Article 10a(13), Directive 2009/29/EC.
229. Article 10b(1), Directive 2009/29/EC.
230. Article 9, Directive 2009/29/EC.
231. Article 10a(1), Directive 2009/29/EC.

hardly any discretion is left to the Member States, though of course the clearness and certainty provided by the free allocation rules can only be determined after those rules have been adopted. It nevertheless appears that Member States may not enjoy an appreciable degree of discretion to interpret the Commission Regulation on the allocation of emission allowances. Any violation of State aid rules would then be attributable directly to the secondary law itself and should therefore be subject to a 230 EC Treaty procedure.

With regards to auctioning, arguably more discretion may be left to Member States. Despite the significant level of uncertainty[232] that still prevails with regard to the future regulatory framework of the EU ETS, Member State influence on auctioning can be inferred from the EU ETS amendment that requires Member States to submit reports to the Commission on each and every auction that is being conducted. In Article 10(4) it is stated that Member States must report on each auction regarding the proper implementation of the auctioning rules, in particular with respect to fair and open access, transparency, price formation and technical and operational aspects.

Before engaging into a close examination of State aid involvement in the case of auctions under the EU ETS amendment an incidental question needs to be examined. It needs to be considered how the Commission in general addresses issues of State aid review in the contexts of auction settings.

4.1   Auctions and Commission Involvement

Auctions under the EU ETS will be subject to a Commission regulation and to the formal criteria of openness, transparency, harmonization and non-discrimination.[233] These criteria seek to ensure that EU ETS auctions yield market prices that per definition will not give rise to any State aid concern. It may therefore be questioned whether a State aid examination is at all required and how the Commission deals with such issues in practice.

With regard to the procurement tenders that are executed in accordance with the European public procurement directives in an open, unconditional and transparent manner – also including auctions – this interesting issue has been addressed. From a conceptual point of view public procurements awarded on the basis of such open, transparent and non-discriminatory procedures are per definition awarded on the basis of normal market conditions. This perspective has been used by the Commission to presume that the prices determined through such public procurement procedures are reflecting the market value and do therefore not constitute State aid.[234] While this may be a correct assessment in many cases this does not allow for the conclusion that State aid is not involved if public procurement rules

---

232. See on uncertainty in the framework of the EP legislative resolution Peeters, M. & Weishaar, S., (2009).
233. Article 10(4), Directive 2009/29/EC.
234. European Commission (1997). See also Tosics, N. & Gaál, N., (2007), 6.

have been observed.[235] This presumption has been criticized in the context of a pre-emptive State aid assessment that the Commission conducts on the basis of public procurement designs, public procurement procedures ability to always guarantee market prices and the possibility that advantages could be accruing to third parties.[236]

Thus not only in situations where no public tender was held but also in cases where public procurement procedures have been employed, it has to be examined how the granting of advantages is to be determined in the context of a business contract that involves public authorities. Clearly, the mere award of a profitable contract by a public procurement entity that leads to a higher capacity utilization and other positive side effects to the undertaking does not allow the conclusion that State aid has been granted[237] and is therefore not at issue here. Thus in order to distinguish between an orderly market price and State aid the European Commission employs the so-called private investor test. If a private investor would have entered into the transaction under the same terms and circumstances there is no State aid.[238] In this way the Commission seeks to prevent that an undertaking receives an advantage it would not have received under normal market conditions.

Similarly, also in the framework of EU ETS auctions the mere fact that auctions were held in conformity with the Commission regulation does not entail that no advantage has been granted. To elucidate this point, one can take a look at the first auction conducted in the second trading phase of the EU ETS.[239] The UK auction was held on 19 November 2008. Approximately four million allowances were sold at a clearing price of EUR 16.15 (GBP 13.60) per allowance, raising about GBP 54 million. The auction was more than four times oversubscribed and organized with the support of large corporations (Barclays Capital, BNP Paribas, JP Morgan and Morgan Stanley) that acted as primary participants. This competitive auction has generated a very positive price and raised millions of Euros and could thus be described as an overwhelming success. Taking the market price of EU ETS emission allowances into account, however, the appreciation is dramatically changed. On the 19th of November the spot market price of EUAs at the Leipzig Energy Exchange was EUR 16.42 per allowance[240] while other sources suggest that the market value of an EUA at the time of closing of the auctions was as high as EUR 16.60 per allowance.[241] In light of these figures the auction certainly may have caused much dismay to the Chancellor of the Exchequer who was selling emission allowances for EUR 1–1.8 million below their market value.

---

235. Case C-334/99 *Germany v. Commission* [2003] n.y.r., para. 142.
236. See Tosics, N. & Gaál, N., (2007), 16 ff. and case law contained therein.
237. See Frenz, W., (2007), 105.
238. Case C-482/99 *France v. Commission* [2002] ECR I-4397, para. 70 and Bultmann, P. F., (2004), 29 ff.
239. Defra (2009).
240. See <www.eex.com>. Other sources report that the market price at the time of closure of the auctions was standing at EUR 16.60 implying a loss to the ministry of finance of EUR 1.8 million, see Business Green (2009).
241. Business Green (2009).

Undertakings participating in a competitive auctions have thus been attaining a benefit to the extent that the difference does not reflect their increased transaction costs incurred due to auctioning.[242]

In light of the above it can be concluded that operating State aid review on the basis of a presumption that a test of structural auction criteria is sufficient to determine that no advantage has been accruing to undertakings has to be viewed critically. Due examination on a case by case basis seems to be necessary. The reporting requirement placed upon Member States can be expected to constitute an important element to facilitate the work of the Commission in this respect.[243]

4.2   AUCTIONS IN THE THIRD TRADING PHASE

This section of the text examines in which particular situations auctions in the post 2012 trading period could give rise to State aid concerns under the auctioning regulation drafted by the European Commission in accordance with the EU ETS amendment. It bears mentioning that the legal treatment here cannot be conclusive without sound economic examination of the actual auctions that will be implemented. This section does, however, seek to raise awareness that auctions as well are not by design beyond State aid reservations when they are operated in the context of a sizable secondary market and specific sectors are specifically addressed by the regulatory framework. This section follows the same structure as was employed earlier in this chapter.

4.2.1   Transfer of a Benefit or an Advantage (Notion of Aid)

As outlined in chapter three, one of the very appealing features of auctions is that they are able to lead to bidders' self revelation of preferences and thereby allow the determination of a market price in the absence of adequate knowledge of what the price should be. In the previous treatment of auctions their generic capabilities have been reviewed. In light of the EU ETS amendment, however, one needs to qualify auctions' capabilities with regard to the process of price determination.

As has become apparent from the UK auction described above, the secondary market can indeed play a crucial role in influencing bidders' willingness to pay. The potentially large amounts of emission allowances that will still be allocated for free during the third trading phase may give rise to a situation in which auction prices will be based upon information derived from secondary markets. In such an environment the interaction between both the auction and the secondary market becomes crucial from a State aid perspective in the sense that bidders will not be willing to pay a price for emission allowances that exceeds the secondary market price. If prices in auctions are, however, below the prices on the secondary market, undertakings could

---

242.   The problem of profits steming from arbitrage opportunities has also been recognized in the context of auctions by Hepburn C. et. al. (2006), 146.

243.   Article 10(4), Directive 2009/29/EC.

be deriving benefits from participating in auctions that are undesirable from a State aid perspective.

Undertakings would be able to attain such benefits from auctions if they were able to pay prices in auctions that are below those prevailing on a secondary market as to overcompensate them for the costs they incurred in participating in an auction. Depending on the variety of auction designs permitted by the Commission and the frequency that a particular undertaking engages in auctions that employ the same mechanism designs, it could also be argued that undertakings would regard their preparatory costs for auctions as non-recoupable. Economists refer to such non-recoupable expenses as sunk costs. In such situations it is rational behaviour to disregard these 'lost' costs for the purpose of economic decision making. This in turn could imply that undertakings in such circumstances should be willing to compete more strongly and drive auction prices up to the secondary market level. In this line of thinking a benefit would already accrue to an undertaking if it would not be paying the market price prevailing on the secondary market. A cost differential between the auctioning market and the secondary market would not be rational. In the presence of highly harmonized mechanism designs, however, remaining cost differentials could reflect transaction costs.

It will therefore have to be estimated to what extent the cost differentials between auctioning and secondary markets – that are obviously necessary to lure undertakings that are able to procure emission allowances on the secondary market into participating in auctions – are reflecting mere transaction cost, and to what extent they should constitute a barrier to price equalization on two competitive markets.

To what extent this price gap between the two markets is persistent cannot be answered in the absence of a sound analysis of the emission allowance market. The price differential is unlikely to be long lived if the secondary market is not sufficiently liquid and participants from this market are forced to engage in auctioning in order to meet their demand. Whether the auctioning of the electricity sector paired with an 80% free allocation for non-exposed sectors is sufficient, is an empirical question that cannot be answered in the abstract. Some clarity may be gained from the publication of how many allowances the Commission intends to be auctioned. The estimates have to be published by 31 December 2010.[244]

4.2.2 Aid Favouring a Certain Undertaking over Others (Selectivity Principle)

From the viewpoint of selectivity the assessment of auctions is complicated by the post 2012 regime shift in the EU ETS. Under the first two trading periods auctions were limited to 5% and 10% respectively. These auctions were in principle open to anyone. Thus there is no indication that auctioning in such a context could give rise to selectivity concerns and hence would not constitute State aid within the meaning of Article 87(1) EC Treaty.

---

244. Article 10(1), Directive 2009/29/EC.

Auctioning is getting an ever more important role of in the post 2012 period. As under the NAPs in the first and second trading period auctions will remain open to all market participants. While this again would suggest that there are no concerns from a selectivity point of view, it bears noticing that the electricity sector is not benefiting from benchmarking and will be subject to 100% auctioning.

Targeting a specific sector directly and thereby limiting the scope of beneficiaries has been held to be selective.[245] Similarly the Court held in a different case that the fact that a measure constitutes an advantage not only for medical practitioners specializing in dentistry but also for other operators in the medical field or even for all operators in that field does not mean that measure does not fulfil the condition of selectivity.[246] Also measures addressing natural persons and smaller companies are held to be selective if they exclude large companies.[247] Furthermore – and more importantly for the EU ETS auction schemes post 2012 – the Court held in other contexts that having a large number of undertakings falling under a measure, or that they belong to the same economic sector[248] or different sectors of activity, is insufficient to call into question the measures' selective nature.[249]

While clear indication on the interpretation of the selectivity criterion has not been found, it is believed that selectivity could be viewed to be existing post 2012 if the electricity sector is singled out. A measure would also appear to be selective if some undertakings were to participate in auctioning while others would not. This would parallel the Court's line of argumentation that measures accruing to all exporters are selective because there are undertakings that are not exporting.

It bears mentioning that a measure that is by its nature selective cannot be cured by reference to its general economic nature.[250] Furthermore it has to be observed that besides measures that are by their nature selective, aid that is not aimed at pre-determined recipients but subject to objective allocation criteria, within the framework of an overall budget, to an indefinite number of beneficiaries can still be selective within the meaning of Article 87(1) EC Treaty.[251] Selectivity does thus require that the undertakings participating in auctions can be distinguished from those that may not participate in them even though they are in a comparable legal and factual situation in light of the objective pursued by the EU ETS directive.[252]

---

245. European Commission, (2000), Statsstøttesag Nr. N 653/1999 – CO2-kvoter, SG (2000) D/, 12 Apr. 2000, 5–6.
246. Case C-172/03 Heiser [2005] ECR I-1627, para 41.
247. Case 409/00 *Spain v. Commission* [2005] ECR I-1487, para. 49.
248. Case C-75/97 *Belgium v. Commission* [1999] ECR I-3671, para. 33, and the case-law cited).
249. Case C-172/03 Heiser [2005] ECR I-1627, para. 42, C-409/00 *Spain v. Commission* [2003] ECR I-1487, para. 48, and the case-law cited.
250. Case 409/00 *Spain v. Commission* [2005] ECR I-1487, para. 49.
251. Case T-55/99 *CETM v. Commission* [2000] ECR II-3207, para. 40.
252. Case C-143/99 *Adria-Wien Pipeline GmbH and Wietersdorfer & Peggauer Zementwerke GmbH v. Finanzlandesdirektion fuer Kaernten* [2001] ECR I-8365para 41, Case C-75/97 *Belgium v. Commission* [1999] ECR I-3671, para. (26) 28–31, C-409/00 *Spain v. Commission* [2003] ECR I-1487, para. 49, Case T-55/99 *CETM v. Commission* [2000] ECR II-3207, para. 53–54.

The case law of the Court appears to be very broadly constructed so as to bring various measures within the reach of Article 87(1) EC Treaty. It appears to require a detailed analysis of all relevant provisions providing for derogations from a general scheme in order to determine whether the derogations to a general economic measure are based on the same principles and objectives as the general measure itself.[253] If the logic of the employed system is not apparent or consistent, a derogation from a general economic measure would not be justified and the selectivity requirement of the State aid examination would be satisfied.

In this line of thought one would have to examine if the selectivity principle regarding auctions in the post 2012 EU ETS would be consistent with the objective of the Directive and its amendment. Of particular relevance in this context appear the avoidance of windfall profits and the polluter pays principle. While the former objective could allow in principle a derogation from a general economic measure it could be questioned why this objective is not phrased in a quantified and objectively measurable way. The legislator could for example have opted for having all companies – independent of their sectoral affiliation – that enjoy for example a certain market price elasticity of demand not to benefit from any free allocation. It is after all this degree of price elasticity that lies at the route of an undertaking's ability to pass through costs to consumers and not their sectoral affiliation as such. Similarly the polluter pays principle as a second objective attributable to the EU ETS might be viewed to be inconsistently applied if some undertakings are forced to engage in auctioning while other undertakings benefit from free allocation.

It is recognized that the treatment in this section is far too limited to give a complete overview of the underlying complexities and that a satisfactory appraisal of selectivity under post 2012 auctions will have to be conducted on a case by case basis. Yet it is this section's objective to raise awareness to the potential existence of problems under the selectivity principle that may lead to the finding of State aid in post 2012 auctions.

4.2.3 Granted by the State or through State Resources

The third criterion to be assessed is whether the measure at hand is granted by the State or through State resources. In light of the ECJ's ruling in *Preussen Elektra*, as discussed before, these two elements contained in Article 87(1) EC Treaty are cumulative rather than alternative. At first it will be assessed if advantages through auctioning are granted by the State and subsequently, it is examined if a financial burden is placed upon the Member State. Only if both elements are confirmed, can the measure constitute aid within the meaning of Article 87(1) EC Treaty.

---

253. See Golfinopoulos, C., (2003), 549, Case T-55/99 *CETM v. Commission* [2000] ECR II-3207, para. 53–54.

4.2.3.1   *By the State*

The relationship between institutions that auction emission allowances and the State can take many forms. According to ECJ case law not only direct allocation through a State institution or public bodies but also through private bodies which are either established or appointed to administer aid[254] constitutes an 'emanation' of the State and hence falls within the meaning of 'State' as used in Article 87(1) EC Treaty. Thus any institution granting or administering CO_2 emission allowances does meet this criterion.

The issue of imputability is, however, more interesting. Pursuant to the EU ETS amendment Member States are obliged to auction all allowances that are not allocated for free.[255] Member States will also have to observe the Commission Regulation on timing, administration and other aspects of auctioning that seeks to ensure that auctions are conducted in an open, transparent, harmonized and non-discriminatory manner. To the extent that Member States merely implement legislation, there is no imputability and hence no State aid. It is, however, not yet known how these rules will be designed and what discretion is left to the Member States. It could for example be possible that the European Commission designs two or three auction systems from which Member States can choose – after all auctions perform best if they are tailored to the bidder preferences and some Member States have already expressed that they see auctioning as an allocation mechanism and not as a means to raise revenues.[256]

In such a context it could for example be entirely possible that a Member State selects an auctioning mechanism that is not designed to produce high revenues rather than an auctioning mechanism that does exactly this. The discretion granted to the Member State to select the less profitable auctioning mechanism would imply that Member States would be wilfully forgoing revenue. Since Member States are only under a duty to auction those allowances not allocated for free, their decision to do so in a less profitable way constitutes an act which is at their own discretion. Member States are thus bound by their obligation to respect the EC Treaty, including Article 87(1). In such situations it may well be questioned if advantages accruing to undertakings as a consequence of Member State choices are not imputable to Member States within the meaning of Article 87(1).[257] In light of the CFI's and the Commission's decision on the Dutch NOx system failure to generate profits at a time when Member States are at a discretion to do so should lead to the establishment of State aid.[258]

---

254. Case 78/76 *Steinike v. Bundesamt für Ernährung und Forstwirtschaft* [1977] ECR 595, para. 21.
255. Article 10(1), Directive 2009/29/EC.
256. See Sectorakkord Engergie 2008–2020 (2008), 4.
257. Case 61/79 *Amministrazione delle finanze dello Stato v. Denkavit italiana* [1980] ECR 1205, para. 31, Case T-351/02 *Deutsche Bahn AG v. Commission* [2006] n.y.r., para. 100.
258. See above for a discussion of this case.

4.2.3.2   *Through Member State Resources*

With regard to the second element, the granting of aid through Member State resources, it is clear that auctions directly affect public accounts. After all auctions lead to direct financial transfers. The attainment of a price below the secondary market price will therefore constitute a negative effect on State finances if compared to a situation where the allowances would have been sold directly on the secondary market. The State is thus in fact forgoing revenue it could have been attaining by failing to maximize auction revenues. Whether the price differential between the primary and secondary market is long lived is a question of fact not of law and cannot be determined in the abstract. It requires an analysis of the market liquidity and the likely strategic behaviour of the bidders. In light of the experience at the UK auction it nevertheless appears to be an issue that cannot be lightly disregarded.

4.2.4   It Should be an Undertaking or . . . Production

As discussed previously Article 87(1) EC Treaty limits the scope of incompatible aid by express reference to undertakings or the production of certain goods. Hence part of the legal exercise is to determine whether successful bidders are to be considered to be undertakings. Only if this is assessed in the affirmative, there can be State aid.

From the foregoing legal assessment, it is apparent that the ECJ applies a broad interpretation of the concept of undertaking. Whether a particular bidder fulfils this criterion obviously depends on the recipient. Directive 2003/87/EC defines operators as any person who operates or controls an installation.[259] One would thus expect that any operator – legal or natural person alike – participating in an auction of European Emissions Trading allowances will be found to be engaged in an economic activity and identified as an undertaking. Things are, however, less clear for individuals who might be willing to participate in the auctions. Their participation could be motivated by their environmental concern or hoping to benefit from speculation. While Article 12(1) of the current Directive allows the free transfer of allowances between persons, Article 10(4) obliging the Commission to draft auction regulations to ensure that auctions are conducted in an open, transparent, harmonized and non-discriminatory manner, only speaks of operators. It is therefore not entirely certain if the Commission will in fact allow private individuals to participate in auctions. While private individuals who want to buy greenhouse gas emission allowances to compensate for example, their last holiday flight's emissions may not be identified as undertakings, private investors may still be identified as engaging in an economic activity, particularly if their traded volume is so huge that the alleged benefit would surpass the *de minimis* threshold.

---

259.   See Art. 3(f), Directive 2003/87/EC.

4.2.5   **Distorts or Threatens to Distort Competition**

With regard to auctioning and the actual distortion of competition, it is questionable whether anticompetitive distortions exist in the first place and secondly, whether the aid upon which they are based exceeds the *de minimis* threshold as required by Commission Regulation 1998/2006. Each will be discussed in turn.

4.2.5.1   *Between Incumbent and New Entering Firms*

Auction rules have to be designed in an open and non-discriminatory way.[260] This entails that auctioneers are unable to distinguish between incumbents and new entrants. Consequently distortions of competition between incumbents and new entrants is not an issue under auctioning.

4.2.5.2   *Trading Sectors and Non-trading Sectors*

With regard to the relationship between trading and non-trading sectors the same finding as before can be established. To the extent that the costs incurred from buying emission allowances under auctions are higher than costs of producers of similar or substitutable goods in non-covered sectors, distortions of competition are created.

4.2.5.3   *Anticompetitive Distortions between Competing Firms*
                    of the Same Member State

As was the case under previously discussed allocation mechanisms also under auctioning anticompetitive distortions can arise if only a number of undertakings of a given sector are obliged to participate in the EU ETS while others are not. To the extent that the non-participating entities are not subject to additional and presumably costly legislative requirements, they will enjoy a comparative advantage over those undertakings that are participating in the EU ETS.

4.2.5.4   *Between Trading Sectors*

Unlike other allocation mechanisms auctions do not appear to create distortions between sectors as such. All operators are entitled to participate in auctions on an open and non-discriminatory basis. Distortions of competition may, however, appear if undertakings in the electricity sector are participating more strongly and quickly in auctions than operators of other sectors. The electricity producers will not receive emission allowances for free and are thus solely dependent on auctions and the secondary market. If electricity producers – as large demanders of emission allowances – realize that they are unable to satisfy their needs on the secondary market, they will be inclined to participate in auctions at a very early

---

260.   Article 10(4), Directive 2009/29/EC.

stage. To the extent that operators in other sectors are more reluctant to do so, the benefits from participating will fall more heavily on the side of energy producers. An express assumption underlying this finding is of course that in these auctions there is a price differential between allowances on an auctioning and the secondary market.

4.2.5.5 Does It Exceed the de minimis *Threshold?*

As explained earlier, aid within the meaning of Article 87(1) EC Treaty can only be established if it surpasses the *de minimis* threshold of EUR 200.000 granted to an undertaking during the three consecutive years.[261] In light of the large quantities of emission allowances to be allocated it does appear to be very likely that this threshold could be easily surpassed by large undertakings. As mentioned above, the first English auction under the second trading phase led to a price differential of EUR 45 cents per emission allowance between the secondary and the auctioning market and a reduction of income of EUR 1.8 million. If such price differentials were to be witnessed in other auctions as well during the post 2012 period the threshold might indeed be surpassed.

4.2.6 Community Dimension: Aid Which is Capable of Affecting Trade between Member States

As outlined above, the mere presumption of a distortion of competition suffices to constitute State aid. In the presence of price differentials between primary and secondary markets it would be likely that the *de minimis* threshold would be exceeded. Such distortions of competition could also impact trade between Member States if for example auction prices would differ on some auctions between Member States. This may be the case if there is no full harmonization of auctions that is, complete uniformity of the employed auction design or if bidder preferences are not uniformly distributed among Member States. The later in particular could impact bidding behaviour if some bidders were more reluctant than others to actually engage in auctions in other Member States. While the criteria of Article 10(4) EU ETS amendment are intending to create a high degree of transparency on the auction market it is thinkable that in particular small and medium sized enterprises are reluctant to bid on auctions abroad. In such situations auction prices could diverge from each other. If this was the case, trade between Member States could be affected as production costs would differ.

The above treatment has shown that operators may be able to attain preferential dealings when auctions are held in the presence of sizable secondary markets. By paying less for emission allowances an element of comparative advantage is created that could give rise to State aid concerns. While it is unlikely that auction schemes during the first and second trading period meet the selectivity criteria, this issue is less clear for the third trading phase. Directly addressing the energy sector

---

261. Commission Regulation 1998/2006.

could put the general nature of auctioning schemes into question. While the case law in this field does not appear to entirely rule out the possibility that auctions may violate Article 87(1) EC Treaty with regard to the selectivity principle, it should be borne in mind that the derogations introduced earlier in this chapter may well be applicable. Even though the anticompetitive distortions and hence the required compensatory justification is likely to be more limited than under any of the free allocation mechanisms, it is nevertheless noticeable that the warm welcoming of auctioning rules within the EU ETS may still give rise to Competition law questions.

5   STATE AID AND DISTORTIONS OF COMPETITION

The previous section has examined how State aid legislation deals with allocation mechanisms that could give rise to distortions of competition on the merits and that negatively impact allocative efficiency. Grandfathering and the PSR system, as examples of free allocation mechanisms, and auctions have been reviewed. This section first examines distortions of competition under these allocation formats. Subsequently allocative efficiency considerations are reviewed.

Regarding distortions of competition, it has been shown that grandfathering is liable to constitute State aid while PSR may not – though the differing finding of the Commission has to be noticed. Even if PSR were found to constitute State aid, both free allocation formats could be declared compatible with the common market under one of the existing derogations. Perhaps even more importantly it has been found that the level of anticompetitive distortions differ between them. It is this difference that is of particular relevance from an economic perspective because any distortion of competition on the merits can have detrimental effects upon society and allocative efficiency. Regarding auctioning, it has been found that State aid may still be an issue until the secondary market looses most of its importance and auctioneers are willing to truly compete with each other and are willing to place bids around the secondary market price. Central to this finding is of course that State aid is indeed imputable to government action. Existence of Member State discretion to decide upon the timing and choice of the auction mechanism design is thus an important precondition. The distortions of competition that are naturally associated with the obtaining of undue profits, are – except perhaps for the electricity sector that is in principle barred from free allocation – not structural in nature and can thus be regarded to be limited. The degree of compensatory justification needed to allow auctioning under Article 87(3)(c) is thus expected to be limited too.

In the context of free allocation mechanism it has been found that Competition law as such does not foreclose the found anticompetitive distortions. It therefore has to be examined if Competition law encourages the selection of the least distortive means for allocation in order to mitigate negative repercussions on the product market. In general it is to be observed that the liberty Member States enjoy to employ the allocation mechanism they deem most appropriate is subject to EC law and judicial review. Judges are, however, more deferential in their review when political considerations such as economic policy or equality considerations are concerned.[262] It is recognized that governments do make policy choices and courts should not overrule them merely because they believe another choice would have been better.[263] Therefore it is unlikely that judges will interfere strongly with public considerations. Consequently it is unlikely that judges would require Member States to use auctions rather than one of the two free allocation formats discussed in this chapter on the basis of allocative efficiency considerations. Yet with regard to allocation mechanisms that have similar properties things are less clear. Here the ECJ may be more prepared to voice an opinion if it was asked.

Allocation mechanisms falling within the ambit of State aid can be declared to be compatible with the common market under one of the existing derogations. The examination of the employed compensatory justification criterion entails a balancing act between distortions of competition on one side and environmental benefits on the other. Underlying this compensatory justification are the two concepts of necessity and proportionality.

In order to examine whether Competition law is able to lead to the selection of the least distortive allocation mechanism, proportionality is of particular importance. Here it is examined if the applied measure is proportional or excessive in relation to the objective sought. If one of two comparable allocation mechanisms was less distortive to competition, the proportionality principle would in general work towards the selection of the least distortive system. Whether it indeed is effective does essentially depend on the degree of comparability between the allocation formats, and the arguments underlying the Member State's choice. If they are very political in nature, the Court may refrain from criticizing government's selection.

It therefore follows that the degree of comparability of both PSR and Grandfathering has to be examined. The systems are comparable to the extent that they both do not burden undertakings with direct financial expenditures and both fall with the category of free allocation mechanisms. They do, however, differ in their design[264] and in their distributive effects on undertakings. While the former does not appear to be decisive from a proportionality point of view, the latter is. Grandfathering allows allocations which are tailor made to the specific requirements of firms and thus entails different effects on undertakings that impact the level playing field while the PSR system in contrast pursues a sectoral approach that may include several benchmarks for specific product groups. If the Court would indeed

---

262. See De Burca, G., (1993), 111–112.
263. See De Burca, G., (1993), 111–112.
264. Important design characteristics are that the State assumes a less central role in the granting of allowances and that the PSR system in its generic form as established under the Dutch NOx system does not have a legally enforceable cap. Yet with regard to the EC emission trading system it is clear that a cap is intended.

differentiate between both systems, EC Competition law[265] would not encourage the selection of the less distortive PSR allocation format.

The finding that in practice Competition law as interpreted by the ECJ may not work to select the least distortive allocation mechanism exacerbates the legal problem established in chapter four. It has been discovered that an allocation system that gives rise to less anticompetitive concerns than grandfathering has been excluded from the scope of the Directive – at least in the Commission's practice – by outlawing *ex post* adjustments so that effective application of EC Competition law rules is foreclosed.[266]

It has therefore to be concluded that EC Competition law may not lead to the selection of the allocation mechanisms that is least distortive. This is related to the wording of the current EU ETS Directive and the potential consideration of political decisions by the ECJ. Not only does this lead to the conclusion that effective Competition law application is foreclosed but it also results in the finding that the EU ETS is not designed to directly attain allocative efficiency nor a competitive market equilibrium on the production market. This finding has to be qualified with regard to the post 2012 trading regime where auctioning takes a central role. Here the problem of allocative efficiency is ever less pronounced as the secondary market's bearing on bidder's price determination decreases.

---

265. The polluter pays principle, may however be considered and support the selection of a system that burdens operators employing less environmentally efficient means of production.

266. Case T-374/04 *Germany v. Commission* [2007] n.y.r., dealing with the Commission's rejection of the German NAP provisions addressing *ex post* adjustments has been decided upon by the Court of First Instance on 7 Nov. 2007. The Court established that the Commission erred in law on several occasions, yet the practical implications of this ruling may be limited. See Weishaar, S., (2008).

Chapter 7

Auction Design Challenges

1   INTRODUCTION

This section of the book examines the design challenges the Commission faces in designing appropriate auctions for the EU ETS. In order to examine these challenges and their implications this section builds on the elements raised in Chapter 3 and in part repeats central issues in order to elucidate points made. In the following also the implications of the eventuality that the European Commission does not provide a sufficient degree of harmonization in the design of its auctioning procedures is examined. In such cases Member State discretion could lead to the adoption of differing auction mechanism designs and auctioning schedules. It is, however, not the objective of the chapter to present a technical review of a wide array of auction mechanism designs that the Commission (or any legislator establishing a greenhouse gas auction scheme) could be selecting and to review their respective advantages and disadvantages. Instead of focusing on the specificities of auction mechanism design, this section focus on the decisions to be taken when introducing an auctioning system and its associated implications and complexities.

The starting point for examining the implementation of large scale auctioning under the EU ETS is to define the product that is being auctioned. Under Directive 2003/87/EC each tradable allowance is defined as 1 ton of CO_2 equivalent over a designated period of time.[1] The legislation also extends to other greenhouse gases with an equivalent global-warming potential of 1 CO_2 ton.[2] Because the benefit derived by operators from attaining 1 ton CO_2 equivalent is the same irrespective of the greenhouse gas they purchase, the good is clearly a substitute[3] and can thus be

---

1. Defined in Art. 3(a) of Directive 2003/87/EC.
2. See Art. 3 (j) of Directive 2003/87/EC.
3. If obtaining one good makes the bidder willing to pay more for a second good, the goods are complements, if the bidder is willing to pay less, they are substitutes.

treated as a homogeneous good – that is a similar good – from an auction design point of view.

Since participants to an emission trading auction will generally acquire larger quantities of allowances – particularly when one considers institutional investors such as brokers or banks[4] – multiunit auctions are of core interest. As in the case for single-unit auctions, auction designers strive to reach two goals that may well be assumed to be close to the objectives of the government. They try to ensure an efficient outcome and to maximize government revenues. It is, however, also entirely possible that the European Commission is not seeking to maximize auction revenues. In its non-paper published in August 2008 the Commission states that the primary objective of auctioning is to bring allowances into circulation in a pre-dictable and efficient manner at market price and that it should be ensured that the clearing price reflects the true value of allowances.[5] From this one can, however, not conclusively infer if the reflection of the 'true value of allowances' will lead to the selection of auctioning schemes that force bidders to reveal their true prefer-ences. In this context it bears mentioning that the Netherlands have expressed that they do not view auctions as a source of income but as an allocation mechanism.[6]

Reasons for the Commission to consider taking a more modest approach to competitive pricing in auctions could include that high auction prices may lead to carbon leakage or to fierce resistance of stake holders. While the later does not appear to be evidenced in the Commission proposal, the former consideration of carbon leakage associated with a higher cost burden of auctions seems to be reflected in Article 10a(9)(a) of the Commission proposal and 10a15(a) of the EU ETS amendment. The possibility that auctions lead to substantial increases in production costs and to losses of market share of carbon efficient Community installations to less carbon efficient non-community installations, is expressly mentioned as a relevant factor to be considered for awarding free allocation and also recognized by the literature.[7]

Independent of the preferences the Commission may have, it bears mentioning again that a lot of the insights relating to multiunit auctions come by analogy from single-unit auctions[8] and a sound understanding of how equilibria respond when assumptions about values and information change are not yet answered at any level of generality.[9] This dilemma may be exacerbated in the presence of an inherent uncertainty about the bidders and their preferences that characterizes the EU ETS market. In light of the severe over allocation that characterized the first trading period the ability of administrators to correctly estimate the underlying parameters should not be overestimated.

---

4. Any person within the Community is allowed to hold emission allowances. See Arts 3(g) and 12, Directive 2003/87/EC as well as recital 29 of COM (2008) 16 final of 23 Jan. 2008 and recital 38 of Directive 2009/29/EC.
5. European Commission (2008), 2.
6. See Sectorakkord Engergie 2008–2020 (2008), 4.
7. See Kuik, O., (2005), 14 ff.
8. Ausubel, L. & Cramton, P., (2002), 1.
9. See Börgers, T. & Van Damme, E., (2004), 43.

Related to the task to allocate a large number of emission allowances is the choice between two general multiple auction methods that could be used to allocate greenhouse gas emission allowances. Firstly, sequential auctions in which one allowance would be sold after the other and secondly, simultaneous auction models in which multiple greenhouse gas emission allowances are sold at the same time. In light of the long trading period (eight years) and the large number of emission allowances that has to be introduced into the market it can be expected that the Commission will employ a combination of both types. It may allow in its regulation the organization of repeated auctions in which larger and smaller bundles of allowances will be auctioned.

In this context the Commission in its desired capacity as the 'drafting authority' for auctioning rules and Member States as the relevant authorities who have to apply these rules must consider three elements upon which also bidder's preferences are expected to have important influence. These are the bundling of the allowances, the timing of the auctions and the market structure and transparency, each will be addressed in turn.

2   BUNDLING

Single-unit auctions do not impair allocative efficiency considerations irrespective of goods being complements or substitutes. In the case of multiple-unit auctions, concerns regarding allocative efficiency can be severe if complements are auctioned in inefficiency inducing packages.[10] With regard to the EU ETS the theoretical complexity of bundling of emission allowances is limited to strike a balance between sufficiently small lots that allow a large number of natural or legal persons[11] to participate in auctions and increased costs from having to organize a large number of auctions. Cost effectiveness will thus have to be balanced against the self imposed Commission's duty to draft an auction regulation that allows small and medium sized enterprises covered by the Community scheme full market access.[12] While the Commission will have to take decisions in this area, it is not expected that this will be particularly problematic if sufficiently harmonized auctioning schemes are employed.

3   TIMING

With regard to the second issue, the timing of auctions, it can be stated that sequential auctions are in general easy to implement for auctioneers since they

---

10. For a recent review see Milgrom, P., (2004), Ch. 8.
11. Any person within the Community is allowed to hold emission allowances. See Arts 3(g) and 12, Directive 2003/87/EC as well as recital 29 of COM (2008) 16 final of 23 Jan. 2008.
12. See Art. 10(5), COM (2008) 16 final of 23 Jan. 2008.

merely have to repeat the (electronic) auction over and over again without adaptations of the auction design mechanism. Such auctions can be organized consecutively or with identical or dissimilar time periods between them. One could imagine for example that auctions could be organized every day, or during the course of the last week of every month or that auctions are in particular concentrated in the months prior to the submission of emission allowances. Despite the simplicity of organizing (electronic) repeat auctions for auctioneers, sequential auctions are, however, not very much favoured by bidders. One reason is the strategic complexity of bidding decisions and the possible price variation of sequential auctions of homogeneous goods.[13] Ashenfelter (1989) has termed this 'declining price anomaly'[14] and explained a falling price in subsequent auctions[15] in terms of bidders' necessity to acquire particular quantities, risk aversion[16] and uncertainty.[17] Ashenfelter's observation entails that on average bidders are risk averse and appears to be based on a more or less static supply and demand. In the presence of economic growth and a capped amount of greenhouse gas emissions, this may, however, not be a realistic expectation. The expectable effect of a declining price anomaly might be overpowered by the increasing scarcity of emission allowances to the effect that prices are increasing towards the annual compliance date rather than decrease. Nevertheless, the empirical observation of such effects in auctions underlines the relevance of risk averseness as a bidder characteristic to be taken into account by auction designers.

In order to allow for a well placed scheduling of auctions during the eight year trading period starting in 2013 the European Commission will thus have to estimate the degree of risk aversion of the various bidders and the degree to which parties' information on the actual EU ETS markets differs. It could for example be expected that the large energy produces as prominent actors on the EU ETS market do have a better understanding of the market dynamics than a medium sized paper producer or a small installation in any other sector. In such situations prices paid at auctions could differ considerably.

From an allocative efficiency point of view, auctions ensuring a single market price as for example, uniform-price auctions are preferred over sequential auctions because they mitigate the 'price risk' of paying too much for the same good.[18] Yet this does not appear to be an option that is easily reconciled with the Commission's objective to allow new entrants access to the EU ETS market – according to the EU

---

13. Inefficiencies from synergies and complementariness of goods appear to be smaller if goods are homogeneous.
14. Ashenfelter, O., (1989), 29 ff.
15. See Ashenfelter, O. & Graddy, K., (2002), 34–36 and Table 8 for a review of subsequent research.
16. Risk aversion implies that bidders dislike taking fair bids. McAfee, R. & Vincent, D., (1993) show that risk aversion can create declining prices.
17. See Neugebauer, T. & Pezanis-Christou, P., (2005).
18. Milgrom, P., (2004), 256.

ETS amendment new entrants may benefit from free allocation under the transitional community-wide rules.[19] The underlying idea is as follows: sequential auctions can give rise to declining price anomaly if they are for example held on the same day subsequently to each other but also if they are held at different intervals over a period of time. Uniform price auctions could be employed successfully to mitigate price anomaly if auctions are held on a single day but may be unlikely to contain price anomalies between auctions that are held at different points in time. Holding a single uniform price auction – or very few of them (if this were to be expected to have a positive impact on limiting price anomaly) – could, however, provide new entrants with difficulties to procure additional allowances at those points in time that are most convenient to them.

Also related to the difficulty of measuring bidder's risk aversion and associated price anomaly are the varying amounts of revenues that are paid per emission allowance. The amount of revenues does not only depend on the type of auctioning mechanisms that the Commission would be permitting under the Regulation that it shall adopt by 30 June 2010 but also crucially on the timing of the auction.[20] If the Commission decided that Member States were allowed to auction at differing points in time, average revenues could differ considerably. Similarly, if the Commission were to prescribe each Member States at which particular point in time it were to organize an auction, the Commission would have to find a way to reimburse Member States in such a way as to pay an equal amount per ton of CO_2 emission allowances to each or face likely criticism from Member States.

4   MARKET STRUCTURE AND TRANSPARENCY

Two issues that are very closely linked to both the amount of revenue generation and the decision the Commission has to take to provide for centrally publicized auctions or fully decentralized ones, are market structure and transparency. In the EU ETS market for emission allowances large emitters appear to be relatively strong in terms of market size. The largest 7% of installations account for 60% of the total emissions while the 1400 smallest installations only account for 0.14%.[21] To the extent that these large installations belong to a limited number of undertakings or that their operators are particularly active in auctions because they are continuously adjusting their stocks to their predicted needs – such as energy producers may be doing – such operators my become influential. They might even be able to recognize their interdependence and thereby set small and medium sized bidders at a comparative disadvantage in the sense that large operators pay a relatively lower price than they otherwise would have been paying.

---

19.   See Art. 10a(6), COM (2008) 16 final of 23 Jan. 2008 and Art. 10a(7), Directive 2009/29/EC.
20.   See Art. 10(4), Directive 2009/29/EC.
21.   COM (2008) 16 final of 23 Jan. 2008, 5.

If a few undertakings would recognize their interdependence this may give rise to inefficiency[22] inducing 'demand reduction' strategies.[23] In multiunit auctions dominant players recognize the interdependence of their bidding strategy and competitor's bidding behaviour. A strategy of self restricting the quantity demanded while bidding the minimum price to indicate interest in a number of units can generate large consumer surpluses. The inherent inefficiency stems from the fact that users with the highest value for a good do in fact prefer not to attain it: large bidders win too little and small bidders win too much. Salmon (2003)[24] points out that if such behaviour is strictly unilateral, this does not amount to collusion. However, if it does involve strategic considerations exemplified by trigger strategies, such behaviour would amount to (tacit) collusion.[25]

Bidders' ability to use backward induction[26] to limit their capital investments will depend on their degree of interdependence that they are recognizing. If bidders fail to attain low price outcomes they can employ signalling[27] tactics to 'negotiate' mutually acceptable allocations. In a simultaneous auction environment with a limited number of participants and known limit prices of fringe firms, Grimm, Riedel and Wolfstetter (2001)[28] cite the German GSM Spectrum Auctions as an example of how effectively signalling can be used to reach an almost immediate mutually acceptable strategic demand reduction. While signalling and demand reduction certainly are difficult points of certain (multiunit) auctions, there is one factor[29] that is relevant in the context of the EU ETS that complicates effective collusion: a large number of bidders.[30] In light of the above it appears desirable to not only limit the information that bidders are able to exchange through the auctioning process but also to allow for processes that enable a large number of bidders to participate in the auction. Both transparency and sufficiently low transaction

---

22. Inefficiency is created by 'differential bid shading', that is, when bidders with identical marginal values reduce their bids by different amounts so that awarding the bidder who values the item most is impossible. See Ausubel, L. & Cramton, P., (2002), 4.
23. For examples see Weber, R., (1997), and Ausubel, L. & Cramton, P., (2002).
24. Salmon, T., (2003), 5.
25. Distinguishing between tacit collusion and pure strategic firm behaviour is complicated if not impossible. In a multiunit auction both bidders could, for example, independently decide to pursue a 'demand reducing' strategy. The outcome would be identical to tacit and indeed outright collusion.
26. Bidders are assumed to imagine how the auction will be developing and to derive from this a mutually acceptable offer at the beginning of the auction.
27. By for example using the financially inconsequential digits of their bids, parties can signal their identity or indicate the market for which they are retaliating.
28. Grimm, V., Riedel, F. & Wolfstetter, E., (2001).
29. Other factors that are relevant particularly in the context of heterogeneous products are synergies and externalities. For an analysis of externalities in single-unit auctions, the interested reader is referred to Caillaud, B. & Jehiel, P., (1998).
30. Brusco, S. & Lopomo, G., (1999) show that the collusion becomes more difficult as the number of bidders relative to the number of items rises. Weber, R., (1997) presents a vivid example of how difficult it can be to reach a mutually acceptable allocation of heterogeneous permits when the number of participants is large.

costs can be expected to have a positive impact on the number of bidders that can credibly be expected to participate in an auction.

In setting up a Community-wide Regulation on the application and implementation of auctions, the Commission also has to consider how it ensures market transparency. Enhanced transparency can influence the amount of auction proceeds that depend on the degree of actual and potential competition. Bidders may be more willing to fiercely compete with each other if they cannot determine the identity and the quantity of bidders during or before the auction. If participation in an auction does not entail high transaction costs, such as for example expenses for designing a bidding strategy, a wide publication of an auction can credibly attract a larger number of potential bidders, including small and medium sized undertakings. This in turn will impact the bidding behaviour of participants. Since transparency will only lead to behavioural change if costs of participation in an auction are low enough, also and in particular to small and medium sized undertakings, the Commission has to weigh transparency and transaction cost considerations against benefits stemming from mechanism design variety. In particular in decentralized auctions in which auctioneers can more easily determine bidder preferences and fine tune the auction mechanism design to the expected bidders auctions may be expected to realize their full potential.

There is thus a trade off between the Commission prescribing a 'one size fits all' fully harmonized and community-wide publicized auction system through its regulation and granting Member States discretion to tailor auction mechanisms to the assumed preferences of the expected bidders that sets different behavioural incentives. It is in the context of tailor made solutions in which auctions are developing their full potential. Besides this important considerations it should be borne in mind that the Commission's regulation on auctioning has to be compatible with the principle of subsidiarity and that the regulation may bring the Commission in conflict with the conferral of power principle as enshrined in Article 5 EC Treaty.

5   CONCLUDING REMARKS

The EU ETS amendment for amending Directive 2003/87/EC attributes a strong role to auctioning. It is to become the core allocation format to bring emission allowances onto the market. Auctions are not only expected to successfully address the problems of windfall profits and new entrants but also lead to the generation of auction revenues. They should be used to enhance economic growth and solidarity and also be used for environmental purposes. While the Commission has until 30 June 2010 to adopt a Regulation on the timing, administration and other aspects of auctioning to ensure that it is conducted in an open, transparent and non-discriminatory way, a number of problematic issues relating to the introduction of large scale auctioning in the EU ETS can already be identified.

The design of auction mechanisms takes relevant properties of the bidders into account in order to design efficient auctions that yield the desired objectives.

Auction theory is best developed with regard to single unit objects that are sold and gained insights may not be directly transferable to multiunit auctions as likely to be employed under the EU ETS. Independent of any theoretical challenges that may still have to be addressed, it appears clear that the European Commission will have to make assumptions about bidders' characteristics and preferences. Given the inherent uncertainty relating to the characteristics of the market operators, it can be expected to be difficult for the Commission to design an optimal auctioning system, and that the realization of the superior allocation properties commonly attributed to auctions may not be a self fulfilling prophecy.

In particular the Commission will have to decide whether it wishes to set up regulation to create strongly harmonized and simple auctioning rules that reduce costs associated with comprehending auctioning mechanisms and limited strategic complexity or whether it wants to grant Member States a larger degree of discretion. A very substantial harmonization of auctioning rules can reduce transaction costs sufficiently and allow for a large number of bidders to be credibly expected to participate in any given auction. This in turn will enable in particular small bidders to be better able to engage in auctions without being confined to conduct standard market operations via exchange houses or other traders. In contrast to a strong degree of harmonization, the decentralized approach allows for more appropriate auction mechanism designs and to realize the full potential of auctions. It also allows Member States to raise the funds they wish to raise through auctioning and to avoid problems related to subsidiarity and competences of the Commission. Depending on the degree of discretion afforded to Member States and the diversity of auction mechanism designs that would actually be employed, it could effectively exclude small firms from participating in certain auctions without incurring substantial costs. Whether this is decisive does, however, also depend on the frequency with which small installations are expected to participate in auctions. Operators who are already weary about participating in the EU ETS market may be even more reserved when it comes to participating in auctions. Such operators may then predominantly rely on secondary markets.

Chapter 8
Final Conclusion

Agreement among scientists is increasing that the greenhouse gas concentration in the atmosphere is rising and that the societal costs associated with greenhouse gas emissions are substantial.[1] Policymakers in Europe are recognizing the need for immediate emission reductions and are debating how this could be achieved. Under the leadership of the German Presidency the Council of the European Union has agreed to take more drastic steps towards achieving the strategic objective of limiting the global average temperature increase to not more than 2 degrees above pre-industrial levels[2] and international preparations a follow up to the Kyoto protocol, the Copenhagen Climate Change Conference in December 2009, are under way. A core element of the European effort in its European Climate Change Program is the European Emissions Trading System that will help Member States and thus also the EU to make progress towards their objectives.

The Emissions Trading instrument provides a framework for cost effective and efficient emission reductions. Even though this constitutes a perfectly legitimate field of State involvement from an economic point of view, this does not free governments from the (economic) obligation to design a system that operates in the most efficient and least distortive way. This book examines in which way the legal text of the present Directive limits Member States' discretion to select an allocative efficient and environmentally effective allocation mechanism and identifies the existing competition law challenges of state involvement under particular consideration of the Directive's amendment of 23 April 2009. Since an important element for the attainment of allocative efficiency and the market equilibrium is in particular the free working of competition, anti-competitive distortions are examined. Such distortions are of relevance if an efficient allocation is not attained as a direct result of initial allocation of emission allowances but has to be established on

---

1. See IPCC, (2007b), 17.
2. Council of the European Union, (2007), 10.

the secondary Emissions Trading market. This research therefore seeks to assist policy makers to recognize to what extent different allocation mechanisms (grand-fathering, PSR and auctions) impact emissions trading, and to what extent the present competition law framework can support competition on the merits within the EU ETS.

While law essentially tries to set incentives (rules) so that people behave in a desirable manner, economists are able to provide the analytical framework to determine whether legislators and administrative authorities succeed in setting the incentives right so that they actually achieve what they are striving for in a cost efficient way. It can therefore be seen that considerable synergy effects can be expected to be derived from a joint 'Law and Economics' approach.

While much has been written about the benefits and drawbacks of emission trading systems, the impact of allocation mechanisms on allocative efficiency with regard to the EU ETS has not sufficiently been addressed. This book addresses this aspect in a coherent Law and Economics framework through the application of Competition law, auction theoretic and industrial economic insights.

Initial allocation is expected to trigger adaptation processes on the market that will eventually lead to allocative efficiency of the market. Beholding the development of the market price of EU ETS emission allowances for the first trading period (2005–2007) it can be observed that the market prices were strongly positive and increasing over a period of sixteen months and only started to decline after publication of governmental reports which indicated the presence of a very strong over allocation of emission allowances.[3] The market failed to realize this despite a strong over allocation of probably 160 million ton CO_2 (amounting to an over allocation of around 8%).[4] The market price decreased from above EUR 30 to well below EUR 10 cents per emission allowance unit. Another insightful element is that the information asymmetry from which bureaucrats are suffering at both national and European level was indeed so strong that they were unable to detect such flaws and thereby put the environmental effectiveness of the first trading period into doubt. In has therefore been proposed that an allocation mechanism that introduces an initial endowment in the neighbourhood of a market equilibrium and therefore requires relatively less working on the market, could have appealing properties.

In order to determine which allocation formats are most desirable from an allocative efficiency and environmental point of view, this text first established a coherent typology of emission allowance allocation mechanisms for the purpose of emission trading models. Through the subsequent analysis of how the various assignment mechanisms deal with issues of price determination, allocative efficiency and environmental considerations in both a static and dynamic economy,

---

3. For a review of the current EU ETS system see Weishaar, S., (2007d).
4. Data based upon Table 1 of IP/07/1094 of 13 Jul. 2007. The data is not conclusive because it is based upon data from those twenty-two Member States that do have an electronic emission allowance registry. The reason why the expectations of market participants did not adjust is subject to further research and beyond the scope of the text.

this research identifies differences and proposes a ranking with regard to their desirability.

With regard to the static closed economy it has been shown that only auctions – employed at a massive scale to effectively overpowering the secondary market – are capable to solve the problem of price determination and to attain allocative efficiency and even potentially a market equilibrium. In the presence of collusion or declining price anomaly, however, the equilibrium will differ from the perfect competitive outcome. In contrast to auctions, normative distribution considerations prevail over allocative efficiency in administrative allocation mechanisms. Since bureaucrats are expected to be less well informed about the true valuations of market participants than undertakings themselves, administrative mechanisms depend upon efficient secondary markets to ensure allocative efficiency. With regard to the free administrative allocation mechanisms it was found that relative standard base mechanisms award 'early movers' and thus a societal group that is expected to have higher valuation for emission allowances. It has therefore been postulated that such mechanisms are allocatively superior to grandfathering schemes.

Relaxing the strict assumptions towards a dynamic open economy which allowes for strategic firm behaviour and international trade but not for a sufficiently long time period to take account of effects of research and development, the above findings are confirmed with only one qualification. Due to an exacerbated risk of collusion and a declining price anomaly attributable to a larger degree of uncertainty, auctions perform less well in the dynamic open economy than in a static closed economy. Nevertheless it can be concluded that only the auctioning mechanism – provided that it is employed at large scale – enables a timely attainment of allocative efficiency and that the other allocation formats have to rely more heavily upon cost effective secondary markets to safeguard allocative efficiency.

In light of allocative efficiency it can thus be concluded that auctions appear to be most desirable. With regard to administrative allocation mechanisms PSR systems are more desirable than grandfathering allocation schemes.

With regard to the second element which was considered in light of the object and purpose of the Directive, environmental effectiveness, this ranking has to be re-examined. Within the framework of a static closed economy setting it has been found that initial allocation mechanisms cannot have a differential impact upon the environment as long as a strict dichotomy between the quantity setting decision and the distribution issue of CO_2 emission allowances prevails. This does, however, not hold true for open dynamic economies as considered in this book where pollution havens have been assumed. The ability to raise costs to operators and thus to internalize negative externalities is constrained by transportation costs and the profitability of relocating production facilities.

In light of the above auction systems, generating the highest cost burden that induces carbon leakages, perform worst from an environmental perspective. Their performance is only partially mitigated by short-term positive environmental effects stemming from declining price anomaly. The relocation inducing burden under financial administrative systems is less severe than under auctions but not as low as under free administrative allocation systems. Because there are no incentives to postpone investments in environmentally friendly technology and low stranded costs, PSR systems are more environmentally friendly than grandfathering mechanisms. This will be the case in a cap and trade system as well as in the absence of a cap as long as the overall emission quantity is not increased, that standards reflect abatement costs and industry's competitive position and that monitoring costs and uncertainty costs are sufficiently low. It can be concluded that indeed initial allocation mechanisms have a bearing on the environment.

Since irrespective of the existence of an emission cap, the upper limit of the costs to the internalization of environmental costs is thus given by operators' ability to evade such costs, the financial impact of allocation mechanisms may merit more attention than the discussion of capped trade and trade without cap may suggest.[5] It could therefore be possible that from a social welfare point of view sacrifices regarding allocative efficiency may be outweighed by benefits stemming from the internalization of negative externalities. It may thus be possible that the selection of a cost effective administrative allocation mechanism may be more desirable than the selection of auctions that potentially attain a market equilibrium at the expense of potentially limited environmental benefits[6] (stronger carbon leakages).

Regarding the post 2012 regime it is noticeable that the legal framework does provide for auctioning and specific rules for exposed sectors in order to alleviate the cost burden placed upon such sectors. While this differential internalization of negative externalities distorts the relative prices between the goods, it may be expected that the carbon leakage problem that is associated with the introduction of an emissions trading system can be restricted.

Building upon the findings of chapter three, it has been examined whether auctions, grandfathering and PSR systems are compatible with the EU ETS Directive or whether the present Directive constraints the application of desirable allocation methods. It has been found that auctions are compatible with the Directive provided that they are being held in due time before the start of the trading period as to ascertain which installations will be holding which allowances and that not more than 5% during the first (2005–2007), and 10% during the second trading period (2008–2012) are auctioned. It is not expected that this amount is sufficient to fully realize the superior allocative efficiency properties of this allocation mechanism. Its overall social welfare effect is dependent upon the degree to which the limited cost burden and the associated limited inducement of relocation is able to compensate society through internalization of negative external effects. With regard to administrative allocation mechanisms grandfathering schemes were in principle compatible with the Directive provided that they are not infringing upon any of its criteria or stipulations. This is not the case for PSR systems. Even though they may be constructed in such a way as to be compatible with the envisaged absolute cap of

---

5. This underlines the importance to consider enhanced energy efficiency of production, global coverage and environmental protectionism.
6. The magnitude of such effects and the relationship between cost burden and investments in research and development is subject to further research.

the EU ETS, they are excluded from the scope of the Directive on the grounds that they violate a strict interpretation of the Directive – as seems to be favoured by the Commission – with regard to Criterion 10 of Annex III, Article 11(1), and 11(4) concerning *ex post* adjustments. The Court of First Instance's (CFI's) ruling on *ex post* adjustments (T-374/04 of 7 November 2007) came to late to allow for significant effects upon the second trading phase.

It is therefore postulated that the EU ETS in its present form constraints the application of allocative efficient allocation formats and is unable to ensure the timely attainment of a market equilibrium because it forecloses a meaningful application of auctions. It may however attain its environmental objective through ensuring a cost effective system that induces less relocation than is to be expected under auctioning. Even though it has been shown that a PSR system may be allocatively superior to grandfathering, it has been excluded from the application under the Directive. By contrast the amendment to Directive 2003/87/EC fundamentally changes the present allocation format away from grandfathering towards auctioning and harmonized benchmarking. At this stage the details are not yet known and need to be fleshed out by future Regulation which has to be adopted by the Commision by 30 June 2010 (auctioning) and 31 December 2010 (benchmarking) – a further analysis is thus obstructed.

The remainder of the text examines in which ways permitted allocation mechanisms can gives rise to anti-competitive concerns. Distortions of competition stemming from the trading systems such as the EU ETS are countered through the application of Competition law.

A Competition law and industrial economics analysis has shown that the EU ETS is affecting competition on the merits in many ways. State measures impact firms' propensity to collude and abuse while the granting of subsidies is tantamount to a comparative advantage that is rationalized on environmental grounds. In the remainder of the book it has been examined whether the core provisions of Competition law could be employed to contain anti-competitive and thus social welfare reducing effects originating in State measures taken within the framework of the EU ETS that give rise to collusion and abuse or to State aid concerns. Since auctions are a market based instrument and if employed as the sole allocation mechanism are not easily liable to generate such distortions, only grandfathering and PSR systems are considered for the first and second trading period. They are, however, examined in the context of the EU ETS amendment. Though excluded under the present Directive PSR systems are considered because they have been shown to be allocatively superior to grandfathering systems.

With regard to impacts upon the propensity to collude and abuse, it has been concluded that the European Court of Justice (ECJ's) approach of the joint application jurisprudence under Article 81 is not designed to address anti-competitive State measures which facilitate collusion originating from the EU ETS. As for Article 82, there might be the theoretical possibility if the Court were to extent the absolute competition approach applied in certain contexts under Article 86 to Article 82. Despite such theoretical possibilities and the economic insight that State measures impacting undertakings' propensity to collude and abuse will reduce social welfare, it is viewed to be unlikely that such measures are being contained if the Court were asked.

The ECJ's ability to address 'legal gaps' is very much determined by the EC Treaty. Since the Court bases its jurisprudence upon the seemingly vague concept of *effet utile*, it is not surprising that there are natural limits to it. Unless the Court was convinced that the State measures altering undertakings' propensity to collude and abuse would endanger the effectiveness of the Treaty, the author does not expect extensions of the jurisprudence. It appears more likely that the Court would be welcoming a rise in cartelization and abuse proceedings under Competition law articles directed against undertakings as an indirect means to mitigate anti-competitive effects originating in State measures. Even though the former is more desirable from a socio-economic and policy point of view, the latter is more in line with the legal jurisprudence.

Regarding the post 2012 EU ETS regime the role of the joint application jurisprudence also appears to be rather limited. While the implementing legislation of allocations are not yet known, it bears mentioning that the role of Member States is reduced and that consequentially their propensity to induce anti-competitive measures is similarly affected. In situations where the *effet utile* of the EC Treaty is endangered, it may be more promising for the Court to examine if the measures the Commission sets at the disposition of the Member State are in compliance with the amended Directive and primary law itself rather than to call upon the joint application jurisprudence.

With regard to comparative advantages granted by the State, it has been shown that Article 87 EC Treaty can be employed to address such distortions stemming from free allocation mechanisms. The overall finding is that grandfathering can be liable to constitute State aid while the PSR system – despite Commission's finding – may not. Even if the ECJ would find that the two systems fell within the meaning of State aid, both could be declared compatible with the common market under one of the applicable derogations. While the grandfathering method and the PSR system can give rise to repercussions in the economies of the respective Member States and the European Union as a whole, the overall sectoral distortions and hence the required compensatory justification is more limited under the PSR system than under the grandfathering system. PSR systems are therefore preferable to grand-fathering systems under the proportionality test and consequently also the preferred mode of allocation. In light of these findings, the resulting legal problem is evident. An allocation system that gives rise to less anti-competitive concerns than grand-fathering has been excluded from the scope of the Directive so that effective appli-cation of EC Competition law rules is foreclosed. It has therefore to be stated that the current EU ETS does not minimize anti-competitive distortions of allocation mechanisms. Consequently it can be argued that the EU ETS is not designed to directly attain a competitive market equilibrium on the production market.

Also for the period post 2012 State aid may be most relevant to assess the EU ETS regime. Regarding auctioning, State aid issues will have to be considered in those cases where undertakings stand to gain from undue profits obtained during orderly market operations at auctions. Auctioneers may require a premium in order to compensate them for higher transaction costs and uncertainty for participating in auctions rather than to procure emission allowances on the secondary market. Undue advantages may be accruing to auction participants up to the point where the secondary market looses its function to serve as an important point of reference for the upper limit of the market price – only if bidders are willing to compete so fiercely that their bids do indeed approach the secondary market price or even exceed it in those cases where its liquidity is limited, can auctions lead to true price revelation. The establishment of State aid will of course also depend upon the discretion left to the Member States – something that can only be assessed when more information regarding the implementing regulation is available – and the selectivity criterion. The selectivity criterion could be satisfied if the Court would take the express mentioning of the electricity sector in the EU ETS amendment as an indication that auctions do not constitute a general economic measure in the sense that auctions are truly open to all market participants. The later in particular also suggests why auctions during the first and second trading period may not have given considerable concern from a State aid perspective. Even though the degree of anti-competitive distortions caused by auctioning allocation mechanisms appear to be more limited than under the free allocation mechanisms – implying thus that auctioning is assessed favourably under the new environmental guidelines – auctions still merit Competition law review. Auctions thus may not constitute the problem-free solution they are at times proclaimed to be.

Regarding the challenges to EC Competition law, two issues bear mentioning. Firstly, in contrast to the first and the second trading periods, the third trading period – at least as far as auctioning is concerned – limits the expected bearing Member States could have upon undertakings' propensity to collude and abuse. Auctioning could give rise to State aid concerns but the degree of anti-competitive distortions is limited and deplead over time as auctioning becomes an ever more central allocation format. Secondly, as auctions become ever more prominent as an allocation mechanism during the third trading phase, Competition law provisions directed against undertakings could assume a more prominent role. Since collusion and abuse are very difficult to detect in an auction setting that is supposingly very competitive, advanced tools of detection are required.

Competition law analysis needs therefore to be receptive to econometric approaches to both in-auction infringements of Competition law as well as infringements occurring beyond the bidding framework. In particular with regard to collusion (bid rigging) auction theory has gained insights. A number of academic articles within the empirical auction literature analyse bidding patterns of convicted cartels and compares them to non-cartel bidding behaviour.[7] Besides their relevance for identifying collusive bidding behaviour, they postulate three main findings. Firstly that cartel members bid less aggressively than non-cartel members. Secondly, cartel member bids are more correlated than bids of a control

---

7. See for example Porter, R. & Zona, D., (1993), Porter, R. & Zona, D., (1997), Pesendorfer, M., (2000).

group.[8] Thirdly, collusion generates higher prices than the non-collusive control group. By analogy, in the EU ETS one would expect carel members to pay lower prices for their emission allowances.

Other research employs econometric tests to detect collusive bidding. After controlling for demand conditions in the timber industry, Baldwin, Marshall and Richard (1997)[9] test several models to determine if price variations are best explained by variations in supply conditions or by collusion. Porter and Zona (1993)[10] propose that in the absence of phantom biddings, parameters of a regression on the winning bidders should be equal to parameters obtained from a regression on the ranking of all bidders.

For asymmetric independent private and common value auctions, Bajari and Ye (2000)[11] propose a test to determine if in-auction bid-rigging did occur. Using insider information they create a distribution of firms' cost structures and compare them to the distribution of submitted bids. The existence of significant differences is interpreted as evidence of collusion. Despite its intuitive appeal, its practical applicability is severely constrained by data availability problems. Furthermore, if bidder's suspected that procuring entities were well informed about their cost structures, participating in such a tender would be less attractive since bidders' expected profits would be lower. To which extent knowledge about bidders' cost structures outperforms a larger number of participants in terms of procurement entity's revenue is subject to further research. Nevertheless, it is certain that bidder's have an intrinsic self-interest to undermine any attempts to attain information about their true values. For attempts to derive such information from past bidding data, Feinstein, Block, Nold (1985)[12] show that cartels who are aware that information is being extracted, do have an incentive to systematically misinform the procuring entity.[13] In can therefore be concluded that interdisciplinary research leading to a closer co-operation between law and special branches of economics can lead to valuable synergy effects.

The overall conclusion of this book is that the EU ETS does not allow for the timely attainment of a market equilibrium because the use of auctions under the Directive appears to be too limited. The resulting welfare loss may be potentially compensated through the employment of other allocation mechanisms that lead to less carbon leakage. In order to attain allocative efficiency on the production market the EU ETS would have to provide for cost effective exchanges of emissions allowances on a perfectly competitive market. During the post 2012 regime, auctions will be used to allocate large quantities of allowances so that a positive

---

8. See for example Porter, R. & Zona, D., (1993), 528.
9. Baldwin, L., Marshall, R. & Richard, J.-F., (1997).
10. Porter, R. & Zona, D., (1993).
11. See Bajari, P. & Ye, L., (2001a) and Bajari, P. & Ye, L., (2001b). For a general discussion see Bajari, P. & Summers, G., (2002).
12. Feinstein, J., Block, M. & Nold, F., (1985).
13. Feinstein, J., Block, M. & Nold, F., (1985), 452, state three variables which are being applied: mean bids of cartel members, variance of cartel bids and the number of long run market suppliers.

effect on allocative efficiency can be expected. This effect may, however, be reduced by introducing policies directed at mitigating carbon leakage that – as a consequence – distort relative prices between sectors. While the joint application jurisprudence offers the theoretical possibility of addressing anti-competitive repercussions stemming from grandfathering and PSR allocations under the EU ETS, its effective application depends on the seriousness of the case before the Court. With regard to the amendment of the Directive, joint application jurisprudence is not expected to be a central issue. State aid review may still be a relevant issue post 2012 in the context of auctions – if they were held to be selective and be imputable to Member States. With regard to State aid review during the first and the second trading period, it is shown that grandfathering systems could be more liable to constitute State aid than PSR systems. Even though both free allocation mechanisms could be declared compatible with the common market, PSR systems are less distortive and are hence to be preferred under proportionality considerations. In light of the Commission's positions PSR systems are, however, falling beyond the scope of the current Directive so that an effective application of EC Competition law rules is not possible. It can thus be argued that from the above presented Law and Economics perspective the EU ETS is not designed to directly ensure allocative efficiency on the production market – this finding is, of course, without prejudice to the application of other fields of law such as for example, the four freedoms.

That the introduction of auctions is not free from challenges has been described in chapter VII. Given the inherent uncertainty relating to bidder preferences it is found that the superior allocation properties commonly attributed to auctions may not be a self fulfilling prophecy. Furthermore it is suggested that the Commission will have to strike a balance between strong harmonization of the employed auctioning mechanisms that may help to keep transaction costs low and a (limited) variety in mechanism designs that could help to lead to superior allocative results and higher revenues.

Putting the above findings into a broader context, a number of observations appear to be in order. This book has found that Member States' discretion to select the allocative efficient initial allocation mechanisms is constrained by the current Directive 2003/87/EC. The question arises whether social welfare is enhanced despite the negative effects on allocative efficiency and how this will change under the amendment where auction is introduced but where it is entirely possible that the full potential of auctions will not be exploited. In addition to further research a sound understanding of the motivation of policy makers may shed light upon the feasibility to select a more favourable mechanism.

During the first trading periods the European legislator has been selecting the allocation methods and made the choice to forbid auctions to a large degree despite the favourable treatment in economic literature. Auctions have been identified in this body of literature as an appealing allocation formats that could even be used to raise profits for governments and will also be employed in the EU ETS at an increasing scale during the third trading period. Revenue generation is an appealing property that has lead American politicians to introduce large scale CO_2

auctioning schemes.[14] Within Europe it is noticeable that revenue generation is not the declared objective but that a 'fair' price is of more central interest and that some governments do not view auctions as a tool to generate revenue. This is lamentable since profits could be distributed to members of society in such a way as to reduce taxes. Consequentually social welfare could be increased through the reduction of distortions to the economy (double dividend hypothesis).[15] Reasons why the Commission proposal of phasing out free allocation (except free allocation to exposed sectors) has been diluted to attain a strong reduction in free allocation (30% by 2020) could be related to concerns of competitive positions of European enterprises and the lack of political support. Even though a relatively higher cost burden placed upon undertakings negatively impact the environmental effectiveness, the sensitivity of this theoretical finding constitutes an interesting field for further research.

Moreover a comparison to the grandfathering allocation method suggests that the overall social welfare might be better served though the application of a PSR system. This is contingent upon the differences in rent seeking costs, costs associated with negotiations, as well as the effort required to establish and maintain optimal national, European or even world wide PSR benchmarks. A multi-jurisdictional approach to PSR benchmarks may serve to mitigate distortions of competition between covered industries of the participating States. The issue of maintaining an optimal benchmark is of paramount importance for achieving the envisaged environmental goal of ensuring an emissions cap.[16] Research into the possibility of a PSR system to constitute a stepping stone towards a global approach may be particularly interesting. Research in sectoral approaches of relative standard base mechanisms[17] have indicated that sectoral PSR systems may not be welfare enhancing because the benefits stemming from reduced carbon leakage and those accruing to exposed sectors benefiting from PSR are unable to exceed the costs of sheltered sectors.[18] This implies that the loss of those sectors that are not strongly exposed to international trade are larger than the benefits to sectors that are exposed to fierce international competition. It would be particularly interesting to more closely examine the applicability of the PSR system under the amended EU ETS Directive once the detailed allocation regulations are presented. In light of the extremely fast developments of the emission trading system and the mind blowing speed with which legal proposals are hurried through the legislative process, it may be feared that yet again decision makers may make a deliberate policy choice of establishing a system quickly without allowing for due examination of the available policy choices. If this were indeed true, the 'learning by doing' phase, proclaimed to be only applicable to the first EU ETS trading period could thus extend well into the third trading phase.[19]

---

14.   See Arimura, T. H., et al. (2007) and Bogdonoff, S. & Rubin, J., (2007).
15.   Cramton, P. & Kerr, S., (1999).
16.   The issue of over allocation has been duly emphasized by OECD, (2005b). For additional research see OECD (2006).
17.   See OECD (2005a).
18.   Kuik, O., (2005), Ch. 7.
19.   Peeters, M. & Weishaar, S., (2009).

Bibliography

Åhman, M., Burtraw, D., Kruger, J. & Zetterberg, L., (2007) 'A Ten-Year Rule to guide the allocation of EU emission allowances', *Energy Policy*, Vol. 35, 1718–1730.

Alanen, L., (2003) 'The Impact of Environmental Cost Internalization on Sectoral Competitiveness: A New Conceptual Framework', in: Singer, H., Fatti, N., Tandon, R., (2003) *Trade and Environment, Recent Controversies, New World Order Series*, Vol. 21 Part III, 1111–1150.

Antweiler W., Copeland, B. R. & Taylor, M. S., (2001) 'Is free trade good for the environment?' *The American Economic Review*, Vol. 91, No. 4, 877–908.

Arimura, T. H., Burtraw, D., Krupnick, A. & Palmer, K., (2007) 'U.S. Climate Policy Developments', available at: <www.wlu.ca/viessmann/ClimateChange/Arimura.pdf>, 34.

Ashenfelter, O., (1989) 'How Auctions Work for Wine and Art', *The Journal of Economic Perspectives*, Vol. 3, No. 3, 23–36.

Ashenfelter, O. & Graddy, K., (2002) 'Art Auctions: A Survey of Empirical Studies', *NBER Working Paper Series*, Working Paper No. 8997, 2–43.

Austria, (2004) *'National Allocation Plan for Austria, pursuant to Art. 11 of the EZG'*, 31 March 2004, with additions dated 7 April 2004, available at: <http://umwelt.lebensministerium.at/article/articleview/27121/1/16297>.

Ausubel, L., (2002) 'An Efficient Ascending-Bid Auction for Multiple Objects', *University of Maryland Working Paper*, Working Paper 97-06, revised August 2002, 1–25.

Ausubel, L. & Cramton, P., (2002) *'Demand Reduction and Inefficiency in Multi-Unit Auctions'*, University of Maryland, 12 July 2002, 1–30.

Ausubel, L. & Schwartz, J., (1999) *'The Ascending Auction Paradox'*, University of Maryland, 5 July 1999, 1–23.

Babiker, M. H., (2001) 'Subglobal climate-change actions and carbon leakage: the implications of international capital flows', *Energy Economics*, Vol. 23, 121–139.

Babiker, M. H. & Jacoby, H. D., (1999) 'Developing country effects of Kyoto-type emissions restrictions', Report No. 53, Cambridge, MA.: MIT, *Joint program on the Science and Policy of Global Change*, 19.

Bach, A., (1994) 'Case C-185/91, Bundesanstalt für den Güterfernverkehr v. Gebrüder Reiff GmbH & Co. KG; Case C-2/91, Meng; Case C-245/91, OHRA Schadeverzekeringen NV', *Common Market Law Review*, Vol. 31, 1357–1374.

Bacon, K., (2003) 'The Concept of State aid: The Developing Jurisprudence in the European and UK courts', *European Competition Law Review*, No. 2, 54–61.

Bain, J., (1949) 'Price and production policies', in Howard S. Ellis, eds *A survey of Contemporary Economics*, Philadelphia, PA: Blakiston, 129–173.

Bain, J., (1956) '*Barriers to New Competition, their character and consequences in manufacturing industries*', M.A.: Harvard University Press, 329.

Bain, J., (1959) '*Industrial Organization*', New York, Wiley, 678.

Bajari, P. & Summers, G., (2002) 'Detecting Collusion in Procurement Auctions', Revised Version Forthcoming in *Antitrust Law Journal*, 1–31.

Bajari, P. & Ye, L., (2001a) 'Competition Versus Collusion in Procurement Auctions: Identification and Testing', *Stanford University Working Paper*, 1–37.

Bajari, P. & Ye, L., (2001b) 'Deciding Between Competition and Collusion', *Stanford University Working Paper*, 1–38.

Baldwin, R. & Krugman, P., (2000) '*Agglomeration, integration and tax harmonization*', available at: <www.wcfia.harvard.edu/seminars/pegroup/baldwin krugman.pdf>.

Baldwin, L., Marshall, R. & Richard, J.-F., (1997) 'Bidder Collusion at Forest Service Timber Sales', *The Journal of Political Economy*, Vol. 105, No. 4, 657–699.

Ballard C., Shoven, J. & Whalley, J., (1985) 'General Equilibrium Computations of the Marginal Welfare Costs of Taxes in the United States', *American Economic Review*, Vol. 75, 128–138.

Becker, G. S., (1968) 'Crime and Punishment: An Economic Approach', *The Journal of Political Economy*, Vol. 76, No. 2 (March–April, 1968), 169–217.

Belgium, (2006) '*Draft Belgian National Allocation Plan for CO_2-emission allowances 2008–2012 September 2006 plan submitted to the European Commission*' available at: <http://ec.europa.eu/environment/climat/pdf/nap_belgium_final.pdf>.

Bester, H., (2003) '*Theorie der Industrieoekonomik*', second edition, Springer Verlag, Berlin, 268.

Bogdonoff, S. & Rubin, J., (2007) 'The regional Greenhouse Gas Initiative, taking Action in Maine', *Environment*, Vol. 49, No. 2, 9–16.

Bohm, P., (1999) 'International greenhouse gas emission trading – with special reference to the Kyoto Protocol', *TemaNord*, 1999:506, Stockholm: University of Stockholm, 25, available at: <www.environmental-economics.dk/papers/FiDepTQKluwer.pdf>.

Boom, J. T. & Nentjes, A., (2003) 'Alternative design options for emissions trading: a survey and assessment of the literature', in Faure, M., Gupta, J. & Nentjes, A., eds, *Climate Change and the Kyoto Protocol, The Role of Institutions and Instruments to Control Global Change*, Edward Elgar, Cheltenham, 45–67.

Börgers, T. & Van Damme, E., (2004) 'Auction Theory for auction design', in Janssen M. C. W., eds, *Auctioning Public Assets, Analysis and Alternatives*, Cambridge University Press, 19–63.

Böringer, C. & Welsch, H., (2004) 'Contracting and Convergence of carbon emissions: an intertemporal multi-region CGE analysis', *Journal of Policy Modelling*, No. 26, 21–39.

Bovenberg, A. & Goulder, L., (2000) 'Neutralizing the Adverse Industry Impacts of CO_2 Abatement Policies: What Does it Cost?', *NEBR Working Paper Series*, No. 7654, 34.

Brander, L., (2003) 'The Kyoto mechanisms and the economics of their design', in Faure, M., Gupta, J. & Nentjes, A., eds, *Climate Change and the Kyoto Protocol, The Role of Institutions and Instruments to Control Global Change*, Edward Elgar, Cheltenham, 25–44.

Brozen, Y., (1971) 'Bain's concentration and rates of return revisited', *Journal of Law and Economics*, Vol. 14, October 1971, 351–369.

Brusco, S. & Lopomo, G., (1999) 'Collusion via Signalling in Open Ascending Auctions with Multiple Objects and Complementarities', *New York University Leonard N. Stern School of Business Department of Economics Working Paper Series*, Working Paper 99-05, 1–27.

Bultmann, P. F., (2004) Beihilfenrecht und Vergaberecht, Beihilfen und öffentliche Aufträge als functional äquivalente Instrumente der Wirtschaftslenkung – ein Leistungsverleich', Mohr Siebeck. Tübingen, 397.

Burniaux, J. M., (2001) *'International trade and investment leakage associated with climate change mitigation'*, available at: <www.gtap.agecon.purdue.edu/resources/download/503.pdf>, 18.

Burniaux, J. M. & Oliveira Martins, J., (2000) 'Carbon Emission Leakages: A general Equilibrium view', *OECD Economics Department Working Papers*, No. 242, 32.

Business Green, (2009) 'UK successfully auctions four million carbon allowances, But analysts divided over whether €16.15 price should have been higher', James Murray, 19 November 2008, available online at: <www.businessgreen.com/business-green/news/2230789/uk-successfully-auctions-four>, last visited 25.02.2008.

Caillaud, B. & Jehiel, P., (1998) 'Collusion in auctions with externalities', *The RAND Journal of Economics*, Vol. 29, No. 4, 680–702.

Calmette, M. F. & Péchoux, I., (2004) *'Agglomeration, major risky accidents and environmental policies'*, version of April 2004, available at: <www.cireq.umontreal.ca/activites/050930/papers/Pechoux.pdf>.

Camesasca, P. D., (2000) *'European Merger Control: Getting the Efficiencies Right'*, Intersentia – Hart, Antwerps, 498.

Camesasca, P. D. & Van den Bergh, R.J., (2006) *'European Competition Law and Economics: A Comparative Perspective'*, second edition, Sweet and Maxwell, London, 463.

Campins Eritja, M., (2006) 'Reviewing the challenging task faced by Member States in implementing the Emissions Trading Directive: Issues of Member State liability', in Peeters, M. & Deketelaere, K., eds, *EU Climate Change Policy, The Challenge of New Regulatory initiatives*, Edward Elgar, 69–82.

Carraro, C., (1999) 'Environmental conflict, bargaining and cooperation', in van den Bergh, J. C. J. M., eds, *Handbook of environmental and resource economics*, Cheltenham, U.K., Northampton, MA. Edward Elgar, 461–471.

Coase, R. H., (1960) 'The Problem of Social Cost', *Journal of Law and Economics*, Vol. 3, 1–44.

Collins, Norman R. & Preston, Lee E., (1969) 'Price-Cost margins and industry structure', *Review of Economics and Statistics*, 51 (August), 226–242.

COM (2000) 87 final, '*Green Paper on greenhouse gas emissions trading within the European Union*', European Commission, Brussels, 8.3.2000, 28.

COM (2000) 88, '*Communication from the Commission on EU policies and measures to reduce GHG emissions: Towards a European Climate Change Programme (ECCP)*', 13.

COM (2001) 31 final, '*On the sixth environment action programme of the European Community, "Environment 2010: Our future, Our choice"*', Brussels, 24.1.2001, 2001/0029 (COD), 24.

COM (2001) 579 final, '*Proposal for a Council Decision concerning the approval, on behalf of the European Community, of the Kyoto Protocol to the United Nations Framework Convention on Climate Change and the joint fulfilment of commitments there under*', OJ 075 E, 26/03/2002 P. 0017–0032.

COM (2001) 580 final, '*Communication from the Commission on the implementation of the first phase of the European Climate Change Programme*', 26.

COM (2001) 581 final, '*Proposal for a Directive of the European Parliament and the Council establishing a scheme for greenhouse gas emission allowance trading within the Community and amending Council Directive 96/61/EC*', 51.

COM (2003) 403 final, '*Proposal for a Directive of the European Parliament and of the Council amending the Directive establishing a scheme for greenhouse gas emission allowance trading within the Community, in respect of the Kyoto Protocol's project mechanisms*', 22.

COM (2003) 492 final, '*Proposal for a Directive of the European Parliament and of the Council for regulating certain fluorinated gases*', 44.

COM (2003) 830 final, '*Communication from the Commission on guidance to assist Member States in the implementation of the criteria listed in Annex III to Directive 2003/87/EC establishing a scheme for greenhouse gas emission allowance trading within the Community and amending Council Directive 96/61/EC, and on the circumstances under which force majeure is demonstrated*', Brussels, 07.01.2004, 27.

COM (2004) 3982/7 final, '*Commission Decision of 20 October 2004 concerning the national allocation plan for the allocation of greenhouse gas emission allowances notified by France in accordance with Directive 2003/87/EC of the European Parliament and of the Council*', Brussels, 20.10.2004.

COM (2005) 107 final, *'State Aid Action Plan, Less and better targeted State aid: a roadmap for State aid reform 2005–2009'* (Consultation document), Brussels, 7.6.2005, 18.

COM (2005) 549 final, *'Commission Decision concerning the national allocation plan for the allocation of greenhouse gas emission allowances notified by Poland in accordance with Directive 2003/87/EC of the European Parliament and of the Council'*, Brussels, 08.03.2005.

COM (2005) 703 final, *'Communication from the Commission: Further guidance on allocation plans for the 2008 to 2012 trading period of the EU Emission Trading Scheme'*, Brussels, 22.12.2005, 39.

COM (2005) 703 final, 'Further guidance on allocation plans for the 2008 to 2012 trading period of the EU Emission Trading Scheme', 39.

COM (2008) 0016, Committee on the Environment, Public Health and Food Safety to the European Parliament (2008), Draft report on the proposal for a directive of the European Parliament and of the Council amending Directive 2003/87/ EC so as to improve and extend greenhouse gas emission allowance trading system of the Community (COM (2008) 0016 – C6-0043/2008 – 2008/ 0013(COD)), 05.10.2008, 30.

Comanor, W. S. & Wilson, T. A., (1967) 'Advertising market structure and performance', *Review of Economics and Statistics*, Vol. 49, No. 4, 423–40.

Commission Decision (2007) regarding Belgium Nap, of January 2007 concerning the national allocation plan for the allocation of greenhouse gas emission allowances notified by Belgium in accordance with Directive 2003/87/EC of the European Parliament and of the Council, Brussels, 16.01.2007, available at: <http://ec.europa.eu/environment/climat/pdf/be_nap_decision_en. pdf>, 17.

Commission Decision 2006/780/EC, 2006, notified under document number C(2006) 5362, Commission Decision of 13 November 2006 on avoiding double counting of greenhouse gas emission reductions under the Community emissions trading scheme for project activities under the Kyoto Protocol pursuant to Directive 2003/87/EC of the European Parliament and of the Council.

Commission Decision 82/776/EEC, of 22 July 1982 on a Belgian Government aid scheme concerning the expansion of the production capacity of an undertaking manufacturing mineral water, hot spring water and soft drinks, 1982 OJ L 323/37.

Commission Decision 84/499/EEC, of 11 July 1984 on the Proposal by the Netherlands Government to grant aid for the building of a petrol additive production plant in the Rotterdam Europoort area, 1984 OJ L 276/43.

Commission Decision 88/468/EEC, of 29 March 1988 on aids granted by the French Government to a farm machinery manufacturer at St Dizier, Angers and Croix (International Harvester/Tenneco), 1988 OJ L 229/37.

Commission Decision 89/254/EEC, of 15 November 1988 relating to aid which the Belgian Government has granted to a petrochemicals company at Ottignies/ Louvain-la- Neuve (SA Belgian Shell), 1989 OJ L 106/34.

Commission Decision C(2004) 2515/2 final, of 7 July 2004 concerning the national allocation plan for the allocation of greenhouse gas emission allowances notified by Germany in accordance with Directive 2003/87/EC of the European Parliament and of the Council, Brussels, 07.07.2004.

Commission Decision (2007), of 16 January 2007 concerning the national allocation plan for the allocation of greenhouse gas emission allowances notified by The Netherlands in accordance with Directive 2003/87/EC of the European Parliament and of the Council, 19.

Commission Directive 2006/111/EC, of 16 November 2006 on the transparency of financial relations between Member States and public undertakings as well as on financial transparency within certain undertakings, OJ L 318, 17.11.2006, 17.

Commission Guidelines on National Regional Aid, (1998), OJ C 74/9 of 10.3.1998.

Commission Regulation (EC) No. 2216/2004 of 21 December 2004 for a standardised and secured system of registries pursuant to Directive 2003/87/EC of the European Parliament and of the Council and Decision No. 280/2004/EC of the European Parliament and of the Council, OJ L 386, 29.12.2004, 1.

Commission Regulation (EC) No. 69/2001 of 12 January 2001 on the application of Articles 87 and 88 of the EC Treaty to de minimis aid, OJ L 10, 13.1.2001, 30–32.

Commission Regulation (EC) 1998/2006 of 15 December on the application of Articles 87 and 88 of the EC Treaty to de minimis aid, OJ L 379, 28.12.2006, 5–10.

Community Guideline on State aid for environmental protection (2001/C 37/03), OJ C 37 of 03.02.2001, 3–15.

Community Guideline on State aid for environmental protection (2008/C 82/1), OJ C 82, 01.04.2008. 1–33.

Cooter, R. & Ulen, T., (2004) '*Law & Economics*', fourth edition, Pearson, Addison Wesley, Boston, 533.

Copeland, B. R. & Taylor, M. S., (2003) '*Free trade and global warming: a trade theory view of the Kyoto protocol*', Revised version: <http://econ.ucalgary.ca/fac-files/st/KyotoMarch13.pdf>.

Council Decision 2002/358/EC, (2002) '*Council Decision of 25 April 2002 concerning the approval, on behalf of the European Community, of the Kyoto Protocol to the United Nations Framework Convention on Climate Change and the joint fulfilment of commitments there under*', OJ L 130, 15.5.2002, 1–3.

Council Decision 2002/358/EC, (2002) '*Council Decision of 25 April 2002 concerning the approval, on behalf of the European Community, of the Kyoto Protocol to the United Nations Framework Convention on Climate Change and the joint fulfilment of commitments thereunder*' (OJ L 130, 15.5.2002, 1–3).

Council Decision 93/389/EEC, (1993) '*Council Decision of 24 June 1993 for a monitoring mechanism of Community CO_2 and other greenhouse gas emissions*' OJ L 167, 9.7.1993, 31. Decision as amended by Decision 1999/296/EC (OJ L 117, 5.5.1999, 35).

Council Decision 94/69/EC, (1993) *'Council Decision of 15 December 1993 concerning the conclusion of the United Nations Framework Convention on Climate Change'*, OJ EG L 033 of 07/02/1994, 011–012.

Council Decision 2006/513/EC, (2006) *'Council Decision amending Decision 1999/468/EC laying down the procedures for the exercise of implementing powers conferred on the Commission'*, OJ L 200/11. of 17/07/2006.

Council of the European Union (2002) Council Decision 2002/358/CE, *Official Journal of the European Communities*, L 130/1 of 15.05.2002.

Council of the European Union (2007) *'Presidency Conclusions of the European Council held in Brussels on 8/9 march 2007'*, 26.

Council of the European Union (2008) *'Presidency Conclusions of the European Council held in Brussles on 13/14 march 2008'*, 21.

Council of the European Union (2008b) *'Energy and Climate change – elements of a final compromise'* Document number 17215/08, Brusels, 12 December 2008, 18.

Council Regulation (EC) No. 659/1999 of March 1999 laying down detailed rules for the application of Article 92 of the EC Treaty, OJ L 83/1 of 27.03.1999.

Cowling, K. & Mueller, D., (1978) 'The social costs of monopoly power', *Economic Journal*, Vol. 88, December 1978, 727–48.

Craig, P. & De Burca, G., (2003) *'EU Law, Text, Cases and Materials'*, third edition, Oxford University Press, 1241.

Craig, P. & De Burca, G., (2008) *'EU Law, Text, Cases and Materials'*, fourth edition, Oxford University Press, 1148.

Cramton, P. & Kerr, S., (1999) *'Tradable Carbon Permit Auctions, How and why to auction not grandfather'*, Wharton, Financial Institutions Center, Implications of Auction Theory for New Issues Markets, No. 02–19, 1–19.

Cramton, P. & Schwartz, J., (2002) 'Collusive Bidding in the FCC Spectrum Auctions', *Contributions to Economic Analysis & Policy*, Vol. 1, No. 1, Article 11, 1–20.

Cullis, J. & Jones, P., (1998) *'Public Finance and Public Choice'*, second edition, Oxford University Press, 422.

Cyprus (2004) *'National Allocation Plan'*, submission of October 2004, <http://europa.eu.int/comm/environment/climat/pdf/cyprus_nap_en.pdf>, viewed on 04.02.2005.

Cyprus (2007) *'National Allocation Plan for Cyprus (revised final draft) 2008–2012 For Submission to the European Commission by 26th February 2007'*, available at: <http://ec.europa.eu/environment/climat/pdf/nap_cyprus.pdf>.

Dales, J. H., (1968) *'Pollution, Property & Prices'*, University of Toronto Press, reprinted in 1975, 111.

De Burca, G., (1993) 'The Principle of Proportionality and its Application in EC Law', *Yearbook of European Law*, Vol. 13, 105–150.

De Cendra de Larragán, J., (2005) *'EU Greenhouse Gas Emissions Trading and Legal Principles'*, written for the contract research: Emissions trading and equal competition, April 2005, Metro/Maastricht University, available at: <www.rechten.unimaas.nl/metro>, 41.

De Cendra de Larragán, J., (2006) 'Linking the project based mechanism with the EU ETS; the present state of affairs and challenges ahead', in Peeters, M. & Deketelaere, K., eds, *EU Climate Change Policy, The Challenge of New Regulatory initiatives*, Edward Elgar, 98–124.

De Mooij, R. A., (1999) 'The double dividend of an environmental tax reform', in van den Bergh, J. C. J. M., eds, *Handbook of environmental and resource economics*, Cheltenham, U.K., Northampton, MA. Edward Elgar, 293–306.

Debreu, G., (1959) 'Theory of Value: An Axiomatic Analysis of Economic Equilibrium', *Cowles Foundation for Research in Economics at Yale University*, 17, 114.

Decision 280/2004/EC of the European Parliament and of the Council of 11 February 2004 concerning a mechanism for monitoring Community greenhouse gas emissions and for implementing the Kyoto Protocol, OJ L 49, 19.2.2004, 1.

Defra, Enviros (2006) 'Appraisal of Years 1–4 of the UK Emissions Trading Scheme', available at: <www.defra.gov.uk/Environment/climatechange/ trading/uk/pdf/ukets1-4yr-appraisal.pdf>.

Defra (2009) Summary of Feedback on first UK auction in Phase II of the EU ETS System, January 2009, available at: <www.defra.gov.uk/environment/ climatechange/trading/eu/pdf/feedback-summary-160109.pdf>.

Demsetz, H., (1973) 'Industry structure, market rivalry and public policy', in: *Journal of Law and Economics*, Vol. 16, No. 1, April 1973, 1–9.

Demsetz, H., (1974) 'Two systems of belief about monopoly', in Harvey, J., Goldschmid, H., Mann, M., & Weston, J. F., eds, *Industrial Concentration: The New Learning*, 164–184, Boston: Little, Brown 1974.

Demsetz, H., (1976) 'More on collusion and advertising: a reply', in: *Journal of Law and Economics*, Vol. 19, No. 1, April 1976, 205–209.

Directive 2001/81/EC of the European Parliament and of the Council, of 23 October 2001 on national emission ceilings for certain atmospheric pollutants, OJ L 309/22 of 27.11.2001.

Directive 2003/87/EC of the European Parliament and of the Council, of 13 October 2003 establishing a scheme for greenhouse gas emission allowance trading within the Community and amending Council Directive 96/61/EC, OJ L 275/ 32 of 25.10.2003.

Directive 2004/101/EC of the European Parliament and of the Council of 27 October 2004 amending Directive 2003/87/EC establishing a scheme for greenhouse gas emission allowance trading within the Community, in respect of the Kyoto Protocol's project mechanisms, OJ L 338/18 of 13.11.2004.

Directive 2002/77/EC of 16 September 2002 on competition in the markets for electronic communication networks and services, OJ L 249, 17.09.2002, 21–26.

Directive 96/61/EC of the Council, of 24 September 1996 concerning integrated pollution prevention and control, OJ L 257/26 of 10.10.1996.

Document 9702/98 of 19 June 1998 from the Council of the European Union reflecting the outcome of proceedings of the Environment Council of 16–17 June 1998.

EC Commission, (1995) '*Notice on co-operation between national courts and the Commission in the State aid field*', OJ C 312/8.

EC Treaty (1993) '*Consolidated Version of the Treaty Establishing the European Community, as amended in accordance with the Treaty of Nice Consolidated Version (OJ 2002 C 325/1-184) and the 2003 Accession Treaty (OJ 2003 L236/17)*'.

ECCP, (2001) '*Long Report*', June 2001, 172 available at <http://ec.europa.eu/environment/climat/pdf/eccp_longreport_0106.pdf>.

Edmonds, J., Wise, M. & Barns, D. W., (1995) 'Carbon Coalitions: The Cost and Effectiveness of Energy Agreements to Alter Trajectories of Atmospheric Carbon Dioxide Emissions', *Energy Policy*, Vol. 23 (4/5), 309–335.

Edward, D. & Hoskins, M., (1995) 'Article 90: Deregulation and EC Law. Reflections arising from the XVI FIDE Conference', *Common Market Law Review*, Vol. 32, 157–186.

Elbers, C. & Withagen, C., (2004) 'Environmental policy, population dynamics and agglomeration', *Contributions to Economic Analysis & Policy*, Vol. 3, 1–21.

Ethier, W. J., (2002) 'Globalization : trade, technology and wages, in International Trade and Factor Mobility', PIER Working Paper 02-031, 34.

EU ETS amendment, Directive 2009/29/EC of the European parliament and of the Council of 23 April 2009 amending Directive 2003/87/EC so as to improve and extend the greenhouse gas emission allowance trading scheme of the Community, (L 140/63) of 5.6.2009.

European Commission (1997) '*Commission Communication on State aid elements in sales of land and buildings by public authorities*' (97/C 209/03), OJ C 209 of 10/07/1997, 3–5.

European Comission (2003) '*Systeem van verhandelbare emissierechten voor Nox*', C(2003)1761fin, 24.06.2003, 13.

European Commission (2003) '*Steunmaatregelen van de Staten N 35/2003 – Nederland*', C(2003) 1761 fin, 24.06.2003.

European Commission (2004a) '*EU Emissions trading, An open scheme promoting global innovation to combat climate change*', 24. available at <http://ec.europa.eu/environment/climat/pdf/emissions_trading_en.pdf>.

European Commission (2004b) '*State Aid and National Allocation Plans*' letter of the Directorate Generals Environment and Competition to the Member States, ENV C2/PV/amh/D(2004)420149, 17.

European Commission (2006) '*Questions and Answers on Emissions Trading and National Allocation Plans for 2008 to 2012*', MEMO/06/452 of 26.11.2006, 9.

European Commission (2006) MEMO/06/452, '*Questions and Answers on Emissions Trading and National Allocation Plans for 2008 to 2012*', Brussels, 29.11.2006.

European Commission (2007) '*Community Guidelines For State Aid For Environmental Protection, Staff paper, preliminary draft*', 55, available at: <http://ec.europa.eu/comm/competition/state_aid/reform/guidelines_environment_en.pdf>.

European Commission, (2008) 'Auctioning and the carbon market: Non-paper of the Commission services' 1–5.

European Commission, (2000) '*Statsstøttesag Nr. N 653/1999 – CO₂-kvoter*', SG (2000) D/, 12.04.2000.

European Commission, (2001) '*State aid No N 416/2001 United Kingdom Emission Trading Scheme*', C(2001) 3739 fin, 28.11.2001.

European Commission, (2001) '*Steunmaatregel nr. N 550/2000 België Groenestroomcertificaten*', SG(2001) D/290545, 25.07.2001.

European Parliament legislative resolution of 17 December 2008 on the proposal for a directive fo the European Parliament and of the Council amending Directive 2003/87/EC so as to improve and extend the greenhouse gas emission allowance trading system of the Community, COM (2008)16.

Evans, A., (1997) '*European Community Law of State aid*', Clarendon Press, Oxford, 484.

Evans & Peck (2007) 'National Emissions Trading Taskforce, Possible Design for a Greenhouse Gas Trading System', 74.

Fairhurst J. & Vincenzi C., (2003) '*Law of the European Community*', fourth edition, Pearson Longman publishers, 527.

Fankhauser, S., Tol, R. S. J. & Pearce, D. W., (1997), 'The Aggregation of Climate Change Damages: A Welfare Theoretic Approach', *Environmental and Resource Economics*, Vol. 10, 249–266.

Farnsworth, N., (2005) 'Emissions Trading and State aid rules', in van Calster and Deketelaere, K., eds, *Energy and Environmental Law – 2005*, Leuven, Acco, 295–337.

Faure, M. & Skogh, G., (2003) '*The Economic Analysis of Environmental Policy and Law, an Introduction*', Edward Elgar Publishing, 354.

Fisher, F. M. & McGowan, J. J., (1983) 'On the misuse of accounting rates of return to infer monopoly profits', *American Economic Review*, Vol. 73, No. 1, March 1983, 82–97.

Feinstein, J., Block, M. & Nold, F., (1985) 'Asymmetric Information and Collusive Behaviour in Auction Markets', *The American Economic Review*, Vol. 75, No. 3, 441–460.

Frank, R., (1997) '*Microeconomics and Behaviour*', third edition, Irwin McGraw-Hill, Boston, 744.

Frenz, W., (2007) '*Handbuch Europarecht, Beihilfe- und Vergaberecht*', third edition, Springer, 1130.

Friedman, M., (1953) 'The Methodology of Positive Economics', in Milton Friedman, *Essays in Positive Economics*, (Chicago: University of Chicago Press) 3–43.

Fujita, M., Krugman, P. & Venables, A., (1999), '*The Spatial Economy: cities, regions and international trade*', Cambridge: the MIT Press, 367.

G8 Summit Declaration (2007), 'Growth and Responsibility in the World Economy, Summit Declaration of 7th June 2007', *G8 Summit 2007 Heiligendamm*, available at: <www.g-8.de/Webs/G8/DE/Homepage/home.html>, 37.

Gayer, T., (2005) 'Auctioning Pollution Rights', *American Enterprise Institute for Public Policy Research*, AEI Print Index No. 17877, <www.aei.org/news21838/>, viewed on 26.01.2005.

Germany (2004) *'National Allocation Plan for the Federal Republic of Germany 2005–2007'*, Federal Ministry for the Environment, Nature Conservation and Nuclear Safety, Berlin, 31 March 2004, Translation of 07 May 2004, available at: <www.bmu.de/files/pdfs/allgemein/application/pdf/nap_kabi_en.pdf>.

Gestel, R. A. J., Van Backes, Ch. W. & Teuben, R., (2002) 'Wetgeving voor de handel in emissiekredieten. Balanceren tussen Europese eisen en nationale verlangens', *RegelMaat*, Vol. 17, No. 5, 162–174.

Gielen, A. M., Koutstaal, P. R. & Vollebergh, H. R., (2002) *'Comparing Emission Trading with Absolute and Relative Targets'*, Paper presented at the 2nd CATEP Workshop on the Design and Integration of National Tradable Permit Schemes for Environmental Protection, hosted by University College London, 25–26 March 2002, available at: <www.ucd.ie/envinst/envstud/CATEP%20Webpage/Papers/Koustaal.pdf>.

Gielen, D. & Karbuz, S., (2003) *'Challenges in energy and environment modeling: A materials perspective'*, EET/2003/05 Paris, OECD/IEA, available at: <www.iea.org/textbase/papers/2003/eet05.pdf>.

Golfinopoulos, C., (2003) 'Concept of Selectivity criterion in State aid definition following the Adria-Wien Judgement – Measures Justified by the "Nature or General scheme of a System"', European Competition Law Review, Vol. 24, No. 10, 543–549.

Gordon, R. H. & Bovenberg, A. L., (1996) 'Why is capital so immobile internationally? Possible explanations and implications for capital income taxation' *The American Economic Review*, Vol. 86, No. 5, 1057–1075.

Greece (2004) *'National Allocation Plan for the period 2005–2007, December 2004'*, available at: <http://ec.europa.eu/environment/climat/pdf/greece_en.pdf>.

Green, A., (2005) 'Climate Change, regulatory policy and the WTO', *Journal of International Economic Law*, 143–189.

Grimeaud, D., (2003) 'To design and implement climate change measures and the need to strike a balance between environmental protection and international trade law', in Faure, M., Gupta, J. & Nentjes, A., eds, *Climate Change and the Kyoto Protocol, The Role of Institutions and Instruments to Control Global Change*, Edward Elgar, Cheltenham, 68–107.

Grimm, V., Riedel, F. & Wolfstetter, E., (2001) *'Low Price Equilibrium in Multi-Unit Auctions: The GSM Spectrum Auction in Germany'*, Institut für Wirtschaftstheorie I, Humboldt Universität zu Berlin, 1–14.

Gürtzen, N. & Rauscher, M., (2000) 'Environmental policy, intra-industry trade and transfrontier pollution', *Environmental and Resource Economics*, Vol. 17, 59–71.

Gyselen, L., (1989) 'State Action and the Effectiveness of the EEC Treaty's competition provisions', *Common Market Law Review*, Vol. 26, 33–66.

Harbord D. & Hoehn, T., (1994) 'Barriers to Entry and Exit in European Competition Policy, *International Review of Law and Economics*, Vol. 14, 411–435.

Harrision, D. & Radov, D., (2002) *'Evaluation of Alternative Initial Allocation Mechanisms in a European Union Greenhouse Gas Emissions Allowance Trading Scheme'*, NERA, 168.

Hay, D. & Morris D., (1991) *'Industrial Economics and Organization, Theory and Evidence'*, second edition, Oxford University Press, 686.

Heal, G. & Kriström, B., (2002) 'Uncertainty and Climate Change', *Environmental and Resource Economics*, Vol. 22, 3–39.

Helm, C., (2003) 'International emissions trading with endogenous allowance choices', *Journal of Public Economics*, Vol. 87, 2737–2747.

Hepburn, C., Grubb, M., Neuhoff, K., Matthes, F. & Tse, M. (2006) 'Auctioning of EU ETS phase II allowances: how and why?' Climate Policy, Vol. 6, 137–160.

Hildebrand, D., (2002) *'The Role of Economic Analysis in the EC Competition Rules'*, European Monographs, second edition, Kluwer Law International, The Hague, 472.

IP/07/1094 of 13/07/2007, *'Emissions trading: Commission adopts decisions on amendments to five national allocation plans for 2008–2012'*, press release of the European Commission.

IP/05/762 of 20/06/2005, *'Emissions trading: Commission approves last allocation plan ending NAP marathon'*, press release of the European Commission.

IPCC (2001) *'Climate Change 2001: Impacts, adaptation and vulnerability'*, third assessment report, Cambridge University Press, available at: <www.grida.no/climate/ipcc_tar/>.

IPCC (2007a) *'Climate Change 2007: The Physical Science Basis'*, Working Group I Contribution to the Intergovernmental Panel on Climate Change Fourth Assessment Report, summary for policy makers, pp. 18, available at <http://ipcc-wg1.ucar.edu/wg1/Report/AR4WG1_Pub_SPM-v2.pdf>.

IPCC (2007b) *'Climate Change 2007: Impacts, adaptation and vulnerability'*, Working Group II Contribution to the Intergovernmental Panel on Climate Change Fourth Assessment Report, summary for policy makers, pp. 23, available at: <www.ipcc.ch/SPM13apr07.pdf>.

IPCC (2007c) *'Climate Change 2007: Mitigation of Climate Change'*, Working Group II Contribution to the Intergovernmental Panel on Climate Change Fourth Assessment Report, summary for policy makers, pp. 35, available at: <www.ipcc.ch/SPM040507.pdf>.

Jensen, J. & Rasmussen, T. N., (2000) 'Allocation of CO_2 Emission Permits: A General Equilibrium Analysis of Policy Instruments', *Journal of Environmental Economics and Management*, Vol. 40, No. 2, September 2000, 111–136.

Jeppesen, T. & Folmer, H., (2001) 'The confusing relationship between environmental policy and location behaviour of firms: a methodological review of selected case studies', *Annals of Regional Science*, Vol. 35, 523–546.

Jeppesen, T., List, J. & Folmer, H., (2002) 'Environmental regulations and new plant location decisions: evidence from a meta analysis', *Journal of Regional Science*, Vol. 42, 19–42.

Jones, A. & Sufrin, B., (2004) '*EC Competition law: Text, Cases, and Material*', second edition, Oxford University Press, 1303.

Jones, R. W. & Kierzkowski, H., (2001) 'A framework for fragmentation', in Arndt S. W. & Kierzkowski K., eds, *Fragmentation: New Production Patterns in the World Economy*, Oxford, Oxford University Press, 272.

Kagel, J. H. & Levin, D., (2001) 'Behavior in Multi-Unit Demand Auctions: Experiments with Uniform Price and Dynamic Vickrey Auctions', *Econometrica*, Vol. 69, No. 2, 413–454.

Kaldor, N., (1934), 'A Classification Note on the Determinateness of Equilibrium', *The Review of Economic Studies*, Vol. 1, No. 2, 122–136.

Katz, M. L. & Rosen, H. S., (1998) 'Microeconomics', third edition, Irwin/McGraw-Hill, 656.

Kerr, S. & Newell, G., (2003) 'Policy-Induced Technology Adoption: Evidence form the U.S. Lead Phasedown', *The Journal of Industrial Economics*, Vol. LI, No. 3, 317–343.

Kortmann, J. S., (2005) 'Handel in emissierechten: het causale stelsel ten onrechte "uitgesloten"?', *Weekblad voor Privaatrecht, Notariaat en Registratie*, Year 136, Number 6631, 27 August 2005, 625–627.

Koster, M. H., (2005) 'Handel in emissierechten: het causale stelsel "uitgesloten"?', *Weekblad voor Privaatrecht, Notariaat en Registratie*, Year 136, Number 6617, 9 April 2005, 301–305.

Krämer, L., (2003) '*EC Environmental Law*', fifth edition, London, Sweet & Maxwell.

Krishna, V., (2002) '*Auction Theory*', Academic Press, San Diego, 303.

Krouse, C., (1990) '*Theory of Industrial Economics*', Basil Blackwell Publishers, Oxford, UK, 602.

Kruger, J. & Pizer, W. A., (2004) 'The EU Emissions Trading Directive: Opportunities and Potential Pitfalls', Resources for the Future, *Discussion Paper*, No. 04 24, April 2004, 1–59.

Krugman, P. & Obstfeld, M., (1997) '*International Economics, Theory and Policy*', fourth edition, Addison Wesley, 766.

Kuik, O., (2005) '*Climate change policies, international trade and carbon leakage: An applied general equilibrium analysis*', dissertation, Vrije Universiteit Amsterdam, 192.

Kuik, O., Tol, R. S. J. & Grimeaud, D., (2003) 'Linkages between the climate change regime and the international trade regime', in van Ierland, E. C., Gupa, J. & Kok, M. T. J., eds, *Issues in International Climate Policy: Theory and Policy*, Edward Elgar, Cheltenham, 201–222.

Kyoto Protocol to the United Nations Framework Convention on Climate Change, United Nations, 1998, available at <http://unfccc.int/resource/docs/convkp/kpeng.pdf>.

Langrock, T. & Sterk, W., (2004) '*Linking CDM & JI with EU Emission Allowance Trading, Policy Brief for the EP Environment Committee*', EP/IV/A/2003/09/01, Brief number 01/2004, 12th January 2004, 16.

Lecocq, R. & Crassous, R., (2003) 'International Climate Regime beyond 2012, Are Quota Allocation Rules Robust to Uncertainty', *Policy Research Working Paper*, No. 3000, The World Bank Development Research Group, Infrastructure and Environment, March 2003, 1–39.

List, J., McHone, W. W. & Millimet, D. L., (2003) 'Effects of air quality regulation on the destination choice of relocating plants', *Oxford Economic Papers*, Vol. 55, 657–678.

Lyon, R. M., (1982) 'Auctions and alternative procedures for allocating pollution rights', *Land Economics*, Vol. 58, No. 1, 16–32.

Maden, P., (1978), 'Why the Edgeworth Process Assumption Isn't that Bad', *The Review of Economic Studies*, Vol. 45, No. 2, 279–283.

Maeda, A., (2003) 'The Emergence of Market Power in Emission Rights Markets: The Role of Initial Permit Distribution', *Journal of Regulatory Economics*, Vol. 24:3, 293–314.

Maks, H. & van Daal, J., (2002) *'Léon Walras: What cutes know and what they should know'*, La Chapelle en Vercors, Maastricht, 23.

Malta (2004) *'National Allocation Plan'*, submission of 18 October 2004, <http://europa.eu.int/comm/environment/climat/pdf/malta.pdf>, viewed on 04.02.2005.

Malta (2006), *'National Allocation Plan for Malta'*, September 2006, available at: <http://ec.europa.eu/environment/climat/pdf/nap_malta_final.pdf>.

Mamuth, H. A., (1993) *'Markteconomie, Analyse en Evaluatie'*, fourth edition, Utrecht.

Mamuth, H. A., (1992), Markteconomie, Analyse en evluatie, third edition, 585.

Manelli, A. M., Sefton, M. & Wilner, B. S., (1999) *'Multi-Unit Auctions: a Comparison of Strategic and Dynamic Mechanisms'*, 1–20, available at <http://wpcarey.asu.edu/tools/mytools/pubs_admin/FILES/wp99_9.pdf>.

Markusen, J. R., (1975). 'International externalities and optimal tax structures', *Journal of International Economics*, Vol. 5, 15–29.

Marrakech Accords, (2001) *'The Marrakesh Accords & The Marrakesh Declaration'*, adopted by the Conference of the Parties at its seventh session', November 2001, unedited form, 245.

Martin, S., (1993) *'Advanced Industrial Economics'*, Blackwell Publishers, Oxford UK, 660.

Martin, S., (1994) *'Industrial Economics, Economic Analysis and Public Policy'*, second edition, Prentice Hall, 623.

Martin, S., (2002) *'Advanced Industrial Economics'*, second edition, Blackwell Publishers, Amsterdam, 533.

Mason, E., (1939) *'Price and production policies of large-scale enterprise, American Concentration and the Monopoly Problem'*, Cambridge, MA: Harvard University Press.

McAfee, R. & McMillan, J., (1987a) 'Auctions and Bidding', *Journal of Economic Literature*, Vol. 25, 699–738.

McAfee, R. & Vincent, D., (1993) 'The Declining Price Anomaly', *Journal of Economic Theory*, Vol. 60, 191–212.

Merola, M. & Crichlow, G., (2004) 'State aid in the Framework of the EU Position after Kyoto: An Analysis of Allowances Granted under CO_2 Emissions Allowance Trading Directive', *World Competition*, Vol. 27, No. 1, 25–51.

Milgrom, P., (2004) *'Putting Auction Theory to Work'*, Cambridge University Press, 368.

Morgan, T., (1988) 'Theory versus Empiricism in academic Economics', in *Journal of Economic Perspectives* 2 (4), 159–164.

Mundell, R. A., (1957) 'International trade and factor mobility', *American Economic Review*, Vol. 47, 321–335.

Munskgaard, J. & Pedersen, K., (2001) 'CO_2 Accounts for Open Economies: Producer or Consumer Responsibility', *Energy Policy*, Vol. 29, No. 4, 327–334.

Nash, J. R, (2000) 'Too much market? Conflict between tradable pollution allowances and the "polluter pays principle" ', *Harvard Environmental Law Review*, Vol. 24, 465–535.

Neergaard, U., (1998) *'Competition & Competences, The Tension between European Competition law and Anti-Competitive Measures by the Member States'*, DJØF Publishing Copenhagen, 358.

Neugebauer, T. & Pezanis-Christou, P., (2005) *'Bidding Behavior at Sequential First-Price Auctions with(out) Supply Uncertainty: A Laboratory Analysis'*, 13 January 2005, 23, republished in *Journal of Economic Behavior & Organization*, 2007, Vol. 63, No. 1, pages 55–72.

OECD, (1974) *'Recommendation of the Council on the Implementation of the Polluter-Pays Principle'*, 14 November 1974 – C(74)223. <http://webdomino1. oecd.org/horizontal/oecdacts.nsf/Display/9DFDD7AF6065709CC1256FA3004 D0413?OpenDocument>, viewed on 04.02.2005.

OECD, (2005a) 'Exploring options for sectoral crediting mechanisms', COM/ ENV/EPOC/IEA/SLT(2005)1.

OECD, (2005b) 'Sectoral crediting mechanisms: an initial assessment of electricity and aluminium', COM/ENV/EPOC/IEA/SLT(2005)8.

OECD, (2006) 'Sectoral crediting mechanisms for greenhouse gas mitigation: institutional and operational issues', COM/ENV/EPOC/IEA/SLT(2006)4.

Pagoulatos E. & Sorensen R., (1976) 'Foreign trade, concentration, and profitability in open economies', *European Economic Review*, Vol. 8, 1976, 255–267.

Pallemaerts, M., (2004) 'De Europese Gemeenschap als Verdragsluitende Partij bij het Protocol van Kyoto', *Tijdschrift voor Milieurecht*, 122–136.

Pallemaerts, M., & Williams, R., (2006) 'Climate change: The international and European policy framework', in Peeters, M. & Deketelaere, K., eds, *EU Climate Change Policy, The Challenge of New Regulatory initiatives*, Edward Elgar, 22–50.

Paltsev, S. V., (2001) 'The Kyoto Protocol: regional and sectoral contributions to the carbon leakage', *The Energy Journal*, Vol. 22, 53–79.

Parry, I., Williams III, R. & Goulder, L. (1999) 'When Can Carbon Abatement Policies Increase Welfare? The Fundamental Role of Distorted Factor Markets', *Journal of Environmental Economics and Management*, Vol. 37, 52–84.

Peck, S. C. & Teisberg, T. J., (1994) 'Optimal Carbon Emissions Trajectories When Damages Depend on the Rate or Level of Global Warming', *Climatic Change*, Vol. 28, 289–314.

Peepercorn, L. & Verouden V., (2007) 'The Economics of Competition', in Faull, J., & Nikpay, A., eds, *The EC Law of Competition*, Oxford University Press, 3–86.

Peeters, M., (1993) 'Towards a European Market for Tradable Pollution Permits?' *Tilburg Foreign Law Review*, Vol. 2, 117–134.

Peeters, M., (2003) 'Internationale klimaatafspraken en Europese emissiehandel. Een onwaarschijnlijk kort implementatietraject voor een niew fenomeen', Nederlands Tijdschrift voor Europees Recht, Nr. 10, 280–286.

Peeters, M., (2006) 'Enforcement of the EU greenhouse gas emissions trading scheme' in: Peeters, M. & Deketelaere, K., eds, *EU Climate Change Policy, The Challenge of New Regulatory initiatives*, Edward Elgar, 169–187.

Peeters, M., De Cendra de Largán, J. & Weishaar, S., (2007) 'Perspectives on the fundamental choice between "cap and trade" and "credit and trade"', *European Environmental Law Review*, Vol. 16, No. 7 (forthcoming).

Peeters, M. & Weishaar, S., (2009) 'Exploring uncertainties, The "Learning by Doing" period has not come to an end, and will still be at the heart of the EU ETS' Carbon Climate Law Review, No. 1, 88–101.

Peltzman, S., (1977) 'The Gains and Losses from Industrial Concentration', *Journal of Law and Economics* 20, Chicago: University of Chicago Press, October 1977, 229–263.

Peltzman, S., (1979) 'The Causes and Consequences of Rising Industrial Concentration: a Reply', *Journal of Law and Economics*, Vol. 22, Chicago, University of Chicago Press, April 1979, 209–211.

Pesendorfer, M., (2000) 'A Study of Collusion in First-Price Auctions', *The Review of Economic Studies*, Vol. 67, No. 3, 381–411.

Peterson, S., (2003) 'Monitoring, Accounting and Enforcement in Emissions Trading Regimes'. Paper for the OECD Global Forum on Sustainable Development: Concerted Action on Tradable Emissions Permits Country Forum, OECD Paris 17–18 March 2003, 21, available at: <www.oecd.org/dataoecd/11/56/2957646.pdf>.

Pethig, R., (1976) 'Pollution, welfare, the environmental policy in the theory of comparative advantage' *Journal of Environmental Economics and Management*, Vol. 2. 160–169.

Pezzy, J., (1992) 'Analysis of Unilateral CO_2 Control in the European Community and OECD', *The Energy Journal*, Vol. 13, 159–171.

Philipsen, N. J., (2003) '*Regulation of and by Pharmacists in the Netherlands and Belgium, an Economic Approach*', Intersatia 195.

Pigou, A. C., (1949) '*A Study in Public Finance*', third (revised) edition (first published in 1928), Macmillan & Co., London, 285.

Poland (2004) '*National Allocation Plan for CO_2 Emission Allowances 2005–2007 Trading Period*', Warsaw 2004, unofficial translation, available at: <www.mos.gov.pl/she/prace_nad_kpru/NAP_2005-2007.pdf>.

Porter, H. & Zona, D., (1997) 'Ohio School Milk Markets: an Analysis of Bidding', *National Bureau of Economic Research, Working Paper*, No. 6037, 1–33.

Porter, R. & Zona, D., (1993) 'Detection of Bid Rigging in Procurement Auctions', *The Journal of Political Economy*, Vol. 101, No. 3, 518–538.

Portugal (2004) '*National Allocation Plan For CO$_2$ Emissions Allowances (NAP) 2005–2007*', Version to European Commision 04 May, 2004, available at: <www.cele.pt/documentos/Portuguese_NAP.pdf>.

Pring, G., (2006) 'A decade of emissions trading in the USA: Experiences and observations for the EU', in Peeters, M. & Deketelaere, K., eds, *EU Climate Change Policy, The Challenge of New Regulatory initiatives*, Edward Elgar, 188–204.

Quigley, C. & Collins, A., (2003) '*EC State aid Law and Policy*', Hart Publishing, Oxford – Portland Oregon, 394.

Rauscher, M., (1997) '*International Trade, Factor Movements, and the Environment*', Oxford, Clarendon Press, 335.

Reader, M., (1982) 'Chicago economics: permanence and change', *Journal of Economic Literature*, Vol. 20, No. 1, March 1982, 1–38.

Reich, N., (1994) 'The "November Revolution" of the European Court of Justice: Keck, Meng and Audi Revisited', *Common Market Law Review*, Vol. 31, 459–492.

Reinaud, J., (2005) 'Industrial Competitiveness under the European Union Emissions Trading Scheme', *International Energy Agency Information Paper*, February 2005, 91.

Rittaler, J. B. & Schmidt, I. L. O., (1990) '*A Critical Evaluation of the Chicago School of Antitrust Analysis*', Kluwer Academic Publishers, 145.

Rothkopf, M. & Harstad, R., (1994) 'Modeling Competitive Bidding: A Critical Essay', *Management Science*, Vol. 40, No. 3, 364–384.

Salmon, T., (2003) '*Preventing Collusion between Firms in Auctions*', Department of Economics (5th February), Florida State University, 1–25.

Scherer, F. M., (1980) '*Industrial Market Structure and Economic Performance*', second edition, Chicago: Rand McNally, 632.

Scherer, F., (1979) 'The Causes and Consequences of Rising Industrial Concentration', *Journal of Law and Economics*, Vol. 22, Chicago, University of Chicago Press, April 1979, 191–208.

Schinkel, M. – P., (2001), '*Disequilibrium Theory, Reflections towards a Revival of Learning*' University Pers Maastricht, dissertation, 272.

Schipper, E. L. F. & Boyd, E., (2006) 'UNFCCC COP 11 and COP/MOP 1, At Last, Some Hope?', *The Journal of Environment & Development*, Vol. 15, 75–90.

Schmitt-Rady, B., (2006) 'A level playing field? Initial allocation of allowances in Member States' implementing the Emissions Trading Directive: Issues of Member State liability', in Peeters, M. & Deketelaere, K., eds, *EU Climate Change Policy, The Challenge of New Regulatory initiatives*, Edward Elgar, 83–97.

Sectorakkoord Energie 2008–2020 (2008) 'Sectorakkoord Energie 2008–2020, Convenant tussen Rijksoverheid en Energiebranches in het kader van het werkprogramma Schoon en Zuinig', Final verision of 28 October 2008 available at: <www.ez.nl/dsresource?objectid=160723&type=PDF>.

Shubik, M., (1983) 'Auctions, bidding, and markets: An historical sketch', in Engelbrecht-Wiggans, R., Shubik, M. & Stark, J., eds, *Auctions, Bidding, and Contracting*, New York University Press, New York, 33–52.

Shughart II, W. F., (1990) *'The Organisation of Industry'*, Irwin, Inc., 492.

Slovak Republic (2004), *'Draft National Allocation Plan for 2005–2007 for the Directive 2003/87/EC establishing a scheme for greenhouse gas emission allowance trading within the Community and amending Council Directive 96/61/EC'*, June 2004, available at <http://ec.europa.eu/environment/climat/first_phase_ep.htm>.

Smith, A., (1937), *'An Inquiry into the Nature and Causes of the Wealth of Nations'*, Edwin Cannan, ed. New York.

Smith, S. & Yates, A., (2003) 'Optimal pollution permit endowments in markets with endogenous emissions', *Journal of Environmental Economics and Management*, Vol. 46, 425–445.

Springer, K., (2000) 'Do we have to consider international capital mobility in trade models?', *Kiel Working Paper*, No. 964, Kiel, Kiel Institute of World Economics, 49.

Staatsblad van het Koninkrijk der Nederlanden (2005a), *'Besluit van 12 mei 2005, houdende vaststelling van het tijdstip van inwerkingtreding van de wet van 28 april 2005 tot wijziging van de Wet milieubeheer en de Wet op de economische delicten ten behoeve van de invoering van een systeem van handel in emissierechten met het oog op het beperken van de emissies van stikstofoxiden (handel in NOx-emissierechten) (Stb. 233)'*, Nr. 248, 3.

Staatsblad van het Koninkrijk der Nederlanden (2005b), *'Wet van 28 april 2005 tot wijziging van de Wet milieubeheer en de Wet op de economische delicten ten behoeve van de invoering van een systeem van handel in emissierechten met het oog op het beperken van de emissies van stikstofoxiden (handel in NOx-emissierechten)'*, Nr. 233, 12.

Stavins, R. N., (2001) 'Lessons From the American Experiment With Market-Based Environmental Policies', Resources for the Future, *Discussion Paper*, No. 01-53, November 2001, 21.

Stavins, R. N., (2003) 'Experience with Market-Based Environmental Policy Instruments', in *Handbook of Environmental Economics, Environmental degradation and institutional responses*, Vol. I, eds, Karl-Göran Mäler & Jeffrey Vincent, Chapter 9, Amsterdam, Elsevier Science, 355–435.

Stavins, R., N., (1997) 'What can we learn from the grand policy experiment? Positive and normative lessons of the SO2 allowance trading', *Journal of Economic Perspectives*, Vol. 12, 68–88.

STEM (2005), *'De Verdeling van broeikasgasemisierechten in de EU bezien in het licht van vonvurrentieverhoudingen'*, Structurele Evaluatie Milieuwetgeving, Oosterhuis, F., Peeters, M., 1110643/CE6/092/000360, 88.

Stigler, G. J., (1968) *'The Organization of Industry'*, Homewood, Ill., R.D. Irwin, 328.

The Netherlands (2004a) '*National Allocation Plan*', submission of 16 April 2004, <www.novem.nl/default.asp?documentId=114203>, viewed on 04.02.2005.

The Netherlands (2004b) '*Allocatieplan CO_2-emissierechten 2005 t/m 2007, Nederlands nationaal toewijzingsplan inzake de toewijzing van broeikasgasemissierechten aan bedrijven, Bijlage E: samenvatting convenanten energie efficiency*', <www.novem.nl/default.asp?documentId=113926>, viewed on 08.02.2005, 1–5.

The Netherlands (2006) '*Netherlands national allocation plan for greenhouse gas allowances 2008–2012, Plan of the Minister for Economic Affairs and the State Secretary for Housing, Spatial Planning and the Environment*' of 26 September 2006, 65.

Tietenberg, T., (1994) '*Environmental Economics and Policy*', Harper Collins College Publishers, 432.

Tietenberg, T., (2002) 'The tradable permit approach to protecting the commons: What have we learned?' *FEEM working paper*, No. 36.2002, 1–40.

Tietenberg, T., Grubb, M., Michaelowa, A., Swift, B. & Zhang, Z. X., (1999) '*International Rules for Greenhouse Gas Emissions Trading: Defining the Principles, Modalities, Rules and Guidelines for Verification, Reporting and Accountability*', UNCTAD/GDS/GFSB/Misc.6, Geneva: United Nations Conference on Trade and Development (UNCTAD), 125.

Tol, R. S. J., (1999a) 'The marginal costs of greenhouse gas emissions', *The Energy Journal*, Vol. 20, 61–81.

Tol, R. S. J., (1999b) 'Spatial and Temporal Efficiency in Climate Change: Applications of FUND', *Environmental and Resource Economics*, Vol. 14, No. 1, 33–49.

Tol, R. S. J., Downing, T. E., Kuik, O. J. & Smith, J. B., (2004) 'Distributional aspects of climate change impacts', Global Environmental Change, Vol. 14, 259–272.

Tosics, N. & Gaál, N., (2007) 'Public procurement and State aid control – the issue of economic advantage', *Competition Policy Newsletter*, 2007, Number 3.

Tweede Kamer der Staten-Generaal (2004) '*Wijziging van de Wet Milieubeheer en de Wet op de economische delicten in verband met de invoering van een systeem van handel in emissierechten met het oog op het beperken van de emissies van stikstofoxiden (handel in Nox-emissierechten)*', Vergaderjaar 2004–2005, 29766 Nr. 3.

UN (1992) Rio Declaration on Environment and Development, United Nations Conference on Environment and Development, U.N. Doc. A/CONF.151.

UNEP, UNCTAD (2002) '*An emerging market for the environment: A Guide to Emissions Trading*', United Nations Publication, 41.

Uzawa, H., (1962), 'On the Stability of Edgeworth's Barter Process', *International Economic Review*, Vol. 3, No. 2, 218–232.

Van Calster, G., (2006) 'Climate change taxes, emissions trading, and international trade law', in Peeters, M. & Deketelaere, K., eds, *EU Climate Change Policy, The Challenge of New Regulatory initiatives*, Edward Elgar, 205–219.

Van den Bossche, P., (2005) '*The Law and Policy of the World Trade Organization, Text, Cases and Materials*', Cambridge University Press, 865.

Van der Kolk Advies (2006) '*Voorevaluatie NOx emissiehandel, In opdracht van het ministerie van VROM*', Van der Kolk Advies, KPMG Sustainability, DHV, Hofland Milieu Consultant, Juli 2006, available at: <www.vrom.nl/get.asp?file=docs/publicaties/w751.pdf&dn=w751&b=vrom>, viewed on 30.06.2007, 89.

Van der Laan, R. & Nentjes, A., (2001) 'Competitive Distortions in EU Environmental Legislation: Inefficiency versus Inequity', *European Journal of Law and Economics*, Vol. 11, No. 2, 131–152.

Van Tol, I. & Oldenziel, H., (2002) 'NOx-emissie-handel in juridisch perspectief', *milieu&recht*, Vol. 29, juli/augustus, number 7/8, 181–188.

Van Tol, I. & Oldenziel, H., (2006) 'Verhandelbare NOx-emissierechten in de Wet milieubeheer', *milieu&recht*, Vol. 33, No. 4, 206–215.

Vattenfall's newsletter on the CO_2 free power plant project, No. 3, November 2005.

Verhoef, E. T, (1999) 'Externalities', in van den Bergh, J. C. J. M., eds, *Handbook of environmental and resource economics*, Cheltenham, U.K., Northampton, MA. Edward Elgar, 197–214.

Vickrey, W., (1961) 'Counter-speculation, Auctions, and Competitive Sealed Tenders', *The Journal of Finance*, Vol. 16, No. 1, 8–37.

Viscusi, K., Vernon, J. & Harrington, J., (1995) '*Economics of Regulation and Antitrust*', second edition, MIT Press Massachusetts, 890.

Vossestein A., J., (2002), 'Case C-35/99, Arduino, judgment of 19 February 2000, Full Court; Case C-309/99, Wouters et al. v. Algemene Raad van de Nederlandse Orde van Advocaten, judgment of 19 February 2002, Full Court, not yet reported', *Common Market Law Review*, Vol. 39, 841–863.

VROM, (2003) '*Erop of eronder, Uitvoeringsnotitie emissieplafonds verzuring en grootschalige luchtverontreining*', <www.vrom.nl/get.asp?file=docs/milieu/ uitvoeringsnotitie_emissieplafonds_dec2003pdf.pdf>, viewed on 08.02.2005, 41.

VROM, (2005) '*Memorie van Toelichting, Wijziging van de Wet milieubeheer en de Wet op de economische delicten in verband met de invoering van een ssysteem van handel in emissierechten met het oog op het beperken van de emissies van stikstofoxiden (handel in NOx-emissierechten)*', 76, available at: <www.vrom.nl/get.asp?file=Docs/milieu/200506_MvTNOx.pdf>.

Wang, Z. K. & Winters, L. A., (2001) 'Carbon taxes and industrial location: evidence from the multinationals literature', in Ulph, A., eds, *Environmental policy, International Agreements, and International Trade*, Oxford: Oxford University Press, 135–151.

Weber, R., (1997) 'Making More from Less: Strategic Demand Reduction in the FCC Spectrum Auctions', *Journal of Economics & Management Strategy*, Vol. 6, No. 3, 529–548.

Weishaar, S., (2005) '*CO_2 Emission Allowance Allocation Mechanisms, Allocative Efficiency and the Environment: a static and dynamic perspective*', written for the contract research: Emissions trading and equal competition, March 2005, Metro/Maastricht University, available at: <www.rechten.unimaas.nl/metro>.

Weishaar, S., (2007a) 'CO$_2$ emission allowance allocation mechanisms, allocative efficiency and the environment', in Deketelaere, K., eds, *Critical Issues in Environmental Taxation IV*, Oxford University Press, Oxford, 393–424.

Weishaar, S., (2007b) 'CO$_2$ emission allowance allocation mechanisms, Allocative efficiency and the environment: A Static and dynamic perspective', *European Journal of Law and Economics*, Vol. 24, No. 1, 29–70.

Weishaar, S., (2007c) 'The European CO$_2$ Emission Trading System and State Aid an assessment of the grandfathering allocation method and the performance standard rate system', *European Competition Law Review*, Vol. 28, No. 6, 371–381.

Weishaar, S., (2007d), *'The EU ETS: current problems and possible ways to move forward'*, invited paper prepared for the climate change conference 'Comparing North American and European approaches to climate change', held by the Viessmann European Research Centre, Wilfrid Laurier University, Waterloo, Canada on 28.09.2007, 23, available at: <www.wlu.ca/viessmann/html_pages/Climate.htm>.

Weishaar, S., (2008) 'Ex-Post-Korrektur im Europäischen CO$_2$-Emissionshandel: Auswirkungen der Rechtsprechung für Deutschland', Zeitschrift für Europäisches Umwelt- und Planungsrecht, Vol. 3, 148–151.

Weiss, L. W., (1974) 'The concentration-profits relationship and antitrust', in Goldschmid, H., Mann, M., & Weston, J., eds, *Industrial Concentration: The New Learning*, Boston: Little, Brown, 184–233.

Welch, W. P., (1983) 'The political feasibility of full ownership property rights: the case of pollution and fisheries', *Policy Sciences*, Vol. 16, 165–180.

Wishlade, F., (1998) 'Competition Policy or Cohesion Policy by the Back Door? The Commission Guidelines on National Regional Aid', *European Competition law Review*, No. 6, 343–357.

Woerdman, E., (2002) *'Implementing the Kyoto Protocol Mechanisms: Political Barriers and Path Dependence'*, Ph.D. Dissertation University of Groningen, The Netherlands, 594.

Woerdman, E., (2003) 'Developing carbon trading in Europe: does grandfathering distort competition and lead to state aid?', in Faure, M. Gupta, J. & Nentjes, A., eds, *Climate Change and the Kyoto Protocol, The Role of Institutions and Instruments to Control Global Change*, Edward Elgar, Cheltenham, 108–127.

Woerdman, E., (2005) 'Tradable Emission Rights', in Backhaus, J. G., ed., *Elgar Companion to Law and Economics*, Cheltenham: Edward Elgar, 364–380.

Woerdman, E. & Arcuri, A., (2006) 'Tradable Emission Rights and the Polluter-Pays Principle: Do Polluters Pay under Grandfathering?' *Working Paper Series in Law and Economics*, University of Groningen, Faculty of Law, 14.

Woerdman, E., Couwenberg, O. & Nentjes, A., (2006) 'Terechte energieprijsver-
hoging door gratis emissierechten', *ESB*, 08.08.2006, 427–429.

Zwingmann, K., (2007) *'Ökonomische Analyse der EU-Emissionshandelsreichtli-
nie, Bedeutung und Funktionsweisen der Primärallokation von Zertifikaten'*,
in der Reihe: Ökonomische Analyse des Rechts, DUV Gabler Edition Wis-
senschaft, Wiesbaden, 358.

List of Cases

Case 30/59, *De Gezamenlijke Steenkolenmijnen Limburg v. High Authority* [1961] ECR 1.

Case 56/64 and 58/64, *Établissements Consten S.à.R.L. and Grundig-Verkaufs-GmbH v. Commission* [1966] ECR 299.

Case 14/68, *Wilhelm v. Bunderskartellamt* [1969] ECR 1.

Joined cases 40 to 48, 50, 54 to 56, 111, 113 and 114-73, *Coöperatieve Vereniging 'Suiker Unie' UA and others v. Commission of the European Communities* [1975] ECR 01663.

Case C-155/73, *Giuseppe Sacchi* [1974] ECR 00409.

Case 173/73, *Italy v. Commission* [1974] ECR 709.

Case 78/76, *Steinike v. Bundesamt für Ernährung und Forstwirtschaft* [1977] ECR 595.

Case C-85/76, *Hoffmann-La Roche* [1979], ECR I-00461.

Case 13/77, *N.V. GB-INNO-B.M. v. Association des détaillantes en tabac* [1977] ECR 2115.

Case 82/77, *Opebaar Ministerie v. Van Tiggele* [1978] ECR 25.

Case 5/79, *Procureur général v. Hans Buys and Han Pesch and Yves Dulieux and Denkavit France SARL* [1979] ECR 3203.

Case 61/79, *Amministrazione delle finanze dello Stato v. Denkavit italiana* [1980] ECR 1205.

Case 730/79, *Philip Morris v. Commission* [1980] ECR 2671.

Joint cases 188/80 – 190/80, *France, Italy and the United Kingdom v. Commission* [1982] ECR I-02545.

Joint cases 213-215/81, *Norddeutsches Vieh- und Fleischkontor v. BALM* [1982] ECR 3583.

Joined cases 177/82 and 178/82, *Criminal proceedings against Jan van de Haar and Kaveka de Meern BV* [1984] ECR 1797.

Case 181/82, *Roussel Laboratoria BV and others v. État néerlandais* [1983] ECR 03849.

Case 232/82, *Duphar BV and others v. The Netherlands State* [1984] ECR 523.

Joined cases 296/82 and 318/82, *Netherlands and Leeuwarder Papierwarenfabriek v. Commission* [1985] ECR 809.

Case 323/82, *SA Intermills v. Commission* [1984] ECR 3809.

Case 123/83, *Bureau national interprofessionnel du cognac v. Guy Clair* [1985] ECR 391.

Case 229/83, *Association des Centres distributeurs Édouard Leclerc and other v. SARL 'Au blé vert' and others* [1985] ECR 1.

Case 231/83, *Henri Cullet and Chambre syndicate des réparateurs automobiles et détaillants de produits pétroliers v. Centre Leclerc à Toulouse and Centre Leclerc à Saint-Orens-d-Gameville* [1985] ECR 305.

Case C-290/83, *Commission v. France* [1985] ECR 439.

Case 209/84 – 213/84, *Lucas Asjes and others, Andrew Gray and others, Andrey Gray and others, Jacques Maillot and others and Léo Ludwig and others* [1986] ECR 1425.

Joint cases 67, 68 and 70/85, *Van der Kooy BV v. Commission* [1988] ECR 219.

Case 259/85, *France v. Commission* [1988] ECR 1573.

Case 310/85, *Deufil GmbH & Co. KG v. Commission* [1987] ECR 901.

Case 311/85, *ASBL Vereniging van Vlaamse Reisbureaus v. ASBL Sociale Dienst van de Plaatselijke en Gewestelijke Overheidsdiensten* [1987] ECR 3801.

Case 66/86, *Ahmed Saeed Flugreisen and Silver Line Reisebüro GmbH v. Zentrale zur Bekämpfung unlauteren Wettbewerbs e.V.* [1989] ECR 803.

Case 136/86, *Bureau national interprofessionnel du cognac v. Yves Aubert* [1987] ECR 4789.

Case 188/86, *Ministère public v. Régis Lefèvre* [1987] ECR 2963.

Case 267/86, *Pascal Van Eycke v. ASPA NV.* [1988] ECR 4769.

Joined cases 62/87 and 72/87, *Exécutif Régional Wallon v. Commission* [1988] ECR 1573.

Case 102/87, *France v. Commission* [1988] ECR 4067.

Case C-142/87, *Belgium v. Commission* [1990] ECR I-959.

Case C-202/88, *France v. Commission* [1991] ECR I-1223.

Case C-303/88, *Italian Republic v. Commission* [1991] ECR I-1433.

Case C-260/89, *Elliniki Radiophonia Tiléorassi AE* [1991] ECR I-02925.

Case C-261/89, *Italy v. Commission* [1991] ECR I-1437.

Case C-305/89, *Italy v. Commission* [1991] ECR I-1603.

Case C-332/89, *André Marchandise, Jean-Marie Chapuis and SA Trafitex* [1991] ECR I-01027.

Case C-339/89, *Alsthom Atlantique SA v. Companie de Construction Mechanique Sulzer SA.* [1991] ECR I-107.

Case C-41/90, *Höfner and Elser v. Macroton GmbH* [1991] ECR I-1979.

Case C-179/90, *Merci convenzionali porto di Genova Spa v. Siderurgica Gabrielli SpA* [1991] ECR I-05889.

Joint cases C-271, 281 and 289/90, *Spain, Belgium & Italy v. Commission* [1992] ECR I-5833.

Case 2/91, *Wolf W. Meng* [1993] ECR I-05751.

Case C-60/91, *José António Batista Morais* [1992] ECR I-02085.

Joined cases C-72/91 and C-73/91, *Sloman Neptun v. Bodo Ziesemer* [1993] ECR I-887.

Joined cases C-159/91 and C-160/91, *Poucet and Pistre* [1993] ECR I-00637.

Case C-185/91, *Bundesanstalt für Güterfernverkehr v. Gebrüder Reiff GmbH & Co. KG.* [1993] ECR I-05801.

Case C-189/91, *Kirsammer-Hack* [1993] ECR I-6185.

Case 245/91, *Ohra Schadeverzekeringen NV.* [1993] ECR I-05851.

Case C-320/91, *Corbeau* [1993] ECR I-02533.

Joint cases C-278/92, C-279/92 and C-280/92, *Spain v. Commission* [1994] ECR I-4103.

Case C-364/92, *SAT Fluggesellschaft mbH v. Eurocontrol* [1994] ECR I-00043.

Case C-387/92, *Banco de Credito Industrial SA (Banco Exterior de España SA) v. Ayuntamiento de Valencia* [1994] ECR I-877.

Case C-393/92, *Almelo and others* [1994] ECR I-01477.

Case C-18/93, *Corsica Ferries Italia Srl. V. Corpo dei Piloti del Porto di Genova* [1994] ECR I-01783.

Case C-153/93, *Delta Schiffahrts- und Speditionsgesellschaft mbH* [1994] ECR I – 02517.

Case C-323/93, *Crespelle* [1994] ECR I-05077.

Joined cases C-329/93, C-62/95 and C-63/95, *Germany, Hanseatische Industrie-Beteiligungen GmbH and Bremer Vulkan Verbund AG v. Commission* [1996] ECR I-5151.

Case C-387/93, *Banchero* [1995] ECR I-04663.

Case C-39/94, *SFEI* [1996] ECR I-3547.

Case C-96/94, *Centro Servizi Spediporto* [1995] ECR I-02883.

Joined cases C-140/94, C-141/94, C-142/94, *DIP* [1995] ECR I-03257.

Case C-241/94, *France v. Commission* [1996] ECR I-4551.

Case C-244/94, *Fédération Française des Soci étés d'Assurance* [1995] ECR I-04013.

Case C-70/95, *Sodemare SA* [1997] ECR I-03395.

Case C-343/95, *Diego Cali* [1997] ECR I-01547.

Case C-35/96, *Commission v. Italy (CNSD)* [1998] ECR I-03851.

Case C-67/96, *Albany International BV v. Stichting Bedrijfspensioenfonds Textielindustrie* [1999] ECR I-05751.

Case C-75/97, *Belgium v. Commission* [1999] ECR I-3671.

Case C-288/96, *Germany v. Commission* [2000] ECR I-8237.

Case C-6/97, *Italy v. Commission* [1999] ECR I-2981.

Case C-38/97, *Autotransporti Librandi Snc di Librandi F. & C.* [1998] ECR I-05955.

Case C-75/97, *Belgium v. Commission* [1999] ECR I-3671.

Joined cases C-115/97 to C-117/97, *Brentjens' Handelsonderneming BV v. Stichting Bedrijfspensioenfonds voor de Handel in Bouwmaterialen* [1999] ECR I – 06025.
Joined cases C-147/97 and C-148/97, *Deutsche Post* [2000] ECR I-00835.
Case C-200/97, *Ecotrade Srl v. AFS* [1998] ECR I-7907.
Case 219/97, *Maatschappij Drijvende Bokken BV v. Stichting Pensioenfonds voor de Vervoer- en Havenbedrijven* [1999] ECR I-06121.
Case C-295/97, *Industrie Aeronautiche e Meccaniche Rinaldo Piaggio SpA v. International Factors Italia SpA* [1999] ECR I-3735.
Case C-404/97, *Commission v. Portugal* [2000] ECR I-4897.
Case C-156/98, *Germany v. Commission* [2000] ECR I-6857.
Joined cases C-180/98 – C-184/98, *Pavel Pavlov and Others v. Stichting Pensioenfonds Medische Specialisten* [2000] ECR I-06451.
Case C-209/98, *Entreprenørforeningens Affalds/Miljøsektion (FFAD) v. Københavns Kommune* [2000] ECR I-03743.
Case C-379/98, *Preussen Elektra v. Schleswag AG* [2001] ECR I-2099.
Case C-35/99, *Manuele Arduino* [2002] ECR I-01529.
Case C-143/99, *Adria-Wien Pipeline GmbH and Wietersdorfer & Peggauer Zementwerke GmbH v. Finanzlandesdirektion fuer Kaernten* [2001] ECR I-8365.
Case C-334/99, *Germany v. Commission* [2003] n.y.r.
Case C-482/99, *France v. Commission* [2002] ECR I-4397.
Case 53/00, *Ferring SA v. Agence centrale des organismes de sécurité sociale* [2001] ECR I-9067.
Case C-218/00, *Cisal di Battistello Venanzio & C. Sas v. Istituto nazionale per l'assicurazione contro gli infortuni sul lavoro (INAIL)* [2002] ECR I-00691.
Case C-409/00, *Spain v Commission* [2003] ECR I-1487.
Case C-501/00, *Spain v. Commission* [2004] n.y.r.
Joined cases C-264/01, 306/01, 354/01, 355/01, *AOK Bundesverband* [2004] ECR I-02493.
Case C-198/01, *Consorzio Industrie Fiammiferi (CIF) and Autorità Garante della Concorrenza e del Mercato CIF* [2003] n.y.r.
Case C-172/03, *Heiser* [2005] ECR I-1627.
Case C-250/03, *Giorgio Emanuele Mauri v. Ministero della Giustizia, Comissione per gli esami di avvocato presso la Corte d'appello di Milano* [2005] n.y.r.
Joined cases C-94/04 and C-202/04, *Federico Cipolla v. Rosaria Fazari and Stefano Macrino, Claudia Capodarte v. Roberto Meloni* [2006], n.y.r.
Case T-459/93, *Siemens SA v. Commission* [1995] ECR II-1675.
Case T-67/94, *Ladbroke Racing Ltd. v. Commission* [1998] ECR II-1.
Case T-214/95, *Vlaamse Gewest v. Commission* [1998] ECR II-717.
Case T-41/96, *Bayer AG v. Commission* [2000] ECR II 3383.
Joined cases T-126/96 and T-127/96, *BFM and EFIM v. Commission* [1998] ECR II-3437.
Case T-46/97, *SIC v. Commission* [2000] ECR II-2125.

Joined cases C-52/97, C-53/97 and C-57/94, *Viscido, Scandella, Terragnolo and Others v. Ente Poste Italiane* [1998] ECR I-2629.

Joint cases T-204/97 and T-270/97, *EPAC v. Commission* [2000] ECR II-2267.

Case T-288/97, *Regione Friuli Venezia Giulia v. Commission* [2001] ECR II-1169.

Joint cases T-298/97, T-312/97, T-313/97, T-315/97, T-600 to 607/97, T-1/98, T-3/98 to T-6/98, T-23/98, *Alzetta Mauro and Others v. Commission* [2000] ECR II-2319.

Case T-613/97, *Ufex v. Commission* [2000] ECR II-4055.

Case T-35/99, *Keller SpA v. Commission* [2002] ECR II-261.

Case T-55/99, *CETM v. Commission* [2000] ECR II-3207.

Case T-152/99, *Hijos de Andrés Molina, SA v. Commission* [2002] ECR II-3049.

Joined cases T-228/99 and T-233/99, *Westdeutsche Landesbank Girozentrale and Land Nordrhein-Westfalen v. Commission* [2003] ECR II-435.

Joined cases T-92/00 and T-103/00, *Territorio Histórico de Álava v. Commission* [2002] ECR II 1385.

Case T-351/02, *Deutsche Bahn AG v. Commission* [2006] n.y.r.

Case T-233/04, *The Netherlands v. European Commission* [2008] n.y.r.

Case T-387/04, *EnBW Energie Baden-Würtemberg AG v. European Commission* [2007] n.y.r.

Case T-28/07, Fels-Werke and Others v. Commission [2007] n.y.r.

Case number 200502867/1, *MOB and Waddenvereniging vs. het college van gedeputeerde staten van Groningen*, of 15 February 2006.

Index

A

Abatement and emission trading, economic
        intuition
    abatement costs, 32–33
    CO_2 emission allowances, 31
    emission trading, 31–33
    initial allocation problem
        allocative efficiency, 34–36
        environmental considerations, 36
        price determination, 33–34
    negative externalities, 30, 31
Administrative allocation mechanism
    financial administrative mechanism,
        69–70, 90
    free administrative mechanism
        grandfathering, 70–71, 90
        relative standard base mechanisms,
            71–73, 91–92
Aid, notion of, 147–148, 161–163, 175,
        192–193, see also State aid for
        environmental protection
Allocation method, see Allocation process
Allocation formats compatibility
    auctions, 101–103
        polluter pays principle, 102
        price determination, 101
    directive establishing the EU ETS, 97
    efficiency, 97
    emission trading system, 101, 103

EU ETS amendment and, 107–108
    commitology procedure, 107
    harmonized allocation, 107
grandfathering, 103
legal perspective, 97
NAPs, Annex III, 98–100
PSR system, 104–106
under Directive 2003/87/EC, 98–101
    allocation mechanism, 100
    EC treaty, 100
    legal requirements, 98
    Member State, 98, 99
Allocation mechanism, 217
    administrative allocation mechanism
        financial administrative mechanism,
            69–70
        free administrative mechanism, 70–73
    auctions as, 60
        auction formats, 62–63
        multiunit auctions, 63–69
Allocation process, 58, 60, see also NAPs
Allocative efficiency
    analysis of allocation mechanisms, 54, 58
    auction benefits, 53
    dynamic open economy
        allocative mechanism and allocative
            efficiency, 79–82
        environmental requirements, 79
        negative externalities, 78
        open economy, 77–79

emission trading systems, 56–59
  allowance allocation mechanism, 57
  CO_2 emission, 56, 58
  CO_2 emission allowances, 56
  Dutch NOx system, 59
  Gross Domestic Product (GDP), 56
  environmental aspects, 73–76
    dynamic interaction of firms and the
        environment, 82–88
    initial allocation mechanisms and the
        environment, 88–91
  initial emission allowance allocation
      system, 54
  relative target system, 54
  revenue recycling, 54
  static closed economy
    allocative mechanism and allocative
        efficiency, 60–73
    closed economy model, 59–60
Allowance prices, 33, 34, 74, 106, 206,
    212, 218
Allowances
  And state aid for environmental protection,
      75, 160, 174
Ascending price auctions, 63, 65–67
Ashenfelter, O., 64, 206
Attainment of allocative efficiency, 92
Auction design challenges
  auction mechanisms designs,
      203, 206, 209
  bundling, 205
  competitive pricing, 204
  emission allowances, 205
  implementation of auction, 203
  market structure and transparency
    bidders' ability, 207
    demand reduction, 208
  objective of, 204
  timing of auction
    declining price anomaly, 206–207
    uniform price auction, 207
  true value of allowances, 204
Auctioning under the EU ETS amendment
  commission involvement and, 190–192
  in third trading phase, 192–199
    aid favouring a certain undertaking,
        193–195
    distort competition, 198–199
    granted by the state or through state
        resources, 196–197

notion of undertaking, 197–198
    transfer of a benefit or advantage,
        192–193
  selectivity principle, 193–195
Auctioning rules, 58, 61, 62, 68, 190, 198,
    200, 205, 210
Auctions
  closed and open, 62–66, 88
  design, 62–64, 67, 80, 192, 193, 199,
      203–210
  Dutch, 62, 63, 179
  economic exchanges, use in, 60, 148,
      153, 154
  emissions trading, *see* Emissions Trading
      systems
  English, 62, 63, 199
  formats, 62–63, 64, 66, 68
  multi-unit, 55, 63–69, 79, 80, 204, 208, 210

B

Balancing test, 159, 161, 173, 186
Barriers to entry, 15, 19, 37–40, 112–115,
    131, 142, 168, 169, 173, 174, 182,
    183, 186, 188,
Belgium, 117, 148, 151, 153, 177–179
Belgium Green Electricity Certificates, 148
Bidders, 61–69, 76, 79, 80, 92, 192, 197, 199,
    204, 206, 207–210, 217, 218
Bundling, 205
Burden sharing agreement, 4, 165, 181

C

Carbon capture and storage (CSS), 7,
    160, 170,
Carbon dioxide emissions, 3, 10, 15, 31,
    32–36, 54, 56, 58, 60, 63–65, 68,
    71–73, 75, 76, 82–86, 88–94, 108,
    131, 138, 150, 161, 164, 168, 181,
    182, 196, 207, 213
  abatement, 9, 15, 30–33, 54, 71, 73, 74, 76,
      84, 85, 88, 90, 91
    basic economic intuition, 15, 19,
        29–36, 84
    operation of, 73
  EU initiative to limit, 2
  trading, *see* Emissions trading systems
Cartelization, 28, 37, 114, 118, 173, 183,
    185, 216

Climate change
European Climate Change Programme, 2, 3, 13
emission trading system, 8, 157, 211
launch of, 2
Comitology procedure, 107, 189
Court of First Instance (CFI), 104, 106, 215
Cramton, P. 53

D

Dales, J.H. 3
De minimis rule, 153, 154, 167, 168, 171–172, 182, 185, 197–199
Denmark, 1, 211
Distort competition, state aid assessment, 181, 199
anticompetitive distortions, trading sectors, 183–184
auctioning and, 198–199
compatible with the common market, 185–187
de minimis threshold, 185
European state aid regulation and, 152
incumbent and new entering firms, 182–183
trading sectors and non-trading sectors, 183
Double dividend hypothesis, 53, 220
Dynamic open economy setting
environmental requirements, 79
negative externalities, 78
open economy, 77–79

E

EUAs, 191
EC Competition law, 14
Economic foundations
abatement and emission trading, economic intuition
abatement costs, 32–33
CO_2 emission allowances, 31
emission trading, 31–33
environmental considerations, 36
negative externalities, 30, 31
problem of initial allocation, 33–36
industrial economics
abuse, 44–45
barriers to entry and price, 39

cartels, 45–48
merger and acquisition, 48–50
oligopoly, 40
policies of firms, 37
static monopoly model, 42–44
structure-conduct-performance, 38
social welfare, efficiency and initial allocation
allocative efficiency, 21, 23, 25, 26
cartelization, 28
edgeworth consumption, 23–25
edgeworth production, 21–23
European emissions trading system, 28
general equilibrium theory, 20
geometric model, 21
pareto optimality, 23, 25
production mix, 25–29
social welfare, 20
Effet utile, 116, 118, 120, 122, 135, 139, 143, 216
Emission cap, 214
Emissions trading systems
advantages and disadvantages of, 11, 53, 65, 77, 112, 113, 126, 145, 149, 150, 153, 167–169, 182, 183, 185, 186, 188, 191, 195, 196
allowance allocation, 11, 15, 55, 56, 57, 59, 71, 98, 115, 138, 182, 188, 212
administrative mechanisms, 89–91, 213
allocative efficiency, 11, 15, 16, 34, 36, 53, 55, 60–73, 76, 78–82, 93, 94, 97, 109, 112, 139, 163, 200–202, 212, 213, 214, 218, 219
assignment mechanisms, analysis of, 55, 212
auctions, 15, 57, 60–69, 69, 76, 79–80, 81, 88–89, 92, 101–103, 109, 145, 191, 196, 198, 200, 201, 204, 205, 215, 217. *See* also auctions below
closed economy settling, comparison in, 54, 55, 60, 79, 88, 91, 92
environmental aspects, 73–76, 82–91, 94
environmental considerations, 15, 33, 36, 55, 77, 155, 212
environmental effectiveness, not influencing, 6, 11, 16, 29, 73, 75, 90, 95, 189, 212, 213, 220

financial administrative mechanisms,
    58, 69–70, 80–81, 88, 89, 90, 213
free administrative mechanisms, 15,
    69, 70, 80–82, 89, 90, 92, 213
grandfathering, 16, 53, 54, 55, 58, 60,
    70–74, 76, 79, 81, 82, 88, 89, 90, 92,
    97, 103, 104, 107, 111–113, 115,
    138, 161, 162, 165, 169–173, 182,
    184, 187, 188, 213, 214
grandfathering schemes, 53, 93, 108,
    163, 169, 170, 213, 214
initial permits, 54
initial problem of, 33–36
issue of, 15, 33, 53, 55, 58, 84, 98,
    145, 153, 189, 216
market distortions, 29, 36, 54
mechanisms, 2, 12, 13, 15, 16, 29, 34,
    53, 54, 55, 57–58,
normative principles, 35
prices, need for, 33–34
relative standard base mechanisms,
    81–82, 89, 91–93, 213
relevant criteria, 34, 145
static close economy, 15, 37, 55,
    59–76, 77, 79, 80, 88, 92, 93,
    94, 97, 213
theoretical framework, 55
auctions
    Ausubel, 67, 68
    closed sealed-bid, 63, 65, 66
    design, 17, 61–64, 67, 80, 101, 192,
        193, 196, 199, 203–210
    effective use of, 55, 67, 68, 208
    formats, 62–63, 64, 66, 68
    open multi-unit, 66–68
    open uniform-price ascending, 66
    pay-as-you-bid, 65, 68
    popularity of, 60
    sequential, 63–65, 68, 79, 80, 205–207
    simultaneous multi-unit, 65, 68, 80
    strategic behaviour, 64, 68, 76
    uniform-price, 64–66, 68, 206
basic economic intuition, 29–36
cap, 5, 7, 8, 54, 56, 58, 73–76, 90–94, 98,
    100–104, 108, 214, 220
cap and trade, 58, 73, 76, 92, 94, 98, 100,
    101, 104, 108, 214
Dutch National Allocation Plan (Dutch
    NAP), 71

Dutch NOx trading system, 172,
    174–176, 178
elements of, 8, 11, 56, 129, 132, 141, 166,
    176, 177, 195, 103
environmental aspects, 73–76, 82–91, 94
European
    cost of, 33
    free exchange of allowances,
    minimal cost, trading at, 34
    performance standard rate system, 15, 16,
        58, 71, 74, 88, 93, 94, 97, 104, 110,
        111, 112, 113, 127, 145, 161, 175,
        176, 182–185, 187, 212, 214, 219,
        220
    quantity setting, 56, 73, 94
    static closed economy
        administrative allocative mechanisms,
            15, 58, 69–74, 80–82, 88–92, 94,
            101, 213, 214
        allocation mechanisms and allocative
            efficiency compared, 60–72, 79–82
        auctions, 59, 60, 61–71, 74, 88, 92, 93,
            213. *See also* Auctions *below*
        environment, impact on, 59, 76, 77
        environmental aspects, 73–76, 82–91, 94
        UK Emissions Trading Scheme, 61
Entrenched market shares, 115
Environmental aspects, allocative efficiency
    dynamic interaction of firms and the
        environment, 82–88
        outsourcing abatement, 85
        outsourcing of production, 86–87
        substitute domestic goods, 87–88
    initial allocation mechanisms and the
        environment, 88–91
        administrative allocation mechanism,
            89–91
        auctions, 88–89
Environmental effectiveness, 6, 11, 16, 29, 73,
    75, 90, 95, 189, 212, 213, 220
EU Emissions Trading system, art. 81 and 82
    anticompetitive distortions, 111
    economic appraisal, 139–142
    economic problem
        barriers to entry, 112–115
        entrenched market shares, 115
    joint application post 2012, 142–143
    legal analysis
        anticompetitive distortions, 116

comissions's decision, 116
in article 86, 117
joined application jurisprudence,
127–136
jurisprudence, development, 117–127
European Climate Change Program (ECCP),
2, 3, 13
European Court of Justice (ECJ), 14, 104, 116,
118, 215
European Emissions Trading System (EU
ETS), 1, 3, 4, 9, 16
action plan, 3
advantages and disadvantages, 11
allocative efficiency, 11
anticompetitive behaviour, 11
burden sharing agreement, 4
CO_2 emissions, 3, 15
Kyoto Protocol, 2, 3
proposal amendment, 4
research methodology
allocative efficiency, 12
anticompetitive distortions, 12, 14
optimal allocation of allowances, 13
research questions
allocative efficiency, 11
endowments allocation, 10–11
externalities, 9–10
legal implications, 8–9
social equity, social welfare, 9
revision, 5–6
trading phases, 1, 3, 4
European state aid regulation
aid favouring a certain undertaking,
149–150
community dimension, 153–154
derogations of article 87(1), 154–160
distort competition, 152–153
existence of state aid, article 87(1),
146–147
granted by the state or through state
resources, 151–152
notion of undertaking, 152
transfer of benefit or advantage, 147–148
Evidence collusion, 218

F

Financial administrative mechanism, 69–70
Free administrative mechanism

grandfathering, 70–71
relative standard base mechanisms, 71–73
Free allocation mechanism, 219
state aid assessment
aid favouring a certain undertaking,
163–164
community dimension, trade and
member states, 172–174
distort competition, 167–172
granted by the state or through state
resources, 164–167
notion of undertaking, 167
transfer of benefit or advantage, 161–163

G

Global CO_2 emissions, 94
Grandfathering, sate aid assessment, 161
community dimension, trade and member
states, 172–174
distorts or threatens to distort competition
competing firms, 170–171
competitive distortions, 167
de minimis threshold, 171–172
incumbent and new entering firms,
168–169
trading sectors and non trading sectors,
169–171
granted by member state resources, 165–166
granted by state, 164–165
notion of aid, 161–163
selectivity principle, 163–164
undertaking or production, 167

I

Industrial economics
barriers to entry and price, 39
economic models
abuse, 44–45
cartels, 45–48
merger and acquisition, 48–50
static monopoly model, 42–44
Oligopoly, 40
policies of firms, 37
structure-conduct-performance, 38
Intangible assets, 148, 151, 175, 178, 188
IPPC (Integrated Pollution Prevention and
Control) Directive, 72, 74

J

Joined application jurisprudence, 216
    in article 81, 127–132
    in article 82, 132–136

K

Kerr, S. 53
Kyoto Protocol, 2, 3

L

Legal analysis, 116
    anticompetitive distortions, 116
    article 86, 117
    comissions's decision, 116
    joined application jurisprudence
        in article 81, 127–132
        in article 82, 132–136
    jurisprudence, development
        delegation of power, 125
        in article 10, 122
        in article 3(g), 120, 123
        in article 5, 120
        in article 81 and 82, 117–119,
            123, 125
        in article 85, 121
    lawfulness of national provisions, 119
    legislations, 119, 120
    national regulation, 118
    reinforcing effects, 124, 126
Legal monopolies, 134

M

Marginal Product of Capital (MPK), 23
Marginal Product of Labour (MPL), 23
Marginal Rats of Technical Substitution
    (MRTS), 23
Milgrom, P., 61, 64, 205, 206
Multiple-unit auctions, 65, 205

N

NAPs (National Allocation Plans), 98, 99,
        100, 102, 105, 107, 115, 127, 134,
        135, 138, 139, 157, 159, 166, 168,
        169, 171–173, 194
Nash, J.R., 101, 102

Netherlands
    environmental and energy taxes, 180, 204
Nitrous oxide, 2, *see* Dutch NOx trading
        system and Emissions Trading
        System

O

Official Journal as Directive 2009/29/EC,
        6, 14
Open auctions, 62, 63
Opportunity costs, 81, 161–163,

P

Performance Standard Rate (PSR), 15–16, 58
    allocation scheme perspective, 105–106
    benchmarks, 220
    Competition law and, 104
    Dutch NOx system, 104
    EU ETS directive, 104
    greenhouse gas efficiency, 105
    State aid and, 104–105
    state aid assessment
        community dimension, trade and
            member states, 185–187
        distort competition, 181–185
        granted by state, 176–178
        granted by state resources, 178–181
        selectivity principle, 175–176
        transfer if a benefit or an advantage, 175
        undertaking or production, 181
Polluter-pays principle, 35
Pigou tax, 30
Precautionary principle, 13
Preussen Elektra case, 178, 180, 195
Price anomaly, 80, 92, 94, 213
Price determination, 33–34
Production Possibility Frontier (PPF), 25, 26

R

Relative performance standards, 15, 58, 75

S

Selectivity principle, 146, 149–150, 163–164,
        175–176, 193–195, 200
Sequential auctions, 64, 65, 205, 206
Signalling, 47, 67, 79, 208

Single-unit auctions, 63, 64, 204, 205
Social costs, 10, 30, 101, 114
Social welfare, efficiency and initial
     allocation
  allocative efficiency, 21, 23, 25, 26
  cartelization, 28
  edgeworth consumption, 23–25
  edgeworth production, 21–23
  European emissions trading system, 28
  general equilibrium theory, 20
  geometric model, 21
  pareto optimality, 23, 25
  production mix, 25–29
  social welfare, 20
Standard base mechanisms, 71–73, 81–82, 91
Standard auctions, 62, 63, 66
State aid,
  distortion and competition, 140, 145, 146,
     149, 152, 153, 154, 162, 163,
     166–168, 172, 174, 181, 182, 184,
     185,186, 188, 198, 199, 200–202
  environmental protection, for, 160, 174
  firms, equal treatment of, 168, 182, 185
  green certificates or tenders as, 177–179,
     190
  renewable energy support schemes, for, 2,
     153, 156
  start-up subsidies, 147
  tax credits, 147, 180
State aid assessment
  auctioning under the EU ETS amendment
     commission involvement and, 190–192
     in third trading phase, 192–199
     selectivity principle, 193–195
  distort competition, 181, 200–202
     anticompetitive distortions, trading
        sectors, 183–184
     compatible with the common market,
        185–187
     de minimis threshold, 185
     incumbent and new entering firms,
        182–183
     trading sectors and non-trading sectors,
        183
  European state aid regulation
     aid favouring a certain undertaking,
        149–150
     community dimension, 153–154
     derogations of article 87(1), 154–160
     distort competition, 152–153

  existence of state aid, article 87(1),
     146–147
  granted by the state or through state
     resources, 151–152
  notion of undertaking, 152
  transfer of benefit or advantage,
     147–148
of free allocation mechanisms
  aid favouring a certain undertaking,
     163–164
  community dimension, trade and
     member states, 172–174
  distort competition, 167–172
  granted by the state or through state
     resources, 164–167
  notion of undertaking, 167
  transfer of benefit or advantage, 161–163
of PSR system
  aid favouring a certain undertaking,
     175–176
  community dimension, trade and
     member states, 185–187
  distort competition, 181–185
  granted by state and state resources,
     176–181
  notion of undertaking, 181
  transfer of benefit or an advantage, 175
State aid for environmental protection, 155,
     160, 174,
Stranded costs, 84, 91, 92, 94, 214
Strategic demand reduction, 67, 208

T

Tax revenues, 180
Transaction costs, 10, 27, 34, 37, 49, 50, 59,
     65, 70, 72, 81, 192, 193, 209, 210,
     217, 219
Transfer rules, 105, 172, 173, 185

W

Welfare effects, 38, 43, 46, 50
Welfare losses, 111, 171, 173
Willingness to pay, 179, 192
Windfall profits, 8, 10, 16, 35, 53, 81, 90, 92,
     162, 195, 209

Z

ZUG, 106

CLIMATE CHANGE LAW, POLICY AND PRACTICE SERIES

1. Clarisse Fräss-Ehrfeld, *Renewable Energy Sources: A Chance to Combat Climate Change*, 2009 (ISBN 978-90-411-2870-6).
2. Stefan Weishaar, *Towards Auctioning: The Transformation of the European Greenhouse Gas Emissions Trading System – Present and Future Challenges to Competition Law*, 2009 (ISBN 978-90-411-3198-0).